Divergence with Genetic Exchange

Divergence with Genetic Exchange

Michael L. Arnold

Department of Genetics, University of Georgia, Athens, Georgia 30602-7223, USA

OXFORD
UNIVERSITY PRESS

OXFORD

UNIVERSITY PRESS

Great Clarendon Street, Oxford, OX2 6DP,
United Kingdom

Oxford University Press is a department of the University of Oxford.
It furthers the University's objective of excellence in research, scholarship,
and education by publishing worldwide. Oxford is a registered trade mark of
Oxford University Press in the UK and in certain other countries

First Edition published in 2016

Impression: 1

Published in the United States of America by Oxford University Press
198 Madison Avenue, New York, NY 10016, United States of America

British Library Cataloguing in Publication Data
Data available

Library of Congress Control Number: 2015939605

ISBN 978–0–19–872602–9 (hbk.)
ISBN 978–0–19–875511–1 (pbk.)

Printed and bound by
CPI Group (UK) Ltd, Croydon, CR0 4YY

To Frances, Jenny, Brian, and Amber

Preface

This book is an investigation into processes associated with evolutionary divergence and diversification. The focus, as the title indicates, is on the role played by the exchange of genes between divergent lineages. This process has been given various monikers, the most recent being "divergence-with-gene-flow." I want to first express my profound relief concerning the degree to which evolutionary biologists in general have embraced a model of diversification that includes some measure of genetic exchange. Many research groups, including my own, have emphasized the important role of gene transfer events in evolution for decades. However, I would be remiss if I did not express some misgivings concerning some of the current research into these processes. As my friends and family members (and especially those people with whom I do not see eye-to-eye) know, I do not hesitate to mention when I think there is a lack of an appreciation of historical precedence. I would argue that often this newest terminology—used in place of what many before us described as natural hybridization, introgression, horizontal gene transfer, viral reassortment, parapatric and/or sympatric divergence, etc.—demonstrates an unfortunate lack of understanding of the rich history of studies of genetic exchange.

Notwithstanding this concern, I once again prefer to emphasize the fact that evolutionary biologists now appear to believe that testing for divergence accompanied by genetic exchange has merit. The order and structure of the various chapters, as with my previous texts, reflect the goal of illustrating the conceptual and biological breadth of genetic exchange-mediated evolutionary processes. I have thousands of Endnote references (10,851 at the time of this writing), most of which discuss web-of-life processes, from which I had to choose the following topics and examples. This means that not only was it necessary to be subjective, but also due to unintentional oversight, I will have minimized some extremely important research. However, as with my most recent two books, I include discussions of viruses, prokaryotes, and eukaryotes. My rationale for discussing clades from all the domains of life remains the same as before. Though the mechanisms by which organisms exchange genomic material differ widely, the outcomes others and I are interested in—adaptive evolution and the formation of new "hybrid" lineages—do not. Furthermore, there remains in some corners a biased outlook on the utility of different organismic groups in defining the "important" evolutionary processes. An interaction I had with a colleague at a Society for the Study of Evolution meeting, I believe, illustrates this point well. In discussing my intention to include organisms from all domains of life in a text like this one, he objected and stated: "Horizontal gene transfer occurs in micro-organisms and has nothing to do with animals. In contrast, natural hybridization affects the latter, but not the former." I addressed this colleague's concern by first reminding him that genetic exchange could produce similar, important evolutionary consequences, regardless of the underlying mechanism causing gene transfer. For example, the evolution of novel adaptations, at least partially via horizontal gene transfer, resulting in the bacterial agent of plague is analogous to the adaptive evolution through introgressive hybridization between annual sunflower species resulting in the formation of several new *Helianthus* species. Finally, I suggested that he keep in mind that natural hybridization *and* horizontal gene transfer have impacted both "higher" and "lower" organisms. In terms of so-called higher lineages, there is evidence for introgressive hybridization between different species of *Homo*, as well as the repeated insertion of retroviral-like elements into the *H. sapiens* genome. On the other hand, the combined effects from natural hybridization and horizontal gene transfer in "lower" eukaryotic clades such as trypanosomes and yeast are now well substantiated.

As mentioned, I am wedded to the idea that we should reflect accurately what our scientific predecessors described, especially when it applies to our chosen field of study. Needless to say, I have not been able to achieve perfection in past books, reviews, and contributions to the primary literature. Neither will I have achieved this in this text. However, in Chapter 1, I attempt to establish what I perceive as some of the most important steps in the history of studies of divergence-with-genetic-exchange. In particular, I highlight work from those on either side of the cultural divide represented by such persons as Lotsy, Anderson, and Stebbins on the one hand and Mayr on the other. I also present a large proportion of definitions used in this book and the genetic exchange literature. I also provide a discussion of theoretical and empirical foundations for the central topic (at least for eukaryotic clades) of hybrid zones. I move on to a discussion of genetic exchange that affects adaptive evolution in plants and animals. I then turn my attention to the origin of plant and animal taxa through hybridization. Finally, I provide exemplars of genetic exchange between lineages from different domains of life.

Chapter 2 reflects a subject on which I have been teaching and writing for over 25 years. I still find it extremely difficult and frustrating. This reaction is because, to un-paraphrase Will Rogers, I have never met a species concept that I liked. Almost certainly this results from the fact that, as much as I want order in the biological world, it is instead extremely messy. This is nowhere more apparent than when we attempt to categorize biological objects into taxonomic categories. I know that humans can attempt this—my first publication (in 1978) was a paper describing a new plant species—but it does not necessarily mean that the organisms typified will fit well within whatever pigeon-hole we construct. In Chapter 2 I emphasize that one source of difficulty for defining species relates to estimates of genetic

exchange. In this regard, I discuss some of the limitations and strengths of various conceptual frameworks upon which evolutionary biologists place their hypothesis testing.

One of the major (and well-justified) complaints by students in graduate courses I have taught, in which I used my previous books, was that there was no meaningful discussion of methodologies used to discern between incomplete lineage sorting (i.e. deep coalescence) and genetic exchange. Thus, though I had discussed ways in which reticulate evolution might be tested for, I had not explicitly provided illustrations of how this had been accomplished relative to the contribution of shared ancestral polymorphisms. In Chapter 3 (and elsewhere), I provide a sample of the methodologies that have begun to be used. An obvious caveat is that by the time this text appears the current methodologies will have likely evolved into many, divergent iterations. In addition to analytical means by which alternative hypotheses may be tested, Chapter 3 also contains a number of other approaches (e.g. examining hybrid zones) by which genetic exchange may be inferred.

I begin Chapter 4 with the following statement, "I often use the analogy of genetic exchange and reproductive isolation as occupying either side of the same coin. In seminars on this topic, I also often show a photo of two North American Rocky Mountain bull elk with their antlers locked in a sparring match to illustrate this analogy. I believe that both the analogy and the photo illustrate the conclusions to be drawn from the accumulated evidence of the past several decades regarding both the observation of the semipermeability of reproductive barriers between divergent lineages (Key 1968) and the processes that cause this semipermeability." This sums up the concept of this chapter. In almost every lineage, genetic exchange is possible, and likely. This is because organisms are highly unlikely to be separated geographically from related lineages until complete reproductive isolation is achieved. This probably should have been obvious in the light of climate change throughout the biological record. In other words, what is the likelihood that two populations somehow isolated geographically would remain so for more than several thousand years or so? This might happen in some regions (i.e. some portion of the tropical latitudes?), and possibly for island migrants from a mainland source, but otherwise the allopatric model of diversification remains unlikely. For example, glacial refugia end and the members of different refugia overlap thus providing opportunities for genetic exchange. Based on the expanding number of genomic data sets, this appears representative for viruses, prokaryotes, and eukaryotes. However, when members of divergent lineages meet and exchange genes, recombination has been shown to be limited to only certain portions of the genome. This is an obvious expectation given that evolutionary history is not only reticulate, but also divergent and diversifying. When discussing barriers to genetic exchange, I follow the conceptualization championed by Dobzhansky in *Genetics and the Origin of Species*; I discuss various barriers in the context of their being either pre- or post-zygotic (i.e. pre- or post-mating according to Dobzhansky 1937).

One of the paradigm shifts that I feel our group has contributed to in a substantial way is that relating to an appreciation of hybrid fitness. Scott Hodges and I argued for the concept of environment-dependent fitness in a 1995 *TREE* paper. Members of our research team subsequently demonstrated the relative fitness of plant hybrids under a number of natural and artificial environmental settings. Noland Martin and I recently (i.e. 2010, *TREE*) revisited the concepts and hypotheses suggested by Arnold and Hodges concerning hybrid fitness, finding support for many of the proposals made 15 years earlier. However, the idea of varying hybrid fitness predated our own work by at least 136 years. Thus, Darwin reported varying fitness among hybrids and parental forms of plants and animals. So, though Chapter 5 in this text reports novel means by which the fitness of hybrids can be estimated, we should keep in mind that many before us predicted these results. However, it is now accurate to state that estimates are available for the fitness effects of specific genomic regions (and in some cases, even specific genes) on the fitness of hybrids under varying environmental conditions. This is an enormous advance in our understanding of the role of genotype × environment interaction in catalyzing the outcome of reticulate evolutionary events.

In Chapter 6, I return to the topic of the outcomes of genetic exchange in terms of the evolutionary and ecological trajectories of individual lineages and entire clades. Though I restrict the examples to animals and plants, I discuss topics as diverse as adaptive radiations and molecular evolution. Examples of both homoploid and allopolyploid hybrid speciation that resulted in sexual or asexual animal taxa are highlighted; animal assemblages characterized by homoploid sexual, parthenogenetic, hybridogenetic, or gynogenetic derivatives are discussed. In addition, studies of allopolyploid sexual and parthenogenetic animal lineages are also presented. Finally, I discuss evidence for an array of genomic changes caused by the whole genome duplications in the animal allopolyploid species. Likewise, I review the overarching role of whole genome duplication (i.e. polyploidy) in plants. In the context of allopolyploidy, I discuss the role of reticulate evolution in adaptive radiations in entire plant clades. Similar conclusions regarding a primary role for hybridization in adaptive radiations of animal assemblages is also discussed in the context of the "hybrid swarm" model of Seehausen.

Introgressive hybridization involving endangered plants and animals continues to be perceived as having mostly negative impacts. Thus, concepts of species "integrity" etc. are often raised. Indeed, the potential loss of biodiversity through genetic assimilation of rare forms by more numerous/prolific congeners is real. However, it is also accurate that more and more instances of intentional introgression into rare forms, to increase genetic variation, have been instigated. In Chapter 7, these and other issues surrounding the conservation of plants and animals involved in divergence-with-gene-flow are discussed. While trying to recognize the real potential for extinction via genetic exchange, I also warn against a simplification of what we now understand concerning evolutionary pattern and process. In particular, I stress that conservation biologists should guard against hiding, from contributors and policy makers, the widespread evidence for non-allopatric divergence. Obfuscation might lead to a loss of credibility for the conservation biologists and thus a diminution of critical conservation priorities.

Debates concerning the evolutionary history of *Homo sapiens* have included many arguments over whether or not lineages related to anatomically modern humans crossed with related species. This includes hybridization within pre-*Homo* clades such as *Australopithecus*, as well as between the proto-*Homo* lineage and proto-*Pan* and proto-*Gorilla*. However, most of the disputes concerning reticulate evolutionary processes have centered on tests of whether or not, as the species migrated out of Africa, *H. sapiens* mated with now-extinct congeners. In Chapter 8, I review both fossil and genomic evidence that falsifies the hypothesis that humans spread into their current geographic range without introgressing with other members of *Homo*. Instead, overwhelming evidence points to a complex evolutionary history involving many genetic exchange events with multiple taxa. I point to the fact that this resembles all other examples in which related primates overlap spatially and temporally. I end the discussion in Chapter 8 with examples of how organisms making up the ecological setting for *H. sapiens* also reflect web-of-life processes. To emphasize the widespread nature of reticulate evolution and the human environment, I discuss examples of plants and animals that provide food, companionship, and drugs.

Chapter 9 is very brief, but summarizes the conceptual framework around which this book was built. Specifically, I draw attention to the underlying pattern of a web of life leading to evolutionary diversification in contrast to the tree of life envisioned by Darwin and others. In particular, I again argue that divergence accompanied by genetic exchange occurred across all domains of life and thus that it has been inaccurate to assign a major role for allopatric diversification. I end this text by suggesting that the "laws" alluded to by Darwin at the end of his *Origin of Species* now must include genetic exchange.

A glossary of definitions can be found after Chapter 9. All terms listed in the Glossary are italicized in the text where they are first used.

This book, like each of the previous three, would not have been possible without help from many friends and colleagues. I would like to first thank Paulo Alves, Jose (Zef) Melo-Ferreira, Jeff Ross-Ibarra, and Ole Seehausen for providing figures and discussion points for various chapters. Reverend Chris Currie graciously provided confirmation (no pun intended) of my translation of the Greek text in the quote used for the frontispiece. I also want to thank my friend and editor, Ian Sherman, for taking another chance on this author. His long-suffering nature during our interactions has always been necessary. Lucy Nash at Oxford University Press answered many questions, sent spreadsheets etc. and provided needed encouragement throughout the writing process. I thank my Head of Department, Allen Moore, for not ending my life or faculty appointment during the gestation of this book, in spite of my brazenly ignoring many faculty meetings. The following people (listed in alphabetical order) graciously spent time and energy in providing such things as data sets, descriptions of relevant work, and computer expertise: Richard Abbott, Becky Ackermann, Dave Brown, Diane Campbell, Miguel Carneiro, Amindine Cornille, Nuno Ferrand, Lila Fishman, Antonio Fontdevila, Dan Garrigan, Peter Grant, Rosemary Grant, Peter Holland, Krushnamegh Kunte, Christian Lexer, Ana Llopart, Joanna Malukiewicz, Noland Martin, Martim Melo, Jess Morgan, Craig Moritz, Jenny Ovenden, Loren Rieseberg, Christian Roos, Glenn-Peter Sætre, Walter Salzburger, Vincent Savolainen, Bob Schmitz, Klaus Schwenk, Rike Stelkens, Andrea Sweigart, Sunni Taylor, Tatiane Trigo, Xiao-Ru Wang, and Dietmar Zinner.

As with each of my previous books, I reserve my greatest appreciation to my wife Frances and our wonderful children, now adults and now including a daughter-in-law—Jenny, Brian, and Amber. I have often felt inadequate, but they have always expressed the assumption that I can accomplish writing projects like this book. Like the three earlier works, I dedicate this book to them.

Contents

CHAPTER 1

Genetic exchange: An historical consideration

"Widely different opinions are held by present-day biologists as to the evolutionary importance of hybridization between species. One of the main reasons for the disagreement seems to be the lack of summarized and codified data bearing upon the problem." **(Anderson 1936)**

"This tree of life notion of evolution attained near-iconic status in the mid-20th century with the modern neo-Darwinian synthesis in biology. But over the past 15 years, new discoveries have led many evolutionary biologists to conclude that the concept is seriously misleading and, in the case of some evolutionary developments, just plain wrong. Evolution, they say, is better seen as a tangled web." **(Arnold and Larson 2004)**

"There are now a variety of methods available for detecting recombination in HIV sequences and estimating recombination rates . . . and it is apparent that recombination plays a significant role in HIV evolution." **(Castro-Nallar et al. 2012)**

"Although the term 'pangenome' has been widely adopted only recently, some of the earlier observations supporting its presence appeared in the late 1980s and have been offered as examples of bacterial HGT. Basically, it appears that the genome complex that characterizes a bacterial species is much larger than can be contained within any single cell." **(Syvanen 2012)**

1.1 The evolutionary role of genetic exchange: Divergent viewpoints

The history of investigations into genetic exchange (i.e., leading to *reticulate evolution*; see the Glossary for definitions of italicized terms) between both closely- and distantly-related organisms has been checkered. On the one hand, scientists such as Lotsy and Anderson wrote treatises (Lotsy 1916; Anderson 1949) on the potential evolutionary effects from the admixture of "germplasms" from related, but distinct, animals and plants (i.e., *introgressive hybridization* or *introgression*; Anderson and Hubricht 1938). Likewise, Stebbins (1959) hypothesized that crosses between divergent lineages (i.e., *natural hybridization*) would form the basis of evolutionary

innovations, stating: "For the major advances to take place, therefore, a population with a high degree of genetic variability must be placed in an environment which is rapidly changing, and which offers to the population new ecological niches to which it can become adapted. Because of the slow rates at which it occurs, mutation can never provide by itself enough variability at any one time to fulfill such conditions. Genetic recombination must, therefore, be the major source of such variability, so that the evolutionary lines most likely to take advantage of a changing environment are those in which recombination is raised to a maximum. This is accomplished most effectively by mass hybridization between populations having different adaptive norms." Stebbins had actually posited

Divergence with Genetic Exchange. Michael L. Arnold.

this hypothesis of rapid, adaptive, and large-scale evolutionary change via natural hybridization five years earlier in a paper coauthored with Anderson (Anderson and Stebbins 1954). In fact, Anderson and Stebbins (1954) argued that natural hybridization likely caused the major evolutionary transitions seen in the biological record.

In contrast to the major role ascribed to genetic admixture from these later workers, Darwin (1859) envisioned the process of hybridization between species as disruptive, as reflected in the sterility of the hybrid offspring (Darwin 1859; pp. 276–277). Yet, even while using terms such as "pure species" and "mongrels" to describe the positive and negative aspects of non-hybrids and hybrids, respectively, Darwin recognized that crosses between less diverged forms were "favourable to the vigour and fertility of their offspring." (Darwin 1859; p. 277). Similarly, 100 years earlier, Linnaeus, the father of the taxonomic nomenclature still in use today, and a creationist, indicated that the "fixity" of specially created forms was not absolute when he stated: "It is impossible to doubt that there are new species produced by hybrid generation" (Linnaeus 1760; as quoted by Grant 1981, p. 245).

If Darwin could be considered, if not an opponent, then at least ambivalent toward the evolutionary significance of genetic exchange, while Stebbins, Lotsy, and Anderson were staunch proponents, then what about the self-named "architects" of the neo-Darwinian synthesis? Given that Stebbins was counted among the scientists who took on the task of explaining Darwinian theory in light of genetics, systematics, paleontology, developmental biology, etc. (Stebbins 1999), it might be concluded that the synthesis saw the incorporation of genetic exchange among the other processes seen as causal for evolutionary change. This was, however, not the case. Some of the reasons that genetic exchange was seen at best as an epiphenomenon were indeed rooted in scientific hypotheses (e.g., concepts and definitions of species and speciation; see Chapter 2). Other motivations apparently had more to do with cultural and philosophical biases. For example, Paterson (1985) pointed to the powerful words chosen to describe types of organisms derived from crosses within or between types: "In English, notice how approbative are words such as 'pure,' 'purebred,'

and 'thoroughbred,' and how pejorative are those like 'mongrel,' 'bastard,' 'halfbreed,' and 'hybrid.'" Even recent treatments have reflected such value judgments, as reflected by the following statement: "when reproductive isolation is complete we consider taxa to be 'good species'" (Coyne and Orr 2004, p. 34).

It is possible to identify recent examples of a negative outlook toward the evolutionary role of genetic exchange (e.g., Schemske 2000; Coyne and Orr 2004; Mayr 2004). Thus, Mayr (2004) concluded, "We have studied the origin of new species in birds, mammals, and certain genera of fishes, lepidopterans, and molluscs, and speciation has been observed to be allopatric (geographical) in most of the studied groups. Admittedly, there have been a few exceptions, particularly in certain families, but no exceptions have been found in birds and mammals where we find good biological species, and speciation in these groups is always allopatric." Yet, it is also accurate to state that there has been a paradigm shift over the past 20+ years in regard to an appreciation of the pervasiveness of viral recombination, *horizontal gene transfer*, introgressive hybridization, and hybrid speciation (both *allopolyploid speciation* and *homoploid speciation*—Arnold 1992, 1997, 2006, 2009; Grant and Grant 1992, 2010; Rieseberg 1997; Seehausen, 2004; Mallet 2005, 2007; Ackermann 2010; Arnold and Martin 2010; Abbott et al. 2013; Mable 2013; Burke et al. 2014; Fuentes et al. 2014; Wang and Wang 2014). Indeed, even while arguing for purely allopatric divergence in a limited number of organismic groups, Mayr (2004) concluded that "numerous other modes of speciation have also been discovered that are unorthodox in that they differ from allopatric speciation in various ways. Among these other modes are sympatric speciation, speciation by hybridization, by polyploidy and other chromosome rearrangements, by lateral gene transfer, and by symbiogenesis. Some of these non-allopatric modes are quite frequent in certain genera of cold-blooded vertebrates, but they may be only the tip of the iceberg."

Mayr (2004) was accurate in his tip of the iceberg analogy for animal taxa. Ironically, this was correct even in groups he suggested as having exclusively allopatric origins. For example, his contention that bird and mammal species originated in geographic isolation has been falsified by a number of synthetic treatments (e.g., see Grant and Grant 1992; Arnold 2009; Arnold et al. 2015). Of course, examples of genetic exchange-accompanied evolution from viral, bacterial, fungal, botanical, and "lower" eukaryotes also abound. Each year our species battles influenza infections from novel variants arising through recombination between divergent viral types (Lam et al. 2013; Latorre-Margalef et al. 2014). Likewise, the dominant, causative agent of bacterial dysentery in human populations, *Shigella sonnei*, was shown to have evolved through both horizontal gene transfer (i.e., HGT) from divergent bacterial lineages and substitution mutations, leading to high levels of resistance to antimicrobials and the ability to destroy other gut bacteria (Holt et al. 2013). In regard to lower eukaryotes, patterns of genomic variation in the human protozoan parasite, *Leishmania*, resulted in the inference of the hybrid origin of *Leishmania infantum* isolated from both vectors (i.e., sand flies) and an infected patient (Rogers et al. 2014). Groups as divergent as mosses, ferns, fungi, and sunflowers evidence the role of reticulate evolution—i.e., horizontal gene transfer, hybrid speciation, and introgressive hybridization—in the origin and adaptive evolution of taxa (Karlin et al. 2009; Whitney et al. 2010; Coelho et al. 2013; Metzgar et al. 2013; Peris et al. 2014; Sigel et al. 2014; Stairs et al. 2014; Wisecaver et al. 2014; Knie et al. 2015). Finally, the generation of hybrid plants can even potentially influence the establishment of associated vegetation (Adams et al. 2011).

Harrison (2012), in discussing the "language of speciation," surmised that new treatments often result in obfuscation because "old definitions have been reconfigured and new terms have been introduced. In some instances, the introduction of new terminology has failed to recognize historical usage, leading to unnecessary ambiguity and redundancy." Significantly, this seems to be occurring in studies of genetic exchange. Even this author's usage of the term "genetic exchange" can be confusing. Originally introduced to encapsulate recombination and admixture between divergent lineages in its many forms (i.e., viral recombination, horizontal gene transfer, introgressive hybridization, and hybrid speciation; Arnold 2006), the term often requires a qualifier in order for the meaning to be clear—e.g., "genetic exchange mediated by introgressive hybridization."

Recently, an entire cottage industry has arisen under the monikers, "divergence-with-gene-flow" and "ecological speciation." Generally reflecting studies of eukaryotic groups (e.g., Galligan et al. 2012; Suárez et al. 2014), a historically more accurate terminology might be "divergence-with-introgressive-hybridization" or simply *"sympatric speciation."* This recognition is in some ways an argument over word usage. However, because there has been such a negative connotation given to natural hybridization in the evolution of eukaryotes, particularly animal groups (Mayr 1963; Schemske 2000) and a limited role for non-allopatric divergence has been argued for since the neo-Darwinian synthesis (Maynard Smith 1966; Futuyma and Mayer 1980; Coyne and Orr 2004), the recognition that divergence-with-gene-flow involves hybridization (e.g., Soria-Carrasco et al. 2014) is crucial if the historical and evolutionary importance of new models are to be appreciated.

In the following sections I will detail examples from organisms representing all domains of life to highlight the widespread effects of events leading to what has been termed the "web-of-life" metaphor. This new metaphor has been proposed as a more accurate model (in contrast to that of the tree of life) by which we can understand the complexity associated with evolutionary diversification (Arnold and Larson 2004; Arnold 2006). In order to illustrate the newest appreciation of the role of reticulate evolutionary change, I will begin by discussing the conceptual framework of divergence-with-gene-flow; I will contrast this paradigm with that of allopatric divergence. I will then turn to a discussion of studies focusing on hybrid zone evolution—both those that established theoretical models and those applying such models in empirical studies—to emphasize their historical role in forming the foundation for ongoing analyses of genetic exchange. To emphasize the role of hybrid zones in creative evolutionary processes in eukaryotes, the next sections will provide examples of both *adaptive trait introgression* and the formation of hybrid taxa in plant and animal clades. The penultimate section will provide examples of genetic exchange between highly divergent organisms, including inter-domain transfers. The chapter will end with a set of general conclusions that follow from the preceding discussion.

1.2 Divergence-with-gene-flow: A process by any other name is still introgressive hybridization and sympatric divergence

To understand that Darwin proposed divergence-with-gene-flow one only has to read the following quote from Mayr (1963; p. 484): "Yet later, under the influence of his work with domestic animals and plants, he increasingly abandoned isolation as an important evolutionary factor (Darwin 1859) and his correspondence with Wagner in 1868 shows how hopelessly confused he had become by then." Darwin (1859) would have indeed observed the type of directional and disruptive selection necessary for divergence to occur with ongoing introgressive hybridization while analyzing data from animal and plant breeding experiments. That he drew the analogy between these findings and what could happen in nature would appear consistent with his overall conclusion that descent with modification was driven by the process of natural selection. In contrast, what seems more surprising was Mayr's (1963; p. 481) conclusion that "geographic speciation" (i.e., *allopatric speciation*) was the prevailing mode, not only for animals, but also for plants. Plants were, by the time of Mayr's writing, well known for sympatric speciation through introgressive hybridization and polyploid formation (e.g., Stebbins 1947, 1950, 1959; Anderson 1949). Recently, Bolnick and Fitzpatrick (2007) have drawn a similar conclusion concerning Mayr's viewpoint stating, "Mayr's biogeographic definition of sympatric speciation excludes any period of geographic isolation, however brief. If we apply an equally stringent definition to allopatric speciation requiring zero gene flow during speciation, we might find that mixed-geography speciation, in which both allopatry and gene flow contribute to divergence, is relatively common."

It can be debated whether or not Darwin or Mayr were "hopelessly confused" concerning the role of sympatric divergence, but if someone *was* confused,

there were two coincident factors that contributed to the confusion; biases arising from their conceptual frameworks and a lack of data to test rigorously their alternative hypotheses. All research scientists have biases that form constraints on how they view the world around them. In evolutionary biology one only has to consider the longstanding discussions concerning, for example, species concepts (e.g., Hey et al. 2003) or the best methodologies to construct phylogenetic frameworks (e.g., Ari et al. 2012) for illustrations of how different paradigms can lead to divergent conclusions. Likewise, until the advent of methods for analyzing large portions of the genomes of organisms, repetitive sequences such as transposable elements were thought to be "junk DNA" (Nowak 1994). Through comparative analyses these elements are now understood to have affected genomic and phenotypic evolution in many, if not all, clades of organisms (e.g., Nowak 1994; Britten 2006).

As with the earlier examples, the conceptual framework (i.e., modern synthesis) formalized during the middle decades of the twentieth century led to the conclusion that species could be defined solely by reproductive isolation and that the origin of reproductive barriers must occur in allopatry (Dobzhansky 1937; Mayr 1942, 1963, 2004). Figure 1.1 comes from Mayr's 1942 work, *Systematics and the Origin of Species*, and reflects his conception of the "stages" that lead to speciation. As can be seen, he proposed that allopatric divergence was key, and that if diverging lineages came back into contact before they were completely reproductively isolated, the process of speciation was not completed. Instead, there would be the formation of a hybrid zone between the partially reproductively isolated forms (Figure 1.1). Mayr (1942, p. 97) reflected this conclusion in the following manner: "Clines [e.g., hybrid zones] indicate continuities, but since species formation requires discontinuities . . . *The more clines found in a region, the less active is species formation*" (italics added by Mayr).

That allopatric speciation was the dominant paradigm to come out of the modern synthesis can also be exemplified by the response of Dobzhansky to a talk on sympatric divergence given by Guy Bush at the 1966 Society for the Study of Evolution meetings. In his talk Bush described his data on the True

Stage 1. A uniform species
with a large range

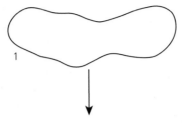

Followed by:
Process 1. Differentiation
into subspecies

Resulting in:
Stage 2. A geographically
variable species with a more
or less continuous array of
similar subspecies (2a all
subspecies are slight, 2b
some are pronounced)

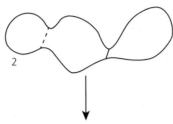

Followed by:
Process 2. a) Isolating action
of geographic barriers between
some of the populations;
also b) development of
isolating mechanisms in the
isolated and differentiating
subspecies

Resulting in:
Stage 3. A geographically
variable species with many
subspecies completely
isolated, particularly near the
borders of the range, and
some of them morphologically
as different as good species

Followed by:
Process 3. Expansion of
range of such isolated populations into the territory of
the representative forms

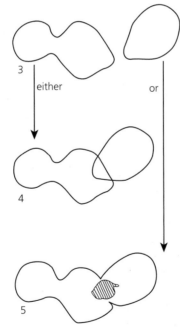

Resulting in either
Stage 4. Noncrossing, that is,
new species with restricted
range
or

Stage 5. Interbreeding, that
is, the establishment of a
hybrid zone (zone of secondary intergradation)

Figure 1.1 Ernst Mayr's (1942, p. 160) conception of the necessary stages for speciation. He hypothesized that the allopatric distribution of the diverging populations was essential and that if the divergent lineages expanded their ranges again before reproductive isolation was finalized, a hybrid zone would form indicating a lack of speciation.

Fruit Fly genus *Rhagoletis* (Bush 1969). At the end of the talk no one asked a question, but Dobzhansky, as moderator of the session stated, "That was an interesting story, but I don't believe it. Sympatric speciation is like the measles; everyone gets it and we all get over it" (Bush 1998). I would suggest that the conclusion drawn by Dobzhansky and others was strengthened by a lack of data to test the alternative hypotheses. In particular, those working on the modern synthesis had no, or very limited, population genetic (not to mention genomic-scale) data sets that could reveal whether or not exchange had occurred during the divergence of closely- or distantly-related lineages.

A recent review of studies that combined the types of genetic and genomic data sets not available during the modern synthesis, with new statistical methodologies, resulted in the following conclusion: "(a) a plurality of studies find low or zero gene flow, while (b) many studies do find evidence of nonzero gene flow (sometimes at quite high levels) between diverging populations or species" (Pinho and Hey 2010). Divergence-with-gene-flow was thus inferred in the evolutionary history of speciation for many of the examples considered by these authors. Furthermore, Pinho and Hey (2010) pointed to the crucial role of linkage disequilibrium (generated by divergent selection; see also Kirkpatrick and Ravigné 2002), if two lineages were to evolve separate trajectories in the face of introgression. In contrast, Via (2012) emphasized the process of "divergence hitchhiking" whereby recombination over large regions of the genomes of hybridizing individuals is reduced by strong divergent selection.

Regardless of the relative roles of linkage disequilibrium or divergence hitchhiking, the earlier models both lead to the prediction that, at least at the earlier stages of divergence, much of the genomes of the hybridizing taxa will likely be "permeable" to introgression. Reflecting this hypothesis, Nosil and Feder (2012) stated, "speciation initiated in the face of gene flow may often begin via divergence in the few specific gene regions directly subject to divergent selection." Thus, while various factors such as direct selection, linkage disequilibrium, trait associations, and genome hitchhiking may lead eventually to reproductively isolated lineages (Smadja and Butlin 2011; Nosil and Feder 2012), they are

sometimes accompanied by adaptive trait introgression and/or *hybrid speciation* (e.g., Arnold 2006; Arnold et al. 2015). In other words, speciation-with-introgression has the potential for producing not only two or more reproductively isolated lineages from a single progenitor through divergent selection, etc., but also novel combinations of adaptive traits and new, hybrid lineages (Lucek et al. 2014). Reviews of the genomic characteristics of speciation, using data generated from *next-generation sequencing approaches* (or "NGS"), led both Sousa and Hey (2013) and Seehausen et al. (2014) to a similar conclusion. The former authors observed that "NGS data have not yet changed our main paradigm of how populations diverge, but they have confirmed that natural selection is sometimes in conflict with gene exchange during the divergence process and that gene flow is a widespread process" while the latter argued that for certain examples "the balance of evidence from NGS data implies introgressive hybridization rather than standing genetic variation as the source of ancient alleles . . . Speciation in these cases might have been facilitated by hybridization that provides genetic material for both adaptation and reproductive isolation in the face of gene flow."

1.3 Hybrid zone studies: Theoretical foundations

In his 1938 *Nature* article, Julian Huxley coined and defined the term cline: "Some special term seems desirable to direct attention to variation within groups, and I propose the word cline, meaning a gradation in measurable characters. This, being technical, seems preferable to such a term as character-gradient or phrases such as 'geographical progression of characters'" A decade later, Haldane (1948) developed the theoretical considerations of clinal variation, especially in the case in which you have two phenotypes favored differentially in two environments, leading to the formation of *natural hybrids* and a *hybrid zone*. He applied his formulae to the case of the deer mouse, *Peromyscus polionotus*. Taking advantage of data from studies by Sumner (1929a, b), Haldane (1948) used his mathematical formulation of cline theory to estimate the effects of the intensity of selection (in this case favoring different coat coloration on lighter or darker soil types; Sumner 1929a, b) and the mean distance migrated per generation on the slope of the clinal changeover across the landscape. Haldane (1948) concluded from his analysis that "a selective advantage of about 0.1% on each side of the boundary [between the lighter and darker coat color phenotypes] would be sufficient to account for the observed cline."

Extensions of Huxley's and Haldane's work continued throughout the twentieth century, particularly in the research of Slatkin, May, Endler, and Barton. However, the conceptual frameworks emphasized by these various workers were quite different. Slatkin (1973), May et al. (1975), and Endler (1973, 1977) emphasized the importance of clinal variation in the habitat leading to differential selection on genotypes (e.g., non-hybrid and hybrid) as a causal factor in structuring the phenotypic and genotypic clines described, for example, by Haldane (1948). In contrast, Barton and his colleagues argued for the overwhelming effect of selection against hybrids, balanced by continual migration of parental individuals into the areas of overlap between the hybridizing taxa, as the cause of hybrid zone structure (e.g., Barton 1979a, b, 1980; Barton and Hewitt 1985). Hybrid zones characterized by the parameters developed by Barton et al. were termed "dynamic equilibrium" or "tension" zones. The assumptions of Barton et al. (e.g., Barton and Hewitt 1985) were that hybrids could be considered uniformly unfit thus resulting in an environment-independent model. Likewise, Howard and Harrison, though emphasizing the more realistic "mosaic" distribution of different habitats, included the assumption of uniformly lower fitness of hybrids across all environments when formulating their "mosaic hybrid zone model" (Howard 1982, 1986; Harrison 1986, 1990).

In contrast to either the tension zone or mosaic models, Endler (1977) and Moore (1977) concluded that differential selection gradients, driven by environment-dependent selection, were the factors responsible for the positioning and overall shape of the clinal variation. In particular, Moore (1977), Moore and Buchanan (1985), and Moore and Price (1993) inferred increased hybrid fitness within hybrid zones (i.e., the "bounded hybrid superiority"

model) as causal in determining hybrid zone structure. Arnold (1997), following on from the bounded hybrid superiority model, suggested that not only could hybrids be more fit in the [often] ecotonal habitats between the hybridizing taxa, but also in parental niches. This latter conclusion was based both on the observation of introgression well outside animal and plant hybrid zones and direct estimates of hybrid fitness (Arnold 1992, 1997; Arnold and Hodges 1995). His "evolutionary novelty" model predicted the occurrence of elevated hybrid fitness across different habitats and thus adaptive trait introgression and the establishment of hybrid species (Arnold 1997, pp. 147–154).

1.4 Hybrid zone studies: Testing the models

One of the most important observations made by Moore and his colleagues was that the application of the mathematical models underlying tension zones and hybrid superiority zones would not distinguish whether the zone being analyzed fitted one paradigm or the other (Moore and Price 1993). In other words, the formulae would result in overlapping predictions, preventing a test of whether a hybrid zone was maintained by uniform, *environment-independent* selection against hybrids or by differential *environment-dependent* selection favoring hybrid and parental genotypes in different habitats (Moore and Price 1993). These authors argued that the most definitive means to distinguish between these two models was to use reciprocal transplant experiments to test directly whether hybrids were more or less fit than parental genotypes across environments. Arnold and Hodges (1995) and Arnold and Martin (2010) reviewed data from reciprocal transplant analyses and other approaches to estimate hybrid fitness, and found that hybrid genotypes do indeed possess a range of fitnesses relative to their progenitors (see Chapter 5).

1.4.1 Hybrid zones and the form of selection: Voles and spruce trees

Two recent examples of hybrid zone analyses (Beysard and Heckel 2014; De La Torre et al. 2014b) can be used to illustrate the strengths and limitations of various approaches for deciphering the form of selection in hybrid zones. These alternative approaches reflect on the one hand, the utilization of cline theory to infer the types of selection occurring (Beysard and Heckel 2014), while on the other hand, the pairing of direct and indirect estimates of fitness of hybrid and parental genotypes to estimate the form of selection (De La Torre et al. 2014b).

In their analysis of hybridization among three divergent lineages of the vole, *Microtus arvalis*, Beysard and Heckel (2014) used a combination of Bayesian admixture models (i.e., "structure": Pritchard et al. 2000 and "newhybrids": Anderson and Thompson 2002) and the cline model of Szymura and Barton (1986) to both define hybrid and parental genotypes and infer the form of selection in the various hybrid zones. Taken together, these analyses led Beysard and Heckel (2014) to conclude that there was both neutral introgression of some markers, and selection against hybrid genotypes. Specifically, they stated, "Narrow coincident and concordant clines are the consequence of strong reproductive barriers between organisms (Barton and Hewitt 1985), but the wider clines in the Central–Eastern hybrid zone . . . may not necessarily be a consequence of weaker reproductive barriers compared with the Western-Central zone" (see Figure 1.2). Though their inferences regarding the extent and directionality of hybridization and introgression are strong, Beysard and Heckel's (2014) conclusions concerning the form of selection

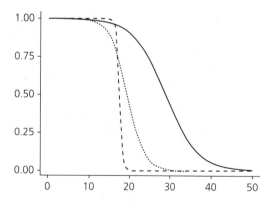

Figure 1.2 Frequency changeover and cline shape across a hybrid zone between divergent lineages of the vole, *Microtus arvalis*, for mtDNA (solid line), autosomal (short dashed line), and Y-chromosome (long dashed line) genes (from Beysard and Heckel 2014).

(i.e., "against male hybrids") must be considered much less certain. As stated earlier, Moore and Price (1993) demonstrated that the application of the models of Barton and his colleagues, or of Moore and his, could not distinguish between selection for or against hybrid and parental genotypes across hybrid zones. Such inferences must come from other lines of evidence.

As with the analysis of the vole hybrid zones, De La Torre et al. (2014b) first estimated admixture and patterns of introgression across zones of hybridization between white spruce (*Picea glauca*) and Engelmann spruce (*Picea engelmannii*). However, unlike the vole study, De La Torre et al. (2014b) also tested the assumptions of the environment-independent and environment-dependent hybrid zone models. First, they asked whether or not there were associations between the genotypic clines and the environmental variability across the parental habitats and the hybrid zones. This analysis revealed strong genotypic clines that were associated with "geographical and climatic gradients" (De La Torre et al. 2014b). In addition to this descriptive analysis, these workers also incorporated data for fitness components (i.e., height, survival, cold hardiness, bud burst, and bud set) across environments for both parental and hybrid genotypes. Taken together, mapping the genotypic clines on the habitats and directly estimating parental and hybrid fitness across the same habitats falsified the hypothesis that the spruce hybrid zones were maintained by hybrid inferiority and continued hybridization (i.e., the assumptions of a tension zone model). Instead, De La Torre et al. (2014b) concluded, "Our results indicate that the *P. glauca* x *P. engelmannii* hybrid zone is maintained by elevational climatic selection gradients resulting from environmental heterogeneity (exogenous selection), where hybrids are fitter than parental species in intermediate habitats."

De La Torre et al. (2014b), like Beysard and Heckel (2014), found abrupt clines in the genotypes associated with parentals and hybrids. Without estimates of genotype × environment associations and genotype × environment fitnesses, they too could have used the clinal variation to conclude that hybrids were uniformly selected against. To reiterate, the steepness of the clinal variation across hybrid zones does not discriminate between the models that assume reduced or elevated hybrid fitness (Moore and Price 1993). Because De La Torre et al. (2014b) collected the data necessary to test the relative fitnesses of hybrids and parental genotypes, they were able to test the assumptions of both the tension zone and bounded hybrid superiority models, resulting in the falsifying of the former.

1.4.2 Hybrid zones and the form of selection: Next-generation sequencing and genome scans in salamanders and butterflies

Though Moore and Price (1993) concluded that the most rigorous tests of hybrid zones would come from reciprocal transplants, they were knowledgeable field biologists and thus understood that such analyses were only likely to occur with some plant systems. Direct estimates of hybrid and parental fitness, across habitats, remain one of the most powerful means by which hybrid zone models can be tested (e.g., Grant and Grant 1992; Arnold and Bennett 1993; Arnold and Martin 2010). However, the pairing of both older and newer analytical methodologies (e.g., Long 1991; Gompert and Buerkle 2011; Nadeau et al. 2013) with the large-scale data sets generated through next-generation sequencing make possible an additional approach for those plant and animal taxa not amenable to reciprocal transplant experiments.

Fitzpatrick et al. (2009, 2010) tested for mosaicism in the genomes of hybridizing salamander species from the genus *Ambystoma*. Their analyses included data from historical records indicating that natural hybridization between the native salamander species (*Ambystoma californiense*) and an introduced form (*A. mavortium*) began during the 1940s. The short time period since the formation of the hybrid zone between these species allowed a test for rapid evolutionary change as reflected in deviations from expected allele frequencies. Such deviations, if present, would be indicative of the impact of selective and/or stochastic factors on different genomic regions in the hybrid animals. The analysis of 64 loci from animals collected inside the hybrid zone detected *mosaic genomes* containing both conspecific and heterospecific alleles. Both stochastic and deterministic processes were inferred as causal in producing the specific hybrid genomes (Fitzpatrick et al. 2009, 2010). However, the "most striking

result" (Fitzpatrick et al. 2009) from the analysis of hybrid *Ambystoma* populations was that three out of the 64 loci were fixed, or nearly fixed, for the non-native alleles. The observation that the frequencies of these three non-native alleles were elevated across all five of the ponds sampled substantiates the inference of adaptive introgression (Fitzpatrick et al. 2009, 2010). At the other 60 loci, one genomic region revealed evidence for selection against heterozygotes, while the remaining 60 markers did not deviate from neutral expectations (Fitzpatrick et al. 2009).

As with the inferences drawn from the scans of the *Ambystoma* genomes, both Martin et al. (2013) and Nadeau et al. (2013) detected genomic regions of various taxa belonging to the butterfly genus *Heliconius* that were more or less likely to be affected by introgression. For example, comparisons of genomic variability in allopatric and sympatric populations of *Heliconius melpomene*, *Heliconius cydno*, and *Heliconius timareta* "revealed a genome-wide trend of increased shared variation in sympatry, indicative of pervasive interspecific gene flow" (Martin et al. 2013). These workers also inferred that this wide-scale introgression began over 1 million years ago and was continuing in current sympatric regions (Figure 1.3). A portion of the "pervasive" interspecific transfer was inferred to involve adaptive trait introgression. Specifically, introgression of genes controlling wing color patterns was detected (Nadeau et al. 2013; Arias et al. 2014). These color patterns are a component of the classic example of Müllerian mimicry found in this species complex and have been shown previously to transfer across hybrid zones between *Heliconius* species (e.g., Pardo-Diaz et al. 2012). Finally, though introgression was inferred to be genomically and temporally extensive, some genomic regions demonstrated limited transfer across all time horizons. In particular, very low rates of introgression of Z-chromosome genes were detected (Figure 1.3; Martin et al. 2013), as expected due to the presence of genes known to cause female hybrid sterility (Jiggins et al. 2001).

1.4.3 Hybrid zones and the form of selection: A genomic outlook

The examples in Section 1.4.2 emphasize the need for additional knowledge concerning the causes of

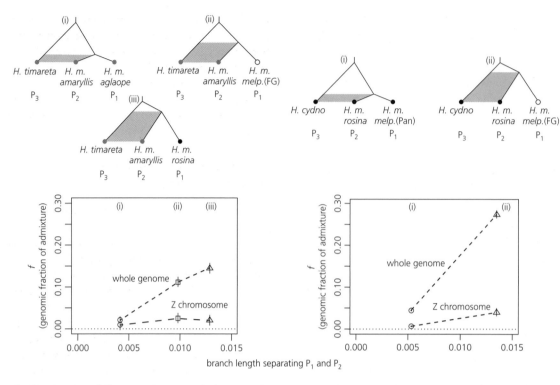

Figure 1.3 Estimated admixture at three time scales between *Heliconius timareta* and *Heliconius melpomene amaryllis* (i-iii) and two time scales between *Heliconius cydno* and *Heliconius melpomene rosina* (i-ii) for the whole genome and Z-chromosome genes. As expected, the Z-chromosome loci show much less introgression across all time periods (from Martin et al. 2013).

the genotypic (and thus phenotypic) structure in hybrid zones. Most of these analyses utilize some form of cline analysis, reflecting the continuing reliance on the theory generated by Huxley and Haldane, as greatly extended by Slatkin, May, Endler, Barton, Moore, and their colleagues. The continual utility of cline theory in exploring the possible forms of selection occurring in cases of introgressive hybridization is obvious. However, it also remains necessary to urge caution when drawing conclusions from such analyses. Furthermore, though some of the previous studies have falsified one model over another (e.g., the tension zone model in the Spruce hybrid zones) at what could be termed a macro-genomic level (i.e., finding indications of hybrid superiority), the availability of extensive genomic data seems to argue for a more refined approach in which individual loci are tested for clinal variation indicative of neutrality or selection for or against hybrids.

Figure 1.4 illustrates the potential illustrative value of such analyses. This figure provides a hypothetical example of mapped loci on a single linkage group across a hybrid zone. If the genome is found to be "semi-permeable" (Key 1968), with introgression being neutral, selected against, or selected for at different loci, we would expect cline shape derived from different loci to falsify or support alternative models (e.g., tension zone versus bounded hybrid superiority; Figure 1.4). Gompert and Buerkle (2009, 2011) have generated methodologies for "genomic cline analyses" from both an analytical (Gompert and Buerkle 2009) and Bayesian (Gompert and Buerkle 2011) approach. These analyses allow descriptions of loci, and genomic regions, that demonstrate patterns of introgression that do or do not differ from neutral expectations (Gompert and Buerkle 2009, 2011), with applications to animal hybrid zones suggesting such differential selection on the loci examined (Gompert and Buerkle 2009, 2011).

Though approaches such as those of Gompert and Buerkle (2009, 2011) allow a dissection of

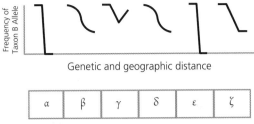

Frequency of Taxon B Allele

Genetic and geographic distance

| α | β | γ | δ | ε | ζ |

Linkage group I containing gene loci α-ζ

Figure 1.4 Hypothetical example of genomic and geographical clinal variation between two Taxa (A and B) at six loci (α-ζ) located on a single linkage group (I). The various clines reflect frequency changeovers both across the spatial distribution of the hybrid zone between Taxa A and B (as reflected by change in frequency of Taxon B alleles) and across the genomic region occupied by the various genes. The steep clinal variation revealed for genes α and ε would usually be interpreted as reflecting negative selection against hybrids (i.e., heterozygotes) thus preventing introgression of the alternate alleles (1) into the genomes of the alternate taxon and (2) across the hybrid zone. The less abrupt changeovers in allele frequencies resolved at genes β, δ, and ζ might best fit a neutral model of diffusion between the genomes of Taxa A and B and across the hybrid zone. Finally, the persistence and increase in frequency of the allele from Taxon B at gene γ across the hybrid zone and genome, reflecting its introgression into Taxon A, would suggest selection favoring hybrid genotypes and adaptive trait introgression at this locus.

patterns of clinal variation across the genome, it still must be asked whether the pattern of sharp transitions reflects reproductive isolation at particular loci or genomic regions (Figure 1.4), or rather elevated hybrid fitness in certain habitats (e.g., within the hybrid zone). Furthermore, although the transfer of alleles from one of the hybridizing forms to another (above the frequency of the neutral expectation) suggests the occurrence of adaptive trait introgression, stochastic processes may also generate such patterns (Figure 1.4; Gompert and Buerkle 2011). It would thus seem necessary to return to the conclusion that testing rigorously for the form of selection affecting both overall hybrid zone structure and individual loci across a hybrid zone will require additional information indicating the affect of environmental variation on the fitness of hybrid and parental genotypes.

In the following sections, the examples chosen will highlight instances where combinations of findings have indeed provided rigorous hypothesis testing for the forms of selection present in cases of divergence-with-gene-flow. In particular, the results

from a wide array of studies will illustrate the similarity of outcomes (e.g., adaptive trait transfer and hybrid lineage formation) found in both plants and animals. Finally, examples will be discussed to demonstrate that genetic exchange can occur not only between evolutionarily related lineages, but also between lineages belonging even to different domains of life.

1.5 Adaptive genetic exchange in animals and plants

One of the challenges to overcome for those who argue that genetic exchange plays a significant role in the evolution of organisms has been the assumption that the vast majority of hybrid genotypes are less fit than within-species progeny. Of course it should be pointed out that evolutionary change, as envisioned from the neo-Darwinian synthesis onward, was seen as resting on very rare events—i.e., mutations that led to increased, rather than decreased, fitness. Therefore, "rarely fit" hybrids, produced at a much higher frequency than point mutations, should logically be considered as evolutionarily very important.

Mayr (1963) set out clearly the predominant viewpoint concerning hybrid fitness that began at least with Darwin, was codified during the modern synthesis, and has continued through the present-day (e.g., Coyne and Orr 2004). Mayr thus concluded, "The majority of such hybrids are totally sterile . . . Even those hybrids that produce normal gametes in one or both sexes are nevertheless unsuccessful in most cases and do not participate in reproduction. Finally, when they do backcross to the parental species, they normally produce genotypes of inferior viability that are eliminated by natural selection" (Mayr 1963, p. 133). In Chapter 5 I will discuss at length evidence that falsifies such a conclusion, however, one prediction proceeding from Mayr's argument is that adaptive genetic exchange (i.e., adaptive trait transfer, adaptive trait introgression) could not occur. By definition, for genetic exchange to be adaptive, some genotypes must demonstrate enhanced fitnesses because of the combination of genes received from their parents, i.e., because they are "hybrids" (e.g., Whitney et al. 2006, 2010; Trucco et al. 2009; Dunn et al. 2013; Baldassarre et al. 2014; Ding et al. 2014a; Fraïsse et al. 2014; Llopart et al. 2014; Vernot and Akey 2014). The following examples will provide evidence to illustrate the role of adaptive trait transfer in both animals and plants.

1.5.1 Adaptive trait transfer: Dogs, wolves, and coat color genes

The evolution of mammalian pigmentation is often described as adaptive (e.g., Sumner 1929a, b), with some of the determining genetic factors belonging to the melanocortin pathway (e.g., Hoekstra et al. 2006). Mutations in the *Melanocortin 1 receptor* (*Mc1r*) gene commonly occupy a causal role in variations in the pigmentation of animals. In contrast to other species complexes, Anderson et al. (2009) defined a melanocortin pathway component, the *K* locus, to be causal in pigment variation in North American gray wolves (i.e., *Canis lupus*). The variation in pelage color across the geographical and ecological range of these animals extends from pale wolves in open tundra regions to darker individuals in forested areas. The habitat associations of paler- and darker-colored wolves in areas with more or less ambient light, respectively, suggested a possible selective constraint on these different

pelages. An investigation of the allelic variability found at the *K* locus supported the hypothesis of positive selection favoring different alleles in different habitats. Specifically, Anderson et al. (2009) detected very low haplotype diversity at this locus, and yet a high frequency of alleles that cause darker pelage in the forested areas. Significantly, Anderson et al.'s (2009) findings pointed to introgression from domestic dogs as the source of the alleles causing darker pelage in North American wolves (Figure 1.5).

The geographical distribution and ecological associations of the melanistic variant in both domestic dogs and wild canid populations supported the hypothesis that the direction of introgression was from domestic dogs into wild canids. First, the *K* locus allelic variant that causes dark pelage is found across dog breeds, including basal lineages from Asia and Africa. This contrasts sharply with the melanistic variant being identified only in wild canid populations in North America, with the exception of some populations of Italian wolves known to have hybridized recently with European domestic dog populations (Anderson et al. 2009). Similarly, the introgression event affecting the coat color in North

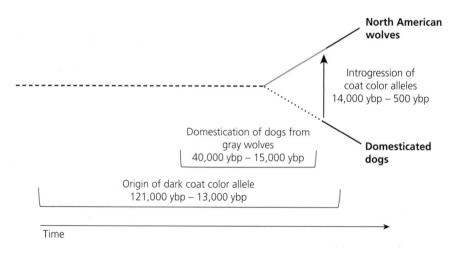

Figure 1.5 Putative events leading to the adaptive trait introgression of the melanistic coat color alleles at the *K* locus from domestic dogs into North American gray wolves. The black, dashed lines before and after the divergence of the wolf and domestic dog lineages reflect when these allelic variants may have arisen. The gray section of the wolf lineage reflects the hypothesis that, if the melanistic *K* locus variants arose before the divergence of dogs and wolves from a common ancestor, they were subsequently lost in the wolf lineage. The dark segment at the tip of the wolf lineage indicates the approximate timing of the introgression of the dark coat color alleles from domestic dogs into North American wolves, after dogs were brought into North America by migrating humans (from Anderson et al. 2009).

American gray wolves would have occurred some time since the former migrated with humans into North America (i.e., *c.* 12,000 ypb; Figure 1.5).

In addition to the correlation of habitats with introgressed and non-introgressed wolf populations, testing the adaptive hypothesis for the pelage trait introgression was also made possible by haplotype analyses in domestic dogs, North American wolves and coyotes, and Italian wolves. Haplotypes associated with dark pelage in all of the domestic and wild canid samples clustered into a well-defined group (Anderson et al. 2009), suggesting a common origin. In contrast, non-melanistic alleles most often grouped together by species. This phylogenetic pattern suggests that the alleles causing darker pelage originated in domestic dogs and subsequently introgressed into not only North American wolves, but also Italian wolves and coyotes (Anderson et al. 2009).

1.5.2 Adaptive trait transfer: Louisiana irises

Genomic and ecological analyses of Louisiana iris species (so named because the species belonging to the complex overlap in the state of Louisiana, and form a wide variety of natural hybrids—Small and Alexander 1931; Viosca 1935; Riley 1938; Anderson 1949) have also tested the hypothesis that introgression in nature sometimes results in the adaptive transfer of traits. For example, though sometimes overlapping near bayous and swamps, *Iris fulva* and *Iris brevicaulis* exhibit habitat differentiation. Thus, *I. fulva* typically grows along the edges of bayous and swamps, with its rhizomes submerged in water, whereas *I. brevicaulis* generally occurs at slightly higher elevations in mixed hardwood forest (Viosca 1935; Cruzan and Arnold 1993; Johnston et al. 2001). Martin et al. (2005, 2006) used analyses in both the greenhouse and natural settings to identify QTL affecting survivorship in different habitats. These analyses involved *I. fulva, I. brevicaulis, I. fulva* first-generation backcross, and *I. brevicaulis* first-generation backcross genotypes. Estimates of fitness of hybrid and parental fitness, across years and habitats, were possible because Louisiana irises are long lived and can be replicated clonally by subdividing rhizomes of a given genotype.

The first analysis examined the genetic architecture associated with the survivorship of hybrid genotypes

in a greenhouse setting (Martin et al. 2005). Although well watered, the *I. fulva* first-generation backcross hybrid genotypes were not exposed to standing water as is typical for *I. fulva* populations (Viosca 1935; Johnston et al. 2001). Inferences from this long-term survivorship analysis falsified a neutral model of introgression and strongly supported adaptive trait introgression between these two iris species. Specifically, the first-generation backcross genotypes in the direction of the dry-adapted *I. brevicaulis* (Viosca 1935) survived at a significantly higher frequency than did those in the direction of the wet-adapted, *I. fulva* (Martin et al. 2005). Furthermore, as expected from the occurrence of adaptive trait transfer, introgression of three genomic regions from *I. brevicaulis* was significantly associated with increased survivorship in the *I. fulva* backcross genotypes.

A second analysis involved genotypes from the same parental and hybrid (i.e., backcross populations), but in this case the genotypes were placed into southern Louisiana in regions occupied by native *I. fulva, I. brevicaulis,* and natural hybrids (Viosca 1935; Arnold 1993; Johnston et al. 2001). The habitats selected for transplantation included hardwood forests (*I. brevicaulis*-like habitats; Viosca 1935; Cruzan and Arnold 1993; Johnston et al. 2001) and bayous (*I. fulva*-like habitats; Viosca 1935; Cruzan and Arnold 1993; Johnston et al. 2001). Yet, within one month, the entire transplant area was inundated with *c.* 1 m of water that remained for several months (Martin et al. 2006). The resulting percentage survivorship reflected a cline among four genotypic classes: *I. brevicaulis* (0%), first-generation backcrosses toward *I. brevicaulis* (5.5%), first-generation backcrosses toward *I. fulva* (9%) and *I. fulva* (27%). Significantly, the *surviving* genotypes of the backcross generation toward *I. brevicaulis* contained significantly more alleles from *I. fulva* than those genotypes from this backcross population that did not survive the flooded conditions. These results supported the hypothesis that adaptive trait introgression had provided some hybrid genotypes with the ability to survive this extreme (i.e., flooded) environment (Martin et al. 2006).

A third analysis by Martin et al., involving pollinator interactions with the same parental and backcross genotypes used previously, also revealed evidence for adaptive trait introgression. In this third study, Martin et al. (2008) defined a complex genetic architecture of QTLs that caused genotypes to be differentially attractive to three pollinator classes (i.e., bumblebees, butterflies, and hummingbirds). The patterns suggesting selection acting on hybrid genotypes were similar to those found in the first two analyses. Specifically, *I. brevicaulis* genotypes were never visited by butterflies, and almost completely avoided by hummingbirds as well (Figure 1.6; Martin et al. 2008). In contrast, backcross hybrids toward *I. brevicaulis* were significantly more attractive to both butterflies and hummingbirds

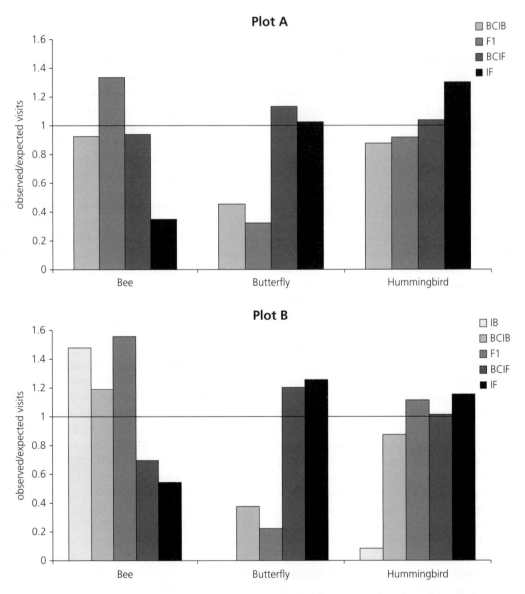

Figure 1.6 Frequency of observed pollinator visits to I. brevicaulis (IB), I. fulva (IF), F₁ hybrids (F₁), backcrosses toward I. brevicaulis (BCIB), and backcrosses toward I. fulva (BCIF). The expected pollinator visits were summed throughout the experiment in two different transplant plots—Plot A and Plot B. A ratio larger than one (horizontal line) indicates a higher proportion of visits relative to that expected, a ratio less than one indicates a smaller proportion of visits relative to that expected, whereas a ratio approaching one indicates that observed and expected visits were relatively equivalent (from Martin et al. 2008).

(Figure 1.6). It therefore appeared that alleles from *I. fulva* introgressed into *I. brevicaulis*, like the coat color alleles from domestic dogs introgressed into North American wolves, provided a selective benefit to the resulting hybrid genotypes (Figure 1.6). These conclusions, also like the analyses of domestic dogs and wolves, agree with genotype × environment associations in *natural* hybrid populations in which *I. brevicaulis*-like hybrid genotypes, introgressed with *I. fulva* genomic segments, occur in flooded habitats (Johnston et al. 2001; Hamlin and Arnold 2014).

1.5.3 Adaptive trait transfer: Darwin's finches

Animal species complexes inferred to have had genetic exchange events during their evolutionary history have rarely been studied using both long-term ecological and genetic analyses. Thus, there are groups for which genomic data have provided clear evidence for adaptive trait introgression (e.g., Fitzpatrick et al. 2010; Pardo-Diaz et al. 2012; Boratyński et al. 2014), yet which do not have multiyear ecological data to test for genotype × environment interactions. There is, however, an example from the zoological literature that does combine both types of information and remains the gold standard for studies of this type—the 40-year research program on the Darwin's finches by Peter and Rosemary Grant and their colleagues.

In hindsight, the fact that the genus *Geospiza* (i.e., Darwin's finches) had a reticulate evolutionary history was recognizable from observations collected by Darwin and others. For example, in his book *The Voyage of the Beagle* (1845), Darwin described variation that would be expected from introgressive hybridization by observing that "The most curious fact is the perfect gradation in the size of the beaks in the different species of *Geospiza*, from one as large as that of a hawfinch to that of a chaffinch, and . . . even to that of a warbler" Approximately 80 years after Darwin's historic voyage, Lowe (1936) recognized the potential role of introgressive hybridization in this group and concluded "in the finches of the Galapagos we are faced with a swarm of hybridization segregates."

Even though there were prior hints concerning the effect of hybridization within the *Geospiza* clade,

the Grants and their colleagues have revealed not only the presence of introgressive hybridization in this bird clade, but also most importantly estimated its relative role in producing and transferring adaptations between the various species. By using long-term ecological, life history, and genetic analyses, they were able to infer the following: (1) habitat shifts due to climatic perturbations resulted in an increase in the fitness of some hybrid genotypes due to environmental selection (Grant and Grant 2010, 2014a); (2) human-mediated habitat modifications also led to increases in hybrid fitness, in some situations leading to a lessening of reproductive isolation between species (Grant and Grant 2010; see also De León et al. 2011); (3) convergence in genotypes and phenotypes occurred over the span of a few decades due to the combinatorial effect of introgression and natural selection (Grant et al. 2004; Grant and Grant 2008); and (4) some hybrid genotypes/phenotypes formed the basis for new, reproductively isolated lineages (Grant and Grant 2009, 2014b). Darwin's finches clearly reflect a species complex marked by web-of-life processes (Farrington et al. 2014; Lamichhaney et al. 2015).

1.5.4 Adaptive trait transfer: Senecio and the origin of floral traits

Studies involving members of the plant genus *Senecio* have provided a rich lode of data for testing models concerning the role of reticulate evolution. A diverse array of outcomes from natural hybridization, both recent and ancient (James and Abbott 2005; Pelser et al. 2010, 2012; Abbott and Brennan 2014), including homoploid and allopolyploid hybrid speciation (James and Abbott 2005; Kadereit et al. 2006), genomic modifications (Hegarty et al. 2011), transgressive phenotypic variation (Brennan et al. 2012), and adaptive introgression of genes affecting floral traits (Kim et al. 2008) have been inferred.

As with the other cases of putative adaptive trait introgression discussed earlier, the transfer between *Senecio* lineages reflects an event that had to pass through a fitness bottleneck (i.e., low fitness of initial hybrid generations) in order to occur. Yet, as pointed out previously (Arnold et al. 1999), the relative rarity of early generation hybrid progeny is not a predictor of the evolutionary importance of genetic exchange. Put another way, an extremely high amount of estimated reproductive isolation (e.g., Ramsey et al. 2003) is *not* inversely related to the likelihood of introgression, hybrid speciation, and/or *adaptive radiations* (e.g., Yatabe et al. 2007).

The earlier conclusion is reflected well by the sequence of events leading to the transfer of adaptive floral traits (i.e., ray florets resulting in large petals attractive to pollinators) from the diploid, *S. squalidus* into the allotetraploid, *Senecio vulgaris*. The production of petals is seen as adaptive in that it promotes outcrossing thus circumventing deleterious effects from inbreeding (Kim et al. 2008). Though the viability of triploid, hybrid progeny produced from crosses between these species is extremely low (<0.02%; Lowe and Abbott 2000), natural introgressive hybridization has resulted in the introduction of a set of regulatory genes named the "*Ray* locus" (Kim et al. 2008). The significance of this reticulate evolutionary event was emphasized in the following conclusion: "Reintroduction of genes that promote outcrossing may therefore allow a self-pollinating species to revert and prevent extinction in the longer term. Our results therefore highlight the interplay between regulatory genes, development, and life history, and show how gene transfers between species may play an important part in the evolution of key ecological and morphological traits" (Kim et al. 2008).

1.6 The origin of hybrid taxa in animals and plants

In Chapter 6, numerous instances of diversification involving genetic exchange will be discussed for organisms as diverse as mammals and viruses (e.g., Larsen et al. 2010; Xu et al. 2013; Melo-Ferreira et al. 2014a). In the present section I will briefly illustrate the potential for reticulate evolution to lead to diversification by discussing a handful of examples from animal and plant species complexes. In particular, illustrations of both homoploid and allopolyploid diversification will be presented, some of which are associated with adaptive radiations. As with adaptive genetic exchanges, the recognition of hybrid lineage formation has been less well appreciated for animal clades compared to plant groups

(Arnold 1997, 2006; Mallet 2007). However, with the increasing interest in "divergence-with-geneflow," and the multiplication of various types of data sets (e.g., morphological data from both contemporary and fossil populations: Ackermann and Bishop 2010, Ackermann et al. 2010, 2014; and molecular data: Monzón et al. 2014; Willis et al. 2014), it has become apparent that zoological examples of reticulate evolution, like those from the botanical literature, are extensive both in number and taxonomic distribution (Arnold 2006, 2009; Arnold et al. 2015). As discussed in Section 1.6.1, both allopolyploid and homoploid hybrid speciation are examples of sympatric speciation, and as such, the reproductive isolation of the portions of the genome of the new hybrid lineage (as indeed with any species) that encode the novel phenotype, sets of adaptations, etc., is key to the stabilization of the new species. The parenthetical portion of the preceding statement is very significant. For biodiversification to occur, the genomic components that make any species look or behave as they do are those that are resistant to recombination during genetic exchange. When recombination does occur in the regions containing "speciation islands" and/or "adaptation genes," a number of outcomes are possible (Arnold 1992); two alternative outcomes are the loss of one of the hybridizing lineages through genetic assimilation and adaptive trait introgression. The following examples illustrate some of the mechanisms by which hybrid species arise and are maintained.

1.6.1 The origin of hybrid animal taxa: Swallowtail butterflies

Butterflies belonging to the New World tropical genus *Heliconius* are a classic example of the role of reticulate evolution in biodiversification and adaptation (see Section 1.4.2 and Jiggins et al. 2008). However, other groups of lepidopterans also demonstrate the occurrence of apparent adaptive, hybrid speciation (e.g., Gompert et al. 2006, 2014; Nice et al. 2013); one such species complex belongs to the North American swallowtail butterfly genus *Papilio*.

Kunte et al. (2011) and Zhang et al. (2013) defined the genetic makeup and adaptive consequences of admixture in the formation of *Papilio appalachiensis* from natural hybridization between *P. glaucus* and

P. canadensis. A hybrid zone, extending between the states of Wisconsin and New York, exists between *P. glaucus* and *P. canadensis* (Winter and Porter 2010). The position and structure of the hybrid zone is apparently maintained by temperature, with *P. glaucus* being adapted to warmer habitats relative to *P. canadensis* (Kunte et al. 2011). What has also become apparent from numerous studies is that genes located on the sex chromosomes of *Papilio* affect many of the adaptive traits that differentiate *P. glaucus* and *P. canadensis*. For example, Hagen and Scriber (1989) demonstrated that genes controlling pupal diapause and female wing color polymorphism were sex-linked (male lepidopterans are ZZ and females are ZW; Hagen and Scriber 1989).

Given the ecological differentiation of the parental species, it was significant that Kunte et al. (2011) found that the formation of the homoploid hybrid species, *P. appalachiensis*, resulted in an admixture of genes located on the Z- and W-chromosomes controlling a suite of adaptive traits. Specifically they demonstrated that "Two sets of traits define *P. appalachiensis'* hybrid phenotype: like *P. canadensis* it inhabits a cold habitat and has a single generation every year, and like *P. glaucus* it mimics a toxic butterfly and its females are dimorphic. The genes responsible for these traits are on two different sex chromosomes. Our genetic data show that *P. appalachiensis* inherited the sex chromosome [i.e., Z-chromosome] associated with the cold habitat from *P. canadensis*, whereas it inherited the sex chromosome [i.e., W-chromosome] associated with mimicry and dimorphism from *P. glaucus*."

Consistent with Kunte et al.'s (2011) conclusions, a transcriptomic analysis by Zhang et al. (2013) detected an enrichment of genes for biological functions including pigmentation, hormonal sensitivity, developmental processes, and cuticle formation, the latter trait likely being important in determining thermal tolerances (Zhang et al. 2013). Also in agreement with the adaptive nature of this homoploid hybrid species formation in *Papilio*, Kunte et al. (2011) proposed a role for climate change during the last interglacial period in North America in the origin of *P. appalachiensis*. Thus, they suggested a model in which, "During one of the late Pleistocene glacial retreats, *canadensis* populations retreated from their southern range while *glaucus* populations advanced northward and upward into the mountains. The changing thermal landscape likely brought the advancing *glaucus* populations into contact with a relict *canadensis* population in the Appalachian Mountains."

1.6.2 The origin of hybrid animal taxa: Sparrows

From a review of known examples, Mallet (2007) concluded, "Homoploid hybrid speciation or recombinational speciation is well known in flowering plants. Speciation takes place in sympatry (by definition, as hybridization requires gene flow). Hybrids must then overcome chromosome and gene incompatibilities, while lacking reproductive isolation via polyploidy . . . Although bisexual polyploids are often barred in animals, there is no reason why homoploid hybrid species would be rarer in animals than in plants." Indeed, since Mallet's (2007) review appeared, a number of new studies of animal, homoploid hybrid speciation have been reported; one example of this process is reflected by European sparrow lineages.

The Italian sparrow, *Passer italiae*, is part of a species complex that also includes the house sparrow, *P. domesticus*, and the Spanish (or Willow) sparrow, *P. hispaniolensis*. The evolutionary history of this complex has long been hypothesized to involve reticulate evolution (see Johnston 1969, and earlier references cited therein). Johnston (1969) presented a detailed morphological analysis of specimens from the entire range of the Italian Sparrow, and selected populations of both the house and Spanish sparrow. As with previous authors, Johnston (1969) recognized the Italian sparrow as having arisen through hybridization between *P. domesticus* and *P. hispaniolensis*; he suggested the taxonomic recognition of the Italian taxon and assigned it the name, *Passer italiae*. Sætre and his colleagues have recently produced a series of studies to test various factors associated with the origin and stabilization of this homoploid hybrid species—including patterns of genetic contributions from the parental species and the relative roles played by various genomic elements in the reproductive isolation of *P. italiae*, *P. domesticus* and *P. hispaniolensis* (Elgvin et al. 2011; Hermansen et al. 2011, 2014; Eroukhmanoff et al. 2013, 2014; Trier et al. 2014).

Single nucleotide polymorphism (SNPs) genotypes were identified in the Italian sparrow species that were diagnostic for either the House or Spanish sparrows. Trier et al. (2014) used the distribution of these SNPs throughout the geographic distribution of *P. italiae*, and across hybrid zones between this species and either House or Spanish sparrows, to estimate levels of reproductive isolation between the homoploid hybrid and its parental taxa. The analyses revealed greatly reduced levels of introgression of certain genomic regions, particularly sex-linked loci, mitochondrial (mtDNA) loci and nuclear loci encoding mitochondrial proteins, suggestive of genomic regions that are reproductively isolated between the hybrid species and its progenitors (Trier et al. 2014). These findings led to the inference of a role for cytonuclear incompatibilities and lower hybrid fitness due to some sex-linked genotypes in reproductive isolation between the three species (Trier et al. 2014). In contrast, the genotypic analysis across the hybrid zones between the sparrow species also detected elevated introgression (compared to neutral expectations) at other loci (Trier et al. 2014) suggestive of adaptive introgression (Toews and Brelsford 2012) and adaptive homoploid hybrid speciation. Consistent with this adaptive inference were findings from beak dimension evolution in the hybrid species leading Eroukhmanoff et al. (2013) to the following conclusion: "The study illustrates how hybrid species may be relatively unconstrained by their admixed genetic background, allowing them to adapt rapidly to environmental variation."

1.6.3 The origin of hybrid animal taxa: Cichlids

The cichlid fish clades that inhabit the Great Lakes of eastern Africa are a model system for studies of adaptive radiations (Henning and Meyer 2014). In the last decade they have also become a model for dissecting the effects of widespread reticulate evolutionary processes associated with rapid, adaptive diversification (e.g., Seehausen 2004; Koblmüller et al. 2007; Genner and Turner 2012; van Rijssel et al. 2015). Thus, Schwarzer et al. (2012b), after analyzing nuclear and mtDNA genotypes from riverine and lake species, concluded, "Hybridization of multiple lineages across changing watersheds

shaped each of the major haplochromine radiations in lakes Tanganyika, Victoria, Malawi and the Kalahari Palaeolakes, as well as a miniature species flock in the Congo basin (River Fwa). On the basis of our results, introgression occurred not only on a spatially restricted scale, but massively over almost the whole range of the haplochromine distribution." In Chapter 6, I will discuss at length the reflection of reticulation in the adaptive radiations of entire clades and lake systems, as indicated by the results of Schwarzer et al. (2012b) and other studies. In the current context, the reticulate evolutionary history of individual lineages helps elucidate the role of homoploid hybrid speciation in this assemblage (Salzburger et al. 2002).

Over the past two decades, Seehausen and his colleagues have developed the African Rift Lakes cichlids into a model system for understanding the processes underlying biodiversification (e.g., Seehausen et al. 1997, 2008; Turner et al. 2001; Carleton et al. 2005; Selz et al. 2014a). Throughout their analyses, Seehausen and colleagues have focused on factors leading to adaptation and speciation. Included in their findings have been inferences of the role of homoploid hybrid speciation and the mechanisms that might underlie reticulate evolution within the cichlid clade (Figure 1.7). For example, Keller et al. (2013) used a genome-wide set of SNP loci to test the hypothesis that hybridization between species belonging to separate genera (i.e., *Mbipia* and *Pundamilia*) had contributed to the genomic and adaptive diversification of five sympatric species belonging to these genera. Associations between adaptive phenotypes and the loci used in the analysis led to the inference that three of the five sympatric taxa were homoploid hybrids formed from intergeneric crosses (Figure 1.7). Specifically, Keller et al. (2013) proposed that introgressive hybridization "led to reshuffling of gene complexes associated with adaptation and mate choice, resulting in hybrid species with novel combinations of ecology and mating behavior" (Figure 1.7).

To further test for the likelihood of adaptive homoploid hybrid speciation in the cichlid radiations, and the causal role of mate choice differentiation in reproductively isolating newly-formed hybrid lineages, Selz et al. (2014b) carried out a

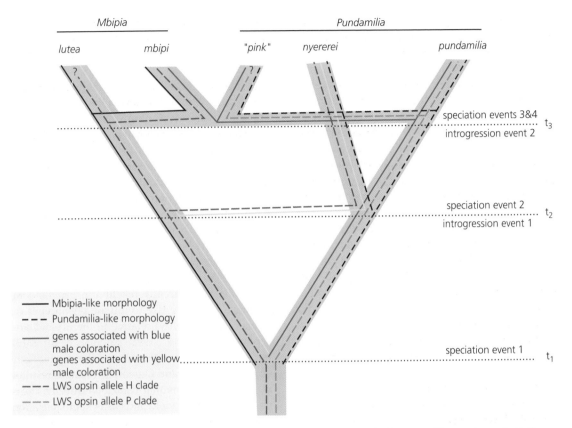

Figure 1.7 A model explaining the reticulate evolutionary history of a clade of African cichlids. The divergence of lineages that resulted in the genera, *Mbipia* and *Pundamilia*, occurs at Speciation event 1 at time point 1 (i.e., t_1). At speciation event 2/t_2 introgressive hybridization from *Mbipia* into *Pundamilia* resulted in the transfer of alleles associated with yellow coloration (i.e., "LWS opsin H clade") catalyzing speciation event 2 that gave rise to *P. nyererei*. A second bout of intergeneric hybridization forms the basis of speciation events 3 and 4 (i.e., t_3) resulting in the origin of *M. mbipi* and *Pundamilia sp.* "*pink*." The "?" at the ends of the *lutea* and "*pink*" lineages indicate that LWS opsin data were not available for these two species (from Keller et al. 2013).

series of crossing experiments between five species of African cichlids. The three F_1 classes synthesized derived from crosses between species from the same lake radiation (two F_1 lineages) or between species from different lakes (one F_1 lineage). The experimental F_1s were used in mate choice experiments in which hybrid and non-hybrid females were given the option to mate with hybrid or non-hybrid males (Selz et al. 2014b). The findings that (1) hybrids tended to mate non-randomly and (2) the expression of hybrid mate choice could lead to isolation of hybrid and parental individuals/lineages suggested that mating behavior could help stabilize homoploid hybrid species (Selz et al. 2014b).

1.6.4 The origin of hybrid plant taxa: Pinus densata

Though plants are paradigms of hybrid lineage formation via allopolyploidy (Stebbins 1947; Soltis and Soltis 2009; Estep et al. 2014), many plant complexes reflect homoploid hybrid speciation as well (e.g., *Iris nelsonii*; Randolph 1966; Arnold 1993; Taylor et al. 2011). One plant species complex that illustrates the latter phenomenon is the Tibetan Plateau endemic, *Pinus densata*. Several different genetical, ecological, and distributional approaches have been used to test whether or not *P. densata* was formed through natural hybridization and, if so, what the factors might be that contributed to its origin and retention as a novel evolutionary unit (Wang and Szmidt 1994; X-R Wang et al. 2001; Song et al. 2002, 2003; Ma et al. 2006; Mao and Wang 2011; B Wang et al. 2011; Gao et al. 2012; Xing et al. 2014).

Variation at both nuclear and cytoplasmic loci, morphological characteristics, and fitness estimates for the parental and hybrid species in different environments have supported the inference that *P. densata* is a homoploid hybrid species generated from crosses between *P. tabuliformis* and *P. yunnanensis* (e.g., Wang and Szmidt 1994; Ma et al. 2006; Xing et al. 2014; Zhao et al. 2014). It has been hypothesized that the stabilization and successful expansion of *P. densata* was facilitated by adaptations to the extreme environments found in its Tibetan Plateau habitat (Ma et al. 2006). Support for this hypothesis comes from the geographic and ecological distribution of the hybrid species and its parents. First, *P. densata* forests occur at high elevations (i.e., forming single-species stands at 2700–4200 m) relative to either of its progenitor species suggesting adaptation to an extreme environment. Second, also suggestive of differential adaptations to alternate habitats: "The geographic distribution of the three pines forms a succession, with *P. tabuliformis* in the north, *P. densata* in the middle, and *P. yunnanensis* in the south" (Ma et al. 2006). Third, geographic information system-based analyses reflected a divergence between the ecological preferences of *P. densata, P. tabuliformis* and *P. yunnanensis* (Mao and Wang 2011). Specifically, the habitat of the hybrid species was diagnosed as "extreme in several environmental dimensions, including mean diurnal range of temperature, vapor pressure, frost frequency, and accumulative heat . . . These environmental factors can induce severe physiological and physical stresses in plants" (Mao and Wang 2011). Taken together, the colonization history, niche breadth, genotypic structure, and geographic distribution of *P. densata* point to divergent environmental selection playing a major role in the formation and survival of this homoploid hybrid species (Mao and Wang 2011; Wang et al. 2011; Gao et al. 2012; Zhao et al. 2014).

1.6.5 The origin of hybrid plant taxa: Cotton

Though a minimum of four separate cotton lineages have been domesticated in different areas of the world, *Gossypium hirsutum* provides the vast majority of the raw material used in cotton clothing (Wendel 1995). Like many other cultivars, *G. hirsutum* reflects the evolution of a domestic species through allopolyploidy and thus is a paradigm for the role of hybrid, polyploid speciation that forms such an important part of plant diversification (Stebbins 1947; Mable 2013; Cannon et al. 2015).

The origin of *G. hirsutum* involved hybridization between diploid lineages that now have allopatric distributions; this species possesses the "A" and "D" genomes present in diploid species from Asia/Africa and the New World, respectively (Wendel 1989, 1995). Not only did the hybridization between Old and New World *Gossypium* c.1–2 million ybp result in the formation of the allotetraploid, *G. hirsutum*, but the bringing together of the different genomes apparently resulted in genetic combinations that gave rise to unique phenotypes (in terms of fiber quality) relative to the diploid progenitors (Jiang et al. 1998; Paterson et al. 2012). In particular, the combination of the diploid genomes from the A- and D-genome lineages resulted in gene interactions that contributed to the novel phenotype of high yield and easily spinnable fibers produced by *G. hirsutum*. The observation of genomic and phenotypic novelty produced through allopolyploid speciation in *Gossypium* led Paterson et al. (2012) to hypothesize that "Emergent features of polyploids may be related to processes that render them no longer the sum of their progenitors and permit them to explore transgressive phenotypic innovations." It would seem that as with adaptive trait introgression and the origin of homoploid hybrid lineages in plants and animals, allopolyploid lineage formation leads to not only biodiversification but also evolutionarily significant alterations to genotypes and phenotypes (Yoo et al. 2014).

1.7 Genetic exchange between divergent lineages

The previous sections have provided the concepts, processes, and examples of genetic exchange between divergent, but relatively closely related organismic lineages. In particular, the occurrence of hybrid zones, the transfer of adaptive traits and the origin of hybrid taxa were based on the ability of the parental lineages to reproduce sexually and to produce at least limited numbers of hybrid offspring. In the following sections, examples of exchanges between highly divergent lineages will be the focus; some of the exchange events occurred between different domains of life. Given this focus, the molecular recombination events are best-termed horizontal gene transfer.

1.7.1 Exchange between divergent lineages: Recombination between viruses and mammals

The mammalian clade provides a rich source of data indicating extensive reticulate evolution, involving hybridization between closely related taxa (e.g., Feulner et al. 2013) and the incorporation of genes from non-mammalian lineages into mammalian genomes (e.g., Gifford 2012). Examples of the latter type of exchange are reflected by studies indicating the integration of viral genes into mammalian genomes (Marco and Marín 2009; Cornelis et al. 2012).

An analysis of genes thought to offer protection to mammals against retroviral infections, in particular those involving the gene *GIN1* (*Gypsy integrase 1*), led to the inference of ancient recombination between retroviral and mammalian genomes. Marco and Marín (2009) thus defined a new member of the *GIN1* gene family named, *CGIN1* that possessed sequence similarity for portions of genes from both retroviruses and mammals. The pattern of sequence similarities suggested strongly that the derivation of *CGIN1* involved a recombination event that fused a mammalian gene with sequences from a retrovirus, and that the timing of this fusion occurred in the ancestor of the marsupial and eutherian clades (Marco and Marín 2009).

As with *CGIN1*, the mammalian gene family, *Syncytins* apparently originated from recombination events between retroviral sequences and their mammalian hosts, with a set of (mammalian) lineage-specific genes found in groups as divergent as opossums, primates, lagomorphs, murids and tenrecs (Cornelis et al. 2012, 2014, 2015). Significantly, though the members of this particular group of genes all contribute to placentation, their derivation appears to have involved independent transfer events from retroviruses into the

various mammalian lineages (Cornelis et al. 2012). The most ancient of these transfers resulted in the Carnivore-specific *syncytin-Car1*; this gene originated before the radiation of Carnivora *c.* 60–85 mya (Cornelis et al. 2012).

1.7.2 Exchange between divergent lineages: Horizontal gene transfer between bacteria and animals

Horizontal gene transfer has been inferred for metazoan lineages as diverse as rotifers and vertebrates (Keeling and Palmer 2008; Boto 2014). Likewise, the diversity of DNA donors to the metazoan recipients is great, including viruses, bacteria, fungi, plants, and other metazoans (Boto 2014). Notwithstanding the diversity of possible donor lineages, the majority of HGT events impacting metazoans involve bacterial genomes (Boto 2014). For example, Mayer et al. (2011) inferred that the transfer of cellulase genes into some parasitic nematode lineages likely involved prokaryotic donors. Likewise, Pauchet and Heckel (2013) found that genes necessary for plant cell wall degradation in the mustard leaf beetle, *Phaedon cochleariae*, were most likely obtained from a species of gammaproteobacteria.

Probably the best-known example of bacteria-to-animal HGT involves the endosymbiont, *Wolbachia*. *Wolbachia* is maternally inherited and infects many arthropod and filarial nematodes, including at least 20% of all insect species (Hotopp 2011). The occurrence of *Wolbachia* in the developing gametes of its hosts provides a mechanism by which horizontally transferred bacterial genes could subsequently be vertically transmitted to host progeny (Hotopp 2011). A survey of genomes retrieved from public databases resolved HGT events involving *Wolbachia* and a divergent array of hosts, including nematode, mosquito, tick, wasp, and several fruit fly (i.e., *Drosophila*) species, with a portion of the transfers involving almost the complete bacterial genome (Hotopp et al. 2007). Some of the transfers resulted in genes transcribed within the host cells, leading to the conclusion that "heritable lateral gene transfer occurs into eukaryotic hosts from their prokaryote symbionts, potentially providing a mechanism for acquisition of new genes and functions" (Hotopp et al. 2007).

1.7.3 Exchange between divergent lineages: Plant genomes receive foreign genes and genomes

Horizontal transfer between highly divergent and thus reproductively isolated plant lineages or between plant and non-plant lineages has become a well-known process (Keeling and Palmer 2008). In an example of plant-to-plant HGT, Christin et al. (2012) inferred a minimum of four, independent transfer events in the acquisition of components of the C_4 photosynthesis pathway by the grass lineage, *Alloteropsis*, from divergent C_4 plant species. Though possibly less frequent than plant-to-plant transfer, HGT between plant and non-plant lineages also occurs and can affect the genomic and adaptive evolutionary trajectories of the recipient lineages. For example, HGT between the marine algae, *Micromonas*, and prokaryotes likely provided the green algal lineage with "unique" genes contributing to its ability to live in diverse environments (Worden et al. 2009). Similarly, the process of HGT-generated adaptations has also been inferred for fungal/plant interactions. Richards et al. (2009) concluded that ancient HGT of genes from fungi into plant genomes "added phenotypes important for life in a soil environment . . . genetic exchange between plants and fungi is exceedingly rare, particularly among the angiosperms, but has occurred during their evolutionary history and added important metabolic traits to plant lineages."

A well-recognized category of plant-to-plant HGT events involves the transfer of mitochondrial DNA (Bergthorsson et al. 2003). Mechanisms that may promote such transfer events include a physical association between a plant parasite and its host (Keeling and Palmer 2008), the movement of mRNAs between parasitic and host species (Kim et al. 2014), and natural grafting (Stegemann et al. 2012). Regardless of the transfer mechanism underlying the exchange of mtDNA, multiple evolutionarily significant outcomes have been described. For example, HGT can result in a large proportion of the mitochondrial genome being transferred. Xi et al. (2013) found that a minimum of 24–41% of genes sampled showed evidence of being involved in transfer among species belonging to the

Family, Rafflesiaceae. In addition, when the transfer of mtDNA genes occurs, one of the consequences can be the formation of gene chimeras. Hao et al. (2010) detected this process, describing their findings in the following passage: "Here we report the discovery, in two lineages of plant mitochondrial genes, of novel gene combinations that arose by conversion between coresident native and foreign homologs. These lineages have undergone intricate conversion between native and foreign copies, with conversion occurring repeatedly and differentially over the course of speciation, leading to radiations of mosaic genes involved in respiration and intron splicing." Finally, Rice et al. (2013) detected the HGT-mediated incorporation of six-genome equivalents of foreign DNA into the mitochondrial genome of *Amborella trichopoda*. The sources of the transferred mtDNA included other angiosperms, green algae, and mosses, with the latter two classes of organisms providing entire mitochondrial genomes (i.e., three green algae genomes and one moss genome; Rice et al. 2013).

1.8 Conclusions

The web-of-life metaphor that took shape during the neo-Darwinian synthesis at the hands of Stebbins and Anderson, and even before the synthesis, through Lotsy and others, has recently been given additional credence with the recognition that divergence-with-gene-flow is widespread among all domains of life. The collection of genome-wide data and phenotypic information from both fossils and extant organisms, coupled with the development and application of novel analytical techniques, has provided evidence that genetic exchange is even more extensive than previously thought. This realization is leading to an even greater appreciation of the need to revise methodologies, such as hybrid zone analyses, in order to test rigorously the strength and form of selection acting on hybrid and parental genotypes across environments. This is of particular importance if we are to be able to define which portions of the genomes of diverging organisms must remain reproductively isolated and which may lead to adaptive admixture (i.e., reflecting islands of divergence and adaptive trait introgression, respectively).

Furthermore, as evolutionary biologists come to a greater realization that allopatric speciation is almost certainly the exception rather than the rule—because of the unlikelihood of geographic ranges of diverging lineages remaining separated across repeated climatic oscillations—the role played by recurrent genetic exchange will be tested more and more. The examples of evolution accompanied by genetic exchange provided earlier support these conclusions, but they represent the tip of the iceberg of data demonstrating the validity of the web-of-life metaphor. In the following chapters we will look in detail at key processes that can affect evolution through genetic exchange.

CHAPTER 2

Genetic exchange and species concepts

"No one definition has as yet satisfied all naturalists; yet every naturalist knows vaguely what he means when he speaks of a species . . . From these remarks it will be seen that I look at the term species, as one arbitrarily given for the sake of convenience to a set of individuals closely resembling each other."
(Darwin 1859, pp. 44, 52)

"A species is a group of individuals fully fertile inter se, but barred from interbreeding with other similar groups by its physiological properties (producing either incompatibility of parents, or sterility of the hybrids, or both)."
(Dobzhansky 1935)

"The conflict centres around the two tenets of the BSC: (i) the genomes between species are extensively divergent; (ii) little genetic exchange is possible during incipient speciation. Surprisingly (i) is probably right but (ii) is not."
(Wu 2001)

"The use of a single criterion to delimit species artificially reduces the complexity of evolving lineages . . . Only when a more eclectic approach is taken, by the use of several criteria, can this complexity begin to be ordered and described."
(Marshall et al. 2006)

"The availability of genome sequences for closely related microorganisms has at the same time clarified and complicated our view of species delineation . . . we have shown how subtle genomic differences between coexisting bacteria contribute to their ecological differentiation within an ecosystem."
(Denef et al. 2010)

2.1 Species concepts and genetic exchange: Resolution and conflict avoidance

The concept of genetic exchange explored in this book is intentionally independent of species concepts (e.g., see the definition of natural hybridization in Chapter 1). Yet, most studies of natural hybridization, hybrid speciation, and horizontal gene transfer continue, understandably, to reflect the assumption of a particular definition of species and concept of the process of speciation. This is understandable because studies of genetic exchange

have historically been designed to address "(i) taxonomy or systematics . . . (ii) mechanisms of reproductive isolation and speciation . . . or (iii) natural hybridization as a fundamental evolutionary process that produces consequences that are significant in their own right" (Arnold 1992).

An examination of "speciation" literature reveals a close association between studies of genetic exchange and debates concerning how to define species and the process of speciation (e.g., Zinner and Roos 2014). For example, natural hybridization has been described either as having a major or minor role in plant and animal evolution based

Divergence with Genetic Exchange. Michael L. Arnold.
© Michael L. Arnold 2016. Published 2016 by Oxford University Press.

largely on an underlying concept and definition of species (Anderson and Stebbins 1954; Mayr 1963; Wagner 1970; Arnold 1992, 1997, 2006; Dowling and DeMarais 1993; Wu 2001; Seehausen 2004; Sobel et al. 2010). In this regard, two arguments provided in support of a relatively minor role for genetic exchange (and natural hybridization per se) in affecting evolutionary process and pattern were that (1) the formation of new evolutionary lineages through heterospecific hybridization is insignificant (or non-existent) because species, by definition, are reproductively isolated (i.e., *"Biological Species Concept"* or *"BSC"*; Dobzhansky 1935; Mayr 1942) and (2) new lineages cannot be polyphyletic in origin (i.e., *"Phylogenetic Species Concept"* or *"PSC"*; Hennig 1966; Mishler and Donoghue 1982; Cracraft 1989).

Acceptance of the two conclusions defines evolution through genetic exchange as nonexistent. For example, Willi Hennig is held to be the architect of modern phylogenetic analysis and thus the founder of the PSC. Yet, in *Phylogenetic Systematics* (Hennig 1966), he emphasized repeatedly an application of the BSC. This application led him to the following conclusion, "Is not the species concept that the species includes all individuals that together are capable of producing completely fertile offspring, and must we not then consider groups whose individuals can produce new species by hybridization as partial groups of one species? If we speak of the origin of species by species hybridization, are we not guilty of circular reasoning between premise and conclusion?" (Hennig 1966, p. 208).

At the other end of the continuum from those taxa that "violate" the BSC by reproducing with other taxa (i.e., that fall into the category of "too much sex"—Templeton 1989; Figure 2.1), are those that violate it by having "too little sex" (Templeton 1989; Figure 2.1). In this regard, Hennig (1966) also considered whether asexual organisms posed a significant problem for species definition. He concluded that it was indeed crucial that species be defined by attributes of bisexual reproduction and thus be demonstrable members of a "reproductive community" (Hennig 1966, p. 65).

However, he did not feel that asexuality was an important consideration because such lineages among animal taxa are scarce. As a general statement of the importance of non-sexually reproducing organisms, his conclusion that "the question of the species concept in organisms without bisexual reproduction is no more than a relatively subordinate species problem of systematics" (Hennig 1966, p. 44), summed up well the viewpoint of many of the architects of the neo-Darwinian synthesis (e.g., Mayr 1963). In contrast, it is now well understood that organisms that reproduce at least partially through asexual/clonal mechanisms make up a large proportion of biodiversity (i.e., plants, microorganisms, some animal clades; see Arnold 2006 and Chapter 6 of the present text). Furthermore, these "asexual" microorganisms, plants, and animals often demonstrate the evolutionary effects of genetic exchange. Indeed, some of the highest frequencies of genetic exchange occur among organisms termed "asexual"—i.e., prokaryotic and viral lineages (Figure 2.1). Therefore, their dismissal by workers such as Hennig and Mayr reflected not so much the "species problem," as a problem with their concept

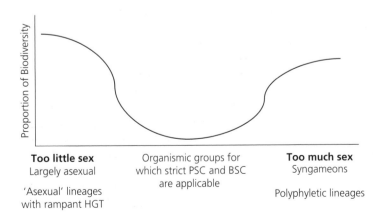

Figure 2.1 Illustration of how the applicability of some species concepts, such as the "biological" and "phylogenetic" are restricted to few organismic groups—i.e., not asexual ("too little sex") and not involved in genetic exchange events ("too much sex"). In contrast, the "cohesion," "prokaryotic", and "genic" species concepts provide conceptual frameworks for studying evolutionary pattern and process regardless of the frequency of asexuality and genetic exchange (adapted from Templeton 1989).

of species. For example, Hey et al. (2003) concluded, "These two ideas—that species are categories that are created . . . by the biologists who study them, and that species are objective, observable entities in nature—have long been in conflict."

Chapter 1 reviewed a small portion of evidence that negates such viewpoints deriving from the strict application of the BSC and PSC, and many other examples will be given in the following chapters. However, one observation that brings into sharp contrast how widespread and significant are web-of-life processes is that of allopolyploid hybrid speciation in plants (e.g., Figure 2.2). Reflecting this significance, Soltis and Soltis (2009) stated: "Recent developments in genomics are revolutionizing our views of angiosperm genomes, demonstrating that perhaps all angiosperms have likely undergone at least one round of polyploidization and that hybridization has been an important force in generating angiosperm species diversity." Though some of these polyploidy events reflect *autopolyploidy*, and thus would derive from whole genome duplication within a lineage, even the origins of these species likely involved natural hybridization between divergent populations (Soltis and Soltis 2009). Furthermore, it is also now well recognized that hybrid speciation leading to polyploid taxa are not merely a plant phenomenon, but are also reflected in animal lineages (e.g., there were at least two whole genome duplication events along the vertebrate lineage; Holland et al. 2008; Van de Peer et al. 2009).

Five species concepts will be discussed in this chapter—BSC, "*Cohesion*" (i.e., CSC), "Phylogenetic," "*Prokaryotic*" (i.e., PrSC), and "*Genic*" (i.e., GSC)—in relation to the web-of-life metaphor. These were chosen because the application of the various concepts has encompassed all domains of life and reflects a number of viewpoints regarding the relative significance of genetic exchange in the evolutionary history of these domains. Conclusions drawn from these five examples, concerning the occurrence and evolutionary importance of divergence-with-gene-flow, are thus a good representation of those found across all species concepts. Specifically, the following discussion will hopefully illustrate (1) how the strict application of some concepts may limit investigations into the evolutionary role of genetic exchange, (2) how, in contrast, a thoughtful application of any given species concept can facilitate such studies, and (3) that we can largely avoid the tension described by Hey et al. (2003) through an appropriate application of definitions. In the following sections, I will borrow heavily from *Natural Hybridization and Evolution* and *Evolution through Genetic Exchange* (Chapter 2 in both texts; Arnold 1997 and 2006, respectively). I believe that my earlier discussions and conclusions remain a useful platform for understanding the issues surrounding the application of these concepts to test the evolutionary role played by genetic exchange between divergent lineages.

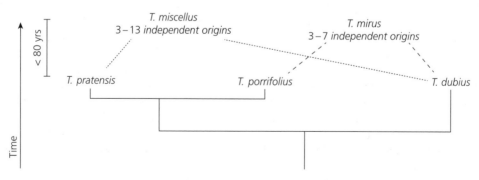

Figure 2.2 The evolutionary origin of allopolyploid hybrid species in the genus *Tragopogon* (i.e., *T. miscellus* and *T. mirus*). The hybrid species formed some time over the past 80 years. *T. miscellus* and *T. mirus* arose from the hybridization of *T. dubius* and *T. pratensis* and *T. dubius* and *T. porrifolius*, respectively. Both allopolyploids originated numerous times, through independent hybridization events (Soltis and Soltis 2009; Symonds et al. 2010).

2.2 Genetic exchange and the biological species concept

It is somewhat inaccurate to represent the BSC as the only concept using the amount of genetic exchange as a metric for defining species. Indeed, most concepts consider the frequency of genetic exchange to be of primary importance for such definitions (e.g., see Sites and Marshall 2004). However, Mayr's (1942) oft-quoted definition that "species are groups of actually or potentially interbreeding natural populations, which are reproductively isolated from other such groups" (based on an earlier description by Dobzhansky 1935, 1937) predicates defining species on the basis of reproductive isolation, with the process of speciation assigned to the development of barriers to reproduction.

Mayr (1942, 1963) was not applying the BSC to organisms in all of the domains life, with the various processes of genetic exchange demonstrated therein. Instead, he focused on eukaryotes and the processes of genetic exchange associated with sexually reproducing organisms (i.e., introgressive hybridization and hybrid speciation). As argued previously (Arnold 2004, 2006), if the process of genetic exchange *sensu lato* is substituted in the place of introgression and hybrid speciation through sexual recombination, estimates of "reproductive isolation" can be extended to include organisms that do not reproduce sexually. However, the substitution of genetic exchange (i.e., resulting from introgression, horizontal gene transfer, and viral recombination) incorporates processes and organisms not encompassed by the BSC (Figure 2.1). Mayr (1963, p. 28) did, however, address the perceived difficulties in applying the BSC to asexually reproducing organisms. He stated, "Various proposals have been made to resolve the difficulty that asexuality raises for the BSC. Some authors have gone so far as to abandon the BSC altogether . . . I can see nothing that would recommend this solution. It exaggerates the importance of asexuality." Though Mayr was speaking mainly of animal taxa, his conclusions portended subsequent findings for "asexual organisms" when he stated, "Indeed clandestine sexuality appears to be rather common among so-called asexual organisms" (Mayr 1963, p. 27). Mayr's contention, that truly asexual organisms are very rare

or nonexistent, has been supported quite well by subsequent studies (e.g., Charney 2012).

Even if the various mechanisms for genetic exchange and the wide array of reproductive systems found across the domains of life are placed under a broadened definition of "reproduction" and thus the BSC, it still begs the question of how to define and study the process of evolutionary diversification. In this regard, it is informative to review the two means by which Mayr (1942, 1963) dealt with the "problem" of natural hybridization within the framework of this species concept. The first approach was to argue that if viable, fertile hybrids were produced then one should consider the hybridizing forms to be subspecies or semispecies—i.e., to define natural hybridization (or genetic exchange) as only possible below the level of species. The second approach reflected a diminution of the importance of natural hybridization based upon the expectation that the vast majority of hybrid genotypes would have reduced fitness (see Section 1.5).

Nothing inherent in the BSC demands that genetic exchange, in all its various forms, cannot occur or is of little evolutionary importance. Individuals from subspecies or semispecies could form fertile hybrids, and this would not represent an explicit violation of the BSC. In fact, subspecies and semispecies are defined under the BSC in part by the presence of ongoing or potential gene flow (Mayr 1942, 1963). Processes such as introgression, HGT, or viral reassortment leading to adaptive evolution are also not explicitly discounted by the BSC. Yet, it has been argued that even crosses between subspecies or semispecies will most likely lead to evolutionary dead ends. Mayr (1963, p. 132) illustrates this latter viewpoint by stating, "In natural populations there is usually severe selection against introgression. The failure of most zones of conspecific hybridization to broaden . . . shows that there is already a great deal of genetic unbalance between differentiated populations within a species." In a similar vein, Mayr (1963) concluded that "mistakes" (i.e., hybrids) were rare and evolutionarily unimportant, at least in animal taxa (see Section 1.5). Standing in sharp relief to this conclusion is the myriad of recent analyses involving microorganisms, plants, and animals that have demonstrated the prevalence and evolutionary import of genetic exchange events (e.g., Grant and Grant 2014a; Nikolaidis et al. 2014; Renaut et al. 2014).

One mechanism by which to illustrate the importance of genetic exchange is to catalog its widespread occurrence in organisms as diverse as bacteria and mammals (e.g., Arnold 1997, 2006, 2009, and the present book). However, in the context of the oft-repeated mantra (see quotes by Mayr and Section 1.5) that genetic exchange is rare and thus evolutionarily unimportant, particularly in animals, it is important to repeat that the rarity of a particular event is not predictive of its potential evolutionary importance. In this regard, the frequent observation of reduced *mean* fertility and viability of F_1, F_2, or BC_1 hybrid classes could lead to the inference that these progeny will play a minor role in the evolution of a given species complex. In contrast to this prediction, there are numerous instances in which, for example, the rarity of viable gametes has not prevented important evolutionary effects from F_1 and later generation hybrid formation (Arnold et al. 1999). A clear example of the validity of this conclusion involves the annual sunflower species *Helianthus annuus* and *Helianthus petiolaris*. Experimental crosses involving these two species produce F_1 individuals that possess pollen fertilities ranging from 0 to 30% (mean *c.* 14%; Heiser 1947). Additionally, experimentally produced F_2 and first backcross generation plants produce a maximum seed set of 1% and 2%, respectively (Heiser et al. 1969). Notwithstanding these extremely low levels of fertility and viability in the initial hybrid generations, natural hybridization between *H. annuus* and *H. petiolaris* has resulted in at least three stabilized hybrid species (Rieseberg 1991) and "rampant" introgressive hybridization (Yatabe et al. 2007).

It can be seen from the discussion that adherents of the BSC have often concluded that genetic exchange, particularly natural hybridization, produces relatively unimportant evolutionary consequences (Coyne and Orr 2004). In contrast, the definition of species in terms of reproductive barriers led Dobzhansky to a model in which genetic exchange played a central role in the process of speciation. However, the assignment of evolutionary importance to web-of-life processes was once again merely a reflection of the assumption of a uniformly maladaptive outcome from genetic exchange.

Dobzhansky (1940, 1970) hypothesized that natural hybridization could lead to the construction of prezygotic barriers to reproduction (i.e., "reinforcement"; Blair 1955). Specifically, reinforcement was posited to occur when individuals from previously allopatric populations overlapped in space and time and mated, resulting in less fit hybrid offspring and thereby to selection favoring those individuals that mated with conspecifics (Dobzhansky 1940; Howard 1993; Noor 1995; Shaw and Mendelson 2013). As seen through the BSC prism, such instances reflect the role of natural hybridization as merely a means for finalizing speciation; reinforcement reflects once again an emphasis on hybridization/genetic exchange as being maladaptive.

As stated earlier, using the degree of reproductive isolation as the litmus test for species and speciation is not unique to the BSC. However, the fact that this species concept was promoted so well during the modern synthesis led to its wide application and thus to an emphasis on cessation of gene flow as the necessary and sufficient component for speciation. In the following section it will be argued that, in a similar way, a strict application of the PSC is problematic for testing the evolutionary role of genetic exchange. However, the application of the PSC does not require such an approach. Indeed, more and more studies have applied a derivation of the PSC in which reticulate evolutionary patterns (rather than simple bifurcating histories) are allowed (e.g., Baum 2007; Halas and Simons 2014). As with the BSC, and indeed any species concept, applying the PSC does not need to limit studies that test the importance of reticulate evolution. Furthermore, if the application of any concept limits the study of observed evolutionary processes, such as genetic exchange, then that species concept needs to be modified or jettisoned in deference to biological reality.

2.3 Genetic exchange and the phylogenetic species concept

The PSC defines species as "an irreducible (basal) cluster of organisms, diagnosably distinct from other such clusters, and within which there is a parental pattern of ancestry and descent" (Cracraft, 1989). The origination of the PSC is most often deemed to belong to Hennig (1966), though Darwin (1859) also presented evolutionary diversification as a bifurcating tree in *The Origin*. Hennig's *Phylogenetic Systematics* is, however, the basis for the modern phylogenetic approach. Inherent in the PSC is the aspect of evolutionary and phylogenetic history. While systematic delimitations are often the ends to studies that apply the PSC, the determination of the pattern of phylogenetic relationships are the means (Sangster 2014). Applying this concept to define species and the process of speciation thus depends upon the identification of the ancestral and derived states of particular characters (i.e., defining evolutionary polarity). Understanding the polarity of character evolution allows a resolution of the pattern of branching within phylogenies and demarcates monophyletic species "boundaries" (i.e., "irreducible clusters"; Cracraft 1989).

As indicated in Section 2.1, the requirement of monophyly in species definition eliminates the possibility of genetic exchange in the origin of new species (Figure 2.1). Cracraft (1989) expressed the viewpoint of Hennig and others when stating, "In the majority of cases, phylogenetic species will be demonstrably monophyletic; they will never be nonmonophyletic, except through error." Hennig (1966, p. 207) recognized that "special complications would arise if new species could also arise to a noteworthy extent by hybridization between species." The possibility of such events were, however, discounted by arguing that "in all cases in which a 'polyphyletic origin of species' has been recognized, the species involved were so closely related that they could just as well be considered races of one species . . . we must consider a polyphyletic origin of species in the category stages below species to be possible, through hybridization for example, but this does not touch the question of the monophyly of the higher taxa" (Hennig 1966, pp. 208–209). As discussed earlier, Hennig (1966) not only used monophyly in his species definition, but also the requirement that different species reflect separate (i.e., non-hybridizing) reproductive communities. Nixon and Wheeler (1990) outlined a similar viewpoint in the following way: "With the strictest application of the PSC, species that show extensive intergradation will be treated as a single species."

Hybrid speciation (both homoploid and allo-polyploid) is only one of the many outcomes that may result from genetic exchange; other outcomes include introgression between the donor and recipient, genetic assimilation of one taxon by another, reinforcement of barriers to reproduction, etc. (see Arnold 1992, 2006; Allendorf et al. 2001; Servedio and Noor 2003 for reviews). Furthermore, the definitions of hybridization and reticulate evolution adopted for this book result in the observation that genetic exchange can have similar effects regardless of taxonomic categories. Yet, the majority of cases of reticulate evolution occur among organisms recognized as belonging to different species or even more distantly related lineages; at the extreme, exchanges can occur between taxa from different domains of life (e.g., Sloan et al. 2014; Wisecaver and Hackett 2014). With the increasing availability of genomic information, the requirement of *strict* monophyly (i.e., all sequences in a particular species, genus etc. are from a single ancestor) in determining "species boundaries" has become problematic. However, the following two examples—one from a fungal clade and another from a mammalian clade—illustrate how application of a phylogenetic approach for delimiting species can result in the recognition of the role played by reticulate events.

Stewart et al. (2014) used sequence information from four nuclear loci, a worldwide population sample, and several phylogenetic/genealogical methodologies to infer species boundaries in the *Alternaria alternata* fungal clade. The significance of defining species in this clade involved not only the formulation of a basic evolutionary understanding, but also the prevention of human mediated transport of a major pathogen of citrus species (i.e., the causative agent of "citrus brown spot"; Stewart et al. 2014). Conclusions drawn from the analysis of *A. alternata* included (1) the recognition of two species that cause citrus brown spot and (2) that one of these species was possibly a hybrid resulting from recombination between different lineages of the alternate species. Most importantly for the present topic was the conclusion that the analysis indicated "that multilocus phylogenetic methods that allow for recombination and incomplete lineage sorting can be useful for the quantitative delimitation of asexual species that are morphologically indistinguishable" (Stewart et al. 2014). In other words, the relaxation of the requirement of monophyly in the determination of "phylogenetic species" allowed a robust inference of evolutionary pattern and process in this fungal assemblage.

As with the *Alternaria* study, Melo-Ferreira et al. (2011, 2012) have documented the utility of testing the hypothesis of reciprocal monophyly for both defining species boundaries and deciphering evolutionary processes associated with speciation. In this case, the phylogenetic relationships of members of the hare genus (*Lepus*) were inferred from 14 nuclear and two mitochondrial loci. Like the inferences derived from *Alternaria*, those for *Lepus* included cases of both reciprocal and non-reciprocal monophyly for different loci (Figure 2.3). In particular, the mtDNA loci reflected the influence of widespread, ancient introgressive hybridization (Alves et al. 2008) leading to paraphyletic relationships throughout the phylogeny (Figure 2.3). Such

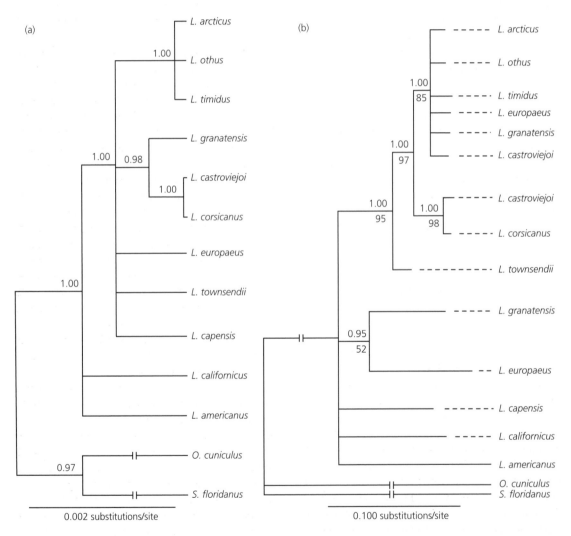

Figure 2.3 An example from the hare genus, *Lepus*, in which extensive genetic exchange-generated paraphyly was detected. (a) Nuclear DNA consensus tree generated from 14 loci (the posterior probability of each clade is given at the base of each clade). (b) mtDNA majority rule consensus tree (numbers above branches indicate the posterior probabilities of each clade and numbers below branches depict the maximum likelihood generated bootstrap supports). Nodes with posterior probability <0.95 were collapsed. In (b) the dashed lines terminating with species designations indicate the presence of multiple haplotypes combined into a single lineage for this illustration (from Melo-Ferreira et al. 2012).

non-concordance between phylogenies derived from different data sets reflects a common signature of genetic exchange and *incomplete lineage sorting* (Arnold 2006; Chapter 3 of the current text). In the analysis by Melo-Ferreira et al. (2012) an explicit test for these two processes resulted in a rejection of a model that lacked genetic exchange, supporting the inference of mtDNA introgression-derived paraphyly. Likewise, an analysis of sequence variation associated with X-chromosome loci (Melo-Ferreira et al. 2011) also revealed a lack of reciprocal monophyly for certain species that was likely caused by past, rather than ongoing, introgressive hybridization (Melo-Ferreira et al. 2014a).

I have argued previously (e.g., Arnold 1997, 2006), that the important evolutionary outcomes from genetic exchange in all domains of life are quite similar—i.e., the origin of new lineages and adaptive evolutionary change. In the next section we will see that, unlike the strict definition and application of either the BSC or the PSC, the cohesion species concept takes into account the role of genetic admixture in evolutionary diversification. One goal of the present text is to highlight the necessity of this type of approach for understanding organismal evolution per se.

2.4 Genetic exchange and the cohesion species concept

Within the framework of the cohesion species concept, species are defined as "the most inclusive group of organisms having the potential for genetic and/or demographic exchangeability" (Templeton 1989). Templeton, as have others, argued that one of the main weaknesses of the BSC was its lack of applicability to either asexual organisms or to individuals belonging to *syngameons* (Figure 2.1). As discussed earlier, Templeton's conclusion was that the difficulty results from either "too little" or "too much" sex. Templeton (1989) proposed the CSC to address problems with a number of species concepts, including the BSC. The CSC was designed to take into account all of the microevolutionary processes thought to contribute to speciation, including the evolution of reproductive barriers (Templeton 1989). "Genetic exchangeability" thus related to the

process of gene flow and "demographic exchangeability" to the processes of genetic drift and natural selection (Templeton 1989). In particular, "genetic exchangeability . . . is the ability to exchange genes during sexual reproduction" and "demographic exchangeability occurs when all individuals in a population display exactly the same ranges and abilities of tolerance to all relevant ecological variables" (Templeton 1989). For the current discussion it is important to ask how the CSC addresses cases of "too much sex" (i.e., introgressive hybridization, hybrid speciation, horizontal gene transfer, and viral reassortment).

Unlike the BSC, PSC, and other similar species concepts, cohesion species are defined on the basis of independent evolutionary/ecological trajectories (Templeton 1989, 2001). The CSC recognizes that species can be definable on one of the criteria, but not the other (Templeton 1989). For example, syngameons are a result of species having greater genetic than demographic exchangeability. Furthermore, the CSC does not require that taxa belonging to such a group be reduced to a subspecific taxonomic category. Indeed, among other implications from this concept, cohesion species may arise as products of a reticulation event between lineages defined as species (Templeton 1981, 2001). Furthermore, cohesion species may participate in some degree of genetic exchange (e.g., a syngameon) with other lineages and yet maintain their species status (Templeton 1981, 1989, 2001, 2004).

It should be pointed out that Templeton (1989) placed those taxa belonging to, for example, a syngameon into a category called "bad species." These species are "bad" because they demonstrate an elevated degree of genetic exchangeability. "Good species" on the other hand are "those that are well defined both by genetic and demographic exchangeability" (Templeton 1989). Yet, in general, rather than proposing hybridization as a problem to be overcome for the process of divergent evolution to proceed, the application of the CSC indicates the viewpoint that reticulate evolutionary processes should be incorporated into an evaluation of evolutionary process and pattern (Templeton 2004). In this regard, the CSC allows the construction of hypotheses that are testable within the widely applied phylogeographic paradigm of Avise (2000b).

By constructing such null hypotheses, it is possible to incorporate geography and distribution of population genetic variation into a genealogical analysis (Templeton 2001, 2004, 2009, 2010a, b; but see Knowles and Maddison 2002, Nielsen and Beaumont 2009 for criticisms of this methodology).

In applying Templeton's approach, the rejection of the null hypotheses—(1) that geographically associated samples are from a single evolutionary lineage and (2) that the lineages discovered by "1" are not exchangeable in terms of genetic or ecological characteristics—results in the inference that more than one evolutionary lineage (i.e., species) is present (Templeton 2001). It could be argued that the PSC also allows a rigorous identification of evolutionary lineages with separate trajectories. However, it is problematic to decide the weight to be given to the characters that identify a "phylogenetic species" (Avise 2000a). In contrast, the CSC allows a statistical test of whether evolutionarily distinct lineages exist, and also whether reticulation, past or present, has impacted said lineages (Templeton 2009, 2010a, b). As such, the application of this conceptual framework allows an encompassing approach for the deciphering of evolutionary pattern and process. In the same way, the following discussion will demonstrate how the prokaryotic species concept (PrSC) incorporates the web-of-life, rather than a BSC- or PSC-based, tree-of-life metaphor.

2.5 Genetic exchange and the prokaryotic species concept

The pervasive effects from horizontal gene transfer in the evolutionary history of prokaryotic taxa are now well established (Boucher et al. 2003; Polz et al. 2013). The appreciation of the fundamental importance of genetic exchange in the diversification of bacterial and archaeal species has been addressed by some systematic treatments. For example, Rosselló-Móra and Amann (2001) reflected the generality of problems inherent in all species concepts as well as the development, in their view, of a scientifically more robust PrSC by stating, "The species concept is a recurrent controversial issue that preoccupies philosophers as well as biologists of all disciplines. Prokaryotic species concept has its

own history and results from a series of empirical improvements."

With regard to the present discussion, one of the salient "improvements" in prokaryotic systematic studies has been the addition of DNA sequence information (Doolittle 1999; Rosselló-Móra and Amann 2001; Cohan 2002; Stackebrandt et al. 2002; Konstantinidis et al. 2009; Richter and Rosselló-Móra 2009; Denef et al. 2010). Workers applying any of the PrSCs utilize this type of data. However, it is of particular significance for studies that incorporate phylogenetic and/or phenetic approaches in defining species (e.g., Woese 1987; Doolittle 1999; Stackebrandt et al. 2002; Papke et al. 2007; Konstantinidis et al. 2009; Richter and Rosselló-Móra 2009; Denef et al. 2010). For example, the original definition of prokaryotic "genospecies" was based on the technique of DNA–DNA hybridization (i.e., DDH); "A bacterial species is essentially considered to be a collection of strains that are characterized by at least one diagnostic phenotypic trait and whose purified DNA molecules show 70% or higher reassociation values . . . " (Konstantinidis and Tiedje 2005).

The reference to "reassociation values" reflected the inference of species status to strains that demonstrated less than 70% DDH during reassociation kinetics experiments (Konstantinidis and Tiedje 2005). While still adhering to the "phylo-phenetic species concept" originally determined through DDH, Richter and Rosselló-Móra (2009) argued for defining prokaryotic species based instead on average nucleotide identity (i.e., ANI). In terms of a pragmatic approach for determining what lineages should be considered species they stated, "We think that it is already time to shift the circumscription gold standards from the traditional DDH to a partial or better full genome sequencing approach. The 95–96% ANI threshold can be readily used as an objective boundary for species circumscription" (Richter and Rosselló-Móra 2009).

In contrast to the phylo-phenetic species concept, Cohan (2002) and Denef et al. (2010) have drawn parallels with concepts applied to eukaryotes when formulating metrics for defining prokaryotic species. For example, Cohan (2002) suggested, "A [prokaryotic] species is a group of organisms whose divergence is capped by a force of cohesion; divergence between different species is irreversible;

and different species are ecologically distinct." Furthermore, Cohan (2002) argued that unlike commonly accepted systematic treatments for bacteria, the properties he proposed as demarcating species were actually present in what prokaryotic systematists termed ecotypes and thus different recognized species were more equivalent to a genus (Cohan 2002). In a similar vein, Denef et al. (2010) analyzed two closely related groups of *Leptospirillum* bacteria that shared 95% ANI between their orthologs. In spite of their high levels of genetic similarity, these two genotypic groups were differentially adapted to various environmental parameters. In summarizing their findings in the context of prokaryotic species delimitation, Denef et al. (2010) concluded, "Our data emphasize the role of sequence and expression variation of shared genes in ecological divergence. As such, we highlight an interesting parallel to higher organisms, in which evolution of gene expression has been suggested as an important factor in species differentiation."

Even though there have indeed been significant advances in the ability to define species of prokaryotes, the occurrence of widespread HGT still represents a dilemma for systematic treatments (Fraser et al. 2007). For example, Doolittle and his colleagues have emphasized repeatedly the significance that must be placed on accounting for genetic exchange in not only prokaryotic origins, but eukaryotic origins as well (Doolittle 1999; Boucher et al. 2003; Nesbø et al. 2006; Papke et al. 2007). Thus, Papke et al. (2007) concluded that the boundaries between different haloarchaeal lineages were porous, leading to the further inference that "no nonarbitrary way to circumscribe 'species' is likely to emerge for this group, or by extension, to apply generally across prokaryotes." Though striking a cautionary note for those who wish to construct prokaryotic phylogenies or, ultimately, the "universal tree of life," the recognition of the role of genetic exchange in organismal origins reflects the richness added by this process to evolutionary pattern and process. One of Doolittle's (1999) conclusions illustrates well the central role played by genetic exchange in the origins, and continued evolution, of the three domains of life. He states, "More challenging is evidence that most archaeal and bacterial genomes (and the inferred ancestral eukaryotic

nuclear genome) contain genes from multiple sources ... Molecular phylogeneticists" may fail "to find the 'true tree,' not because their methods are inadequate or because they have chosen the wrong genes, but because the history of life cannot properly be represented as a tree." In the next section, this conclusion will be highlighted further by a discussion of the genic species concept (GSC).

2.6 Genetic exchange and the genic species concept

As emphasized by Harrison (2012) and in Chapter 1 of the present book, hybrid zone literature has for some decades demonstrated the phenomenon of differential permeability to genetic exchange (à la Key 1968) across different genomic regions (e.g., Hunt and Selander 1973; Gartside et al. 1979; Shaw et al. 1979; Harrison 1986; Santucci et al. 1996; Brumfield et al. 2001; Chavez et al. 2011; Robbins et al. 2014). The observation that some regions of the genomes of related, but taxonomically distinct, lineages are reproductively isolated (i.e., there are few or no hybrids demonstrating recombination in these genomic regions), while genetic exchange occurs in other regions, led Wu (2001) to formulate the GSC. In particular, Wu (2001) proposed that the GSC reflected the reality that speciation occurred before the reproductive isolation of the entire genomes of related taxa. Wu (2001) argued that the process of speciation and the species concept should not be based on the neo-Darwinian construct of whole organism/whole genome reproductive isolation. Instead, he contended that evolutionary biologists should return to the Darwinian conceptual framework of defining the process of speciation in terms of differential adaptation (via natural and sexual selection) at the genic level.

Though Wu's (2001) contention that evolutionary biologists had not recognized the "genic" nature of speciation has been challenged (Harrison 2012), his formal derivation of the GSC is still of great value. For example, the argument that evolutionary biologists—particularly evolutionary zoologists—generally embrace a concept of speciation similar to the GSC is not reflected in the relevant scientific literature from the period of the modern synthesis onward (see Arnold 1997, 2006, and Chapter 1 of the present book). Instead, authors reporting speciation often continue the pattern of describing divergence in such terms as percent fertility or viability of certain hybrid classes (e.g., F_1 or first backcross generations). Though useful for comparing levels of reproductive isolation between related taxa, if speciation is more often than not accompanied by genetic exchange between diverging lineages, these values are at the best confusing and at the worst deceptive for predicting the processes leading to biodiversification. Wu (2001) was thus correct in his contention that there continues to be a need for a concept like the GSC to illuminate the process of speciation.

A number of examples supporting the GSC as a conceptual framework for understanding evolutionary diversification have already been discussed in Chapter 1 and in the previous sections of the current chapter. Likewise, each of the remaining chapters will provide numerous illustrations of the benefits to be derived from divorcing studies of the process of diversification from the whole-genome and whole-organism bias of the BSC and similar conceptual platforms. Indeed, divorce proceedings have actually already begun, as reflected by older and newer treatments summarizing the diversity of mechanisms associated with the process of "speciation" (e.g., Harrison 1990, 2012; Arnold 1992, 1997; Abbott et al. 2013; Seehausen et al. 2014).

2.7 Conclusions

An indication of whether or not a particular investigator defines genetic exchange as evolutionarily significant can often be gathered from how this process has been described. For example, is the process considered, at best, to be an epiphenomenon? The degree to which adherence to a particular species concept reflects or determines the evolutionary importance ascribed to this process can also be assayed by examining the types of analyses carried out by investigators who adopt different species concepts. For example, evolutionary biologists who subscribe to the BSC and PSC often assume that the formation of new species through genetic exchange between different species is vanishingly rare or theoretically impossible. However, these types of conclusions need not be made.

Diversification that is coincident with genetic exchange among lineages can be incorporated into a consideration of evolutionary processes in general. Accepting this approach allows the recognition of the philosophical and scientific tension created by studying a dynamic process (Hey et al. 2003; Pigliucci 2003; Sites and Marshall 2004). This approach also allows the application of conceptually and mechanistically broadly based, rather than narrow, models for defining species. When necessary, any of the species concepts discussed in this chapter may be relaxed to incorporate reticulate evolution. However, models of the type encapsulated by the cohesion, prokaryotic and GSC allow the best opportunity for investigating all of the possible evolutionary processes. Using concepts of this sort, or relaxing the strictures of other concepts, can thus provide rigorous tests of the relative roles of all the potential processes—including genetic exchange—associated with evolutionary diversification in organisms as diverse as bacteria and mammals.

CHAPTER 3

Testing for genetic exchange

"The more imperceptible introgression becomes, the greater its biological significance . . . the wide dispersal of introgressive genes (perceptible only to the most exquisitely precise techniques) would be a phenomenon of fundamental importance." **(Anderson 1949, p. 102)**

"Bifurcating phylogenies are frequently used to describe the evolutionary history of groups of related species. However, simple bifurcating models may poorly represent the evolutionary history of species that have been exchanging genes." **(Machado and Hey 2003)**

"In generations F_0-F_2 . . . two male and five female resident G. fortis that mated with members of the immigrant lineage were genotyped. Their identities were first established by their measurements . . . then assessed with a no-admixture model in Structure." **(Grant and Grant 2009)**

"Combination of the data that the nif cluster is conversed in the 15 N2-fixing Paenibacillus strains and the G+C contents of the nif clusters are higher than those of the average of the entire genomes, we proposed that N2-fixing Paenibacillus strains were generated by acquiring the nif cluster via HGT." **(Xie et al. 2014)**

3.1 Genetic exchange: A testable hypothesis

Since Anderson wrote his quote, there have been technological advances resulting in "exquisitely precise techniques" that provide the means for testing and falsifying the null hypothesis of no genetic exchange during evolutionary diversification. From an optimistic viewpoint we could argue that we are living in a golden age typified by both the raw material (i.e., data sets consisting of many thousands to millions of characters) and the rapid multiplication of methodologies for deciphering signatures of past and contemporary gene flow events (e.g., Huson and Bryant 2006; McBreen and Lockhart 2006; Gauthier and Lapointe 2007; Degnan and Rosenberg 2009; Knowles 2009; Geneva and Garrigan 2010; Choi and Hey 2011; Chan et al. 2013; Ackermann et al. 2014; Andújar et al. 2014; Glöckner

et al. 2014; Robinson et al. 2014; Veeramah and Hammer 2014; Roux and Pannell 2015). Yet, the various approaches, including methods for generating data and the various analytical approaches for deciphering the data, each have their weaknesses and thus necessitate caution in their application (e.g., Wagner et al. 2013; Wisecaver and Hackett 2014; Huber et al. 2015).

Regardless of the type of data being collected to test for genetic exchange (genomic, morphological, behavioral, etc.), there will be limitations in their usefulness. The limitations inherent in all types of data and data analyses indicate the need for a comparative approach (Arnold 2006). In each of the following sections, I will illustrate different methodologies available to test for the occurrence of genetic exchange, beginning with the critical need to be able to distinguish reticulate evolutionary patterns caused by genetic exchange versus those resulting from the coalescent properties of divergence (i.e., due to incomplete lineage sorting).

3.2 Testing the hypothesis: Genetic exchange and incomplete lineage sorting

One conclusion that will be highlighted in the following sections is the need to avoid inferences concerning the presence/absence of genetic exchange based on *a priori* assumptions of the investigator. For example, zoologists often assume that allele sharing between different taxa is due to incomplete lineage sorting, while botanists are more likely to assume that the same pattern of allele sharing in plants reflects introgression. Different investigators studying the same organismic group can even reach contrasting conclusions concerning the role of genetic exchange-mediated evolution (e.g., compare the inferences of Lack 1947 versus those of Grant and Grant 2014a for the Darwin's finches). However, with the collection of multiple classes of data, and the application of an array of analytical methodologies, rigorous inferences are possible regarding the frequency of web-of-life processes.

3.2.1 Genetic exchange and incomplete lineage sorting: Some analytical approaches

The history of phylogenetic investigations is characterized by inferences concerning the contribution of genetic exchange and incomplete lineage sorting in producing non-concordant evolutionary trees (e.g., Hallström et al. 2011; Amaral et al. 2014). However, for inferences to be rigorous they need to be based upon explicit tests to discern between the relative roles of these two processes (Eckert and Carstens 2008; Joly et al. 2009; Bapteste et al. 2012, 2013; Yu et al. 2014). Yu et al. (2013) emphasized the critical need for such tests, stating, "in addition to hybridization, the incongruence among gene trees may be partly caused by incomplete lineage sorting (ILS) . . . Ignoring the presence of ILS could result in an over- or underestimation of the amount of hybridization events and/or wrong inference of the location of these events. Recent studies have documented large extents of ILS in groups of organisms across the tree of life." Kubatko (2009) put

it more succinctly when she argued, "simply observing incongruence in reconstructed phylogenies is insufficient for concluding that hybridization has occurred." Of course, the same can be stated if incomplete lineage sorting is either assumed as the only important factor causing, or inferred simply from the observation of, phylogenetic discordance (e.g., Takahashi et al. 2001; Heckman et al. 2007; Grummer et al. 2014). In this regard, some frequently used phylogenetic methodologies explicitly assume a model in which all discordances are caused by incomplete lineage sorting (e.g., "Best" and "Beast": Edwards et al. 2007; Heled and Drummond 2010; Chung and Ané 2011; Drummond et al. 2012). However, if this underlying assumption is incorrect the "performance" of these models in estimating a species tree can be quite poor (Leaché et al. 2014).

Nakhleh and his colleagues have contributed greatly to both the description of existing models and construction of novel methodologies for estimating the effects of various parameters—including incomplete lineage sorting and genetic exchange—on the topology of phylogenetic trees. Though they and other workers (e.g., Joly et al. 2009; Kubatko 2009; Meng and Kubatko 2009; Gerard et al. 2011; Yu et al. 2011) had proposed important methodological approaches for simultaneously calculating the effects of deep coalescence and genetic exchange (i.e., HGT or hybridization), these methodologies were only applicable to special cases "where the phylogenetic network topology is known and contains one or two hybridization events, and a single allele sampled per species" (Yu et al. 2012).

To overcome the limitations, Yu et al. (2012) proposed methodologies for estimating the probabilities of gene tree topologies, given a phylogenetic network, that could accommodate numerous taxa and genetic exchange events and with no limit on the number of loci sampled per taxon. Though reflecting an improvement over previous methodologies, in terms of allowing the simultaneous analysis of many taxa, genetic exchange events, and loci, Yu et al.'s (2012) methodology still possessed two major weaknesses. The first was the requirement of a phylogenetic network. The second was the fact that the method was computationally very expensive. To address these issues somewhat,

Yu et al. (2013) proposed the use of a parsimony approach (*à la* Maddison 1997). Both simulations and application of this approach to genomic data sets from yeast indicated that this was a computationally efficient technique that accounted for incomplete lineage sorting in genome-wide tests for genetic exchange (Yu et al. 2013). It is important to point to the drawback of the parsimony assumption for this model. However, this methodology requires only gene tree topologies that can be estimated any number of ways and provides a phylogenetic network possessing "inheritance probabilities that correspond to proportions of genes involved in each of the hybridization events inferred" (Yu et al. 2013).

Another methodology, developed originally to test for introgression versus incomplete lineage sorting (i.e., retention of ancestral polymorphisms) between archaic *Homo* species and *Homo sapiens*, has been named "ABBA/BABA" (Green et al. 2010). Durand et al. (2011) explained the meaning of this terminology in the following quote: "For the ordered set {P1, P2, P3, O}, we call the two allelic configurations of interest 'ABBA' or 'BABA.' The pattern ABBA refers to biallelic sites where P1 has the outgroup allele and P2 and P3 share the derived copy. The pattern BABA corresponds to sites where P1 and P3 share the derived allele and P2 has the outgroup allele." Green et al. (2010) defined a statistic (i.e., "D") based on the predictions that if introgression occurred between archaic *Homo* species and *H. sapiens* (1) DNA segments with unusually low divergence between the extinct and extant lineages should be detected and (2) these introgressed segments should show high sequence divergence in comparisons with non-introgressed *H. sapiens* populations. In the absence of introgression, low divergence between archaic and extant lineages could also be caused by low mutation rates since the split from the outgroup species (i.e., in this example, chimpanzee). However, if mutation rate is causative, contemporary human populations should demonstrate low divergence as well, which is the opposite outcome expected from introgression (Green et al. 2010).

Though a useful methodology for inferring the presence of ancient introgression, the D statistic does have the potential to be confounded by ancestral population subdivision (Durand et al. 2011;

Martin et al. 2015). For example, if the ancestral populations of archaic *Homo* lineages had been subdivided and both the archaic species and *H. sapiens* derived from the same subpopulation, then they would be more similar than expected, in the absence of gene flow (Green et al. 2010). However, the D statistic estimates along with other classes of observations can contribute to a robust hypothesis testing exercise. Once again, using the example from Green et al. (2010), there were three separate observations leading to the inference of introgression between archaic and extant lineages (Green et al. 2010; Wall et al. 2013). First, significant D values were detected. Second, there were several haplotypes found in the archaic European species and the European *H. sapiens* genomes that were absent in African populations of *H. sapiens*. Third, significantly more fragments displaying low divergence between the European samples of the archaic lineage and the European *H. sapiens* were detected relative to African *H. sapiens*. Use of the ABBA/BABA approach in concert with other observations (that can be drawn from the same data set used to generate the D statistic) thus reflects a powerful approach to test for genetic exchange.

In the next sections, I will highlight a number of studies using a variety of approaches to infer the relative roles of incomplete lineage sorting and genetic exchange in producing phylogenetic incongruence in plants, fungi, and animals. Some of these examples will reflect the methodologies described while some will illustrate additional means of testing for the effects of these alternative processes.

3.2.2 Genetic exchange and incomplete lineage sorting: Flycatchers

A number of avian clades have provided classic examples of divergence accompanied by genetic exchange (e.g., Darwin's finches: see Chapter 1, Section 1.5.3; sparrows: see Chapter 1, Section 1.6.2; Eurasian crows: Wolf et al. 2010, Poelstra et al. 2013, 2014). As discussed in Chapter 1 for the sparrows and Darwin's finches, the genetic exchange in these cases resulted in hybrid speciation and adaptive trait introgression. Recently, Rheindt et al. (2014) collected morphological (including plumage traits), vocalization, and genomic data for the neotropical tyrant-flycatcher clade. The complex contains two species—the northern, *Zimmerius chrysops* and the southern, *Z. viridiflavus* (Zimmer 1941—placed in genus *Tyranniscus* in this earlier treatment). Analyses of the morphological and vocalization data sets demonstrated that in a region between the northern and southern species a lineage existed (termed "mosaic") that possessed plumage characteristics of *Z. chrysops*, but other morphological and vocalization traits diagnostic for *Z. viridiflavus* (Rheindt et al. 2014).

In order to test for the processes that gave rise to the mosaic lineage, genotyping-by-sequencing was used to generate over 37,000 SNPs. Though showing high similarity across the entire genome to *Z. viridiflavus*, using a subset of SNPs (*c.* 2700) and the ABBA/BABA analytical methodology, Rheindt et al. (2014) were able to discriminate between the contributions of incomplete lineage sorting and genetic exchange in the occurrence of shared alleles. Their analysis implicated low levels of introgression from the northern species into populations of the southern taxon in the origin of the mosaic lineage. Finally, mapping of the flycatcher SNPs onto the zebra finch genome indicated an enrichment of loci that may affect plumage coloration in the genomic regions of the mosaic lineage showing introgression from *Z. chrysops* (i.e., the putative source of the plumage traits; Rheindt et al. 2014).

3.2.3 Genetic exchange and incomplete lineage sorting: Alpine lake whitefish

North American and European clades of lake whitefish belonging to the genus *Coregonus* have become evolutionary model systems, particularly for understanding diversification and adaptive evolution (e.g., Renaut et al. 2011; Winkler et al. 2011; Vonlanthen et al. 2012; Gagnaire et al. 2013; Lundsgaard-Hansen et al. 2014). With regard to the European species complex, both Hudson et al. (2011) and Winkler et al. (2011) inferred a role for genetic admixture in the rapid and parallel adaptive radiations across several alpine lakes. Figure 3.1 reflects the series of events hypothesized by Hudson et al. (2011) to have resulted in the origin of the parallel species

flocks in different lakes. Included in this scenario is the divergence of *Coregonus* lineages in separate glacial refugia, the invasion and formation of a hybrid swarm in habitats following glacial retreat, the dispersal of the introgressed lineages into separate lakes and, finally, the parallel formation of species flocks due to similar classes of ecological selection (Hudson et al. 2011).

The inference of past admixture at the base of the adaptive diversification in the Alpine lake whitefish was supported by a number of observations, including significant discordance between phylogenies constructed alternately from mtDNA or nuclear loci. Though Hudson et al. (2011) concluded that the discordance was due to divergence-with-gene-flow in the establishment of the various species flocks, they likewise recognized incomplete lineage sorting as the "major alternative hypothesis." Yet, the following comparative evidences suggested genetic admixture as the more likely cause of the cytonuclear discordance: (1) previous work led to the inference of two European glacial refugia for whitefish (e.g., Østbye et al. 2005), consistent with the two divergent mtDNA lineages; (2) the lack of phylogenetic resolution (i.e., branching order of the European mtDNA lineages) relative to the North American lake whitefish clade indicated the rapid and recent diversification of the European and North American lineages from a common ancestor; and (3) completed lineage sorting was observed between the North American and European clades. These observations, taken collectively, falsify incomplete lineage sorting as a major contributor to the evolution of the present-day Alpine lake whitefish species flocks, and instead indicate a causative role for introgressive hybridization (Hudson et al. 2011).

3.2.4 Genetic exchange and incomplete lineage sorting: Yeast

Fungi are now understood to reflect the significant role of genetic exchange-mediated evolution. Indeed, the fungal clade contains many member lineages marked by natural hybridization, introgression, or horizontal gene transfer events (Gladieux et al. 2014). The documentation of hybrid speciation and adaptive trait transfer (mediated by introgressive hybridization and HGT) in numerous

Divergence in refugia—
divergent mitochondrial
haplotypes evolve in
North (N) and Central (C)
European refugia

Admixture in hybrid swarm—
nuclear loci from N and C
refugia recombine, genetic
diversity increases

Sympatric speciation—
species within each lake form
clades based on nuclear loci,
mitochondrial haplotypes are
not fixed within lakes

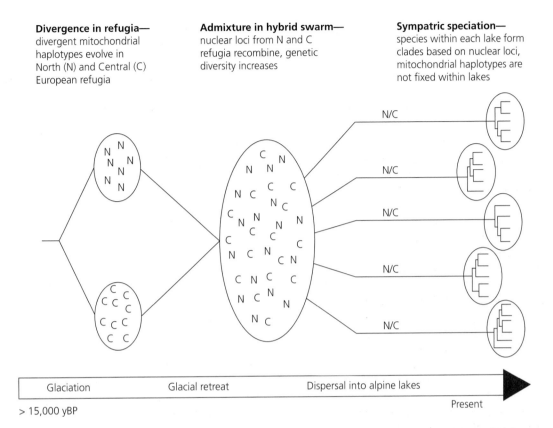

Glaciation   Glacial retreat   Dispersal into alpine lakes

\> 15,000 yBP   Present

Figure 3.1 Putative evolutionary history of the Alpine lake whitefish species complex. The origin of the individual species flocks included introgressive hybridization between multiple, divergent lineages coexisting in glacial refugia, resulting in the formation of hybrid swarms. The signature of introgression was lost from the nuclear genomes of the resulting lineages due to recombination. In contrast, the hybrid nature of the founders of the individual flocks is reflected by the presence of multiple, divergent mtDNA haplotypes both within and among different lakes (Hudson et al. 2011).

fungal lineages has thus led to the inference that "these sources of adaptive change may have made a substantial contribution to evolution, rather than being merely anecdotal" (Gladieux et al. 2014). For example, Morales and Dujon (2012) discussed the evidence for widespread introgressive hybridization between various yeast lineages, including transfers between species of *Saccharomyces*, between *Saccharomyces* and other genera, and between different lineages within the pathogenic genus *Cryptococcus*. These workers also reviewed many examples of hybrid speciation within the *Saccharomyces* sensu stricto species complex. To support their inference of the potential significance of hybrid species formation within this group, they pointed to "the fact that even some type strains (used to define species in classical taxonomy) are actually complex hybrids

emerged from successive ancestral hybridization events" (Morales and Dujon 2012). Likewise, Peris et al. (2012) and Dunn et al. (2012) found that the majority of divergent *S. cerevisiae* strains from natural and artificial (i.e., fermentative) habitats possessed introgressed material from other species in this genus.

Like any other organismic group, testing for the role of reticulate evolutionary processes in the origin, diversification, and adaptation of *Saccharomyces* species and indeed fungi, in general, necessitates analyses that are able to parse out the effects of incomplete lineage sorting and divergence-with-gene-flow. This may be even more critical for a group like fungi, because the inference of genetic exchange-mediated evolution of various lineages and clades has been intractable until recently (Morales and Dujon 2012; Gladieux et al. 2014). However, discerning between coalescence and genetic exchange events has been accomplished in analyses of various members of the genus *Saccharomyces*.

Peris et al. (2014) recently reported the findings of an investigation into the origin of the allopolyploid *S. pastorianus* used in the brewing of lager beer. Though diagnosed as a hybrid species from two independent natural crosses between *S. cerevisiae* and the Patagonian species *S. eubayanus* (Libkind et al. 2011; but also see Bing et al. 2014), it was not known which of the divergent lineages of *S. eubayanus* had contributed to the origin of *S. pastorianus*. In addition, the evolutionary history and genomic structure of natural populations of *S. eubayanus* within Patagonia, and of a recent isolate of *S. eubayanus* outside of Patagonia (i.e., in the state of Wisconsin) were similarly undefined (Peris et al. 2014). Their results included the following: (1) the detection of high genetic diversity, and two divergent lineages, among the South American *S. eubayanus* samples; (2) very low diversity of the *S. eubayanus* genomic contribution (i.e., from a small subpopulation of the "B" lineage) to the allopolyploid *S. pastorianus*; and (3) the inference that the Wisconsin sample of *S. eubayanus* derived from hybridization between the two divergent Patagonian lineages.

Peris et al. (2014) found that an array of methodologies were necessary to diagnose the population genetic structure and evolutionary history of the various *Saccharomyces* lineages. They were thus unable to decipher the "complex and varied reticulation events" using standard phylogenetic methodologies. Instead, robust inferences were only possible by analyzing Bayesian concordance factors and constructing phylogenetic networks (Peris et al. 2014). Combining these two approaches provided an independent topological structure (and model of evolution) derived from each gene with a separate statistical support estimate for the effect of each gene on phylogenetic splits and clades (Peris et al. 2014).

3.2.5 Genetic exchange and incomplete lineage sorting: Pampas grasses

Discordance between phylogenies constructed for the same taxa, but based alternatively on either nuclear or cytoplasmic (i.e., mtDNA for animals and chloroplast DNA [cpDNA] for plants) loci, is common (see Arnold 2006, 2009 for examples). As mentioned until very recently, zoologists and botanists often subjectively assigned such discordances to the alternate processes of deep coalescence and genetic exchange, respectively. Yet, a majority of the time, both zoologists and botanists agreed that to define the "true" evolutionary tree, and thus understand the "true" evolutionary history of a particular group, the non-concordant loci should be somehow minimized, eliminated, or combined with those loci demonstrating concordant phylogenies.

As will be argued throughout this text, this approach may lead to the resolution of a simple bifurcating phylogeny, but such a phylogeny will be demonstrably inaccurate with regard to both the underlying evolutionary processes and indeed even evolutionary relationships. However, this is not a novel conclusion. For example, Pirie et al. (2009) demonstrated the power of defining and then incorporating discordance into phylogenetic analyses in order to thoroughly test evolutionary hypotheses. They argued that "the importance of representing and correctly interpreting differences between gene trees is highlighted by the numerous phylogenetic studies, which have demonstrated gene tree conflict above the species level." In accord with this conclusion, Pirie et al. (2009) utilized the concordance and discordance found among gene trees to discern evolutionary pattern and process within the pampas grass genus *Cortaderia*. Specifically, these workers utilized sequence data from nuclear and cpDNA loci to construct phylogenetic trees both from the independent data sets and from a combination of the nuclear and cytoplasmic data. The general findings were that numerous reticulate events had occurred and that there were "significant discrepancies between inferences [of evolutionary relationships] based on phylogenies representing different proportions of the reticulations" (Pirie et al. 2009). Simulations designed to address the phylogenetic patterns predicted from incomplete lineage sorting versus genetic exchange led to the conclusion that both processes were likely contributing to the discordance in the pampas grass complex (Pirie et al. 2009). However, combining the simulation results, phylogenetic reconstruction methodology, and observations concerning known hybridization in this complex provided a robust inference that web-of-life processes were widespread and evolutionarily significant within the pampas grass clade.

3.2.6 Genetic exchange and incomplete lineage sorting: Orchids

Though genetic exchange and incomplete lineage sorting are considered by many to be the main causes of discordant gene genealogies, Van der Niet and Linder (2008) pointed to the need for testing for other "biological" (e.g., conflation of orthologous and paralogous sequences) and "nonbiological" (e.g., insufficient taxon sampling) factors when defining the evolutionary history of any organismic group. Specifically, Van der Niet and Linder (2008) applied a number of approaches in a study of the orchid genus *Satyrium* in order to diagnose whether the gene trees produced from different loci were significantly discordant and, if so, what the cause(s) of the discordance might be.

The analysis of phylogenetic relationships within *Satyrium* included 63 of the 91 species, from across the range of the genus, including lineages from Madagascar, sub-Saharan Africa, and Asia (Van der Niet and Linder 2008). To infer relationships, sequences were collected from each taxon for both chloroplast and nuclear (i.e., ribosomal RNA or "rDNA") loci. Phylogenetic reconstruction involved either methodologies assuming parsimony or Bayesian approaches that incorporated a number of partitioning schemes for the nuclear and cytoplasmic data sets (Van der Niet and Linder 2008). Though no significant discordances were detected among gene trees generated from the different cpDNA loci, disagreement in phylogenetic placement of taxa was present between trees based on either nuclear or cytoplasmic sequences. Van der Niet and Linder (2008) chose to manually inspect phylogenetic trees derived from the parsimony analyses and thereby determine which clades were producing non-concordance by systematically removing the minimum number of taxa necessary to "solve" the discordant relationships (i.e., to have completely concordant trees). The significance of discordances among the various data sets was also investigated

by individually adding the removed lineages back into the phylogenies and applying Farris et al.'s (1994) incongruence length difference test.

From this and additional analyses, the following conclusions were reached concerning the factors causing the observed phylogenetic discordance: (1) nonbiological, analytical artifacts such as inadequate taxon sampling were not causal; (2) orthology/paralogy conflation did affect the gene trees produced from the nuclear rDNA sequences; (3) incomplete lineage sorting was likely an important factor; and (4) natural hybridization was inferred for some of the discordances detected, particularly in the species occurring in Asia (Van der Niet and Linder 2008). These results once again indicate the necessity of using multiple approaches and explicit tests if a determination of the relative roles of deep coalescence and genetic exchange on phylogenetic discordances is to be accomplished.

3.3 Testing the hypothesis: The fossil record and genetic exchange

Darwin considered the biological record to be a wonderful source of information concerning past and present evolutionary processes as expressed in the following quote: "This wonderful relationship in the same continent between the dead and the living, will, I do not doubt, hereafter throw more light on the appearance of organic beings on our earth, and their disappearance from it, than any other class of facts" (Darwin 1845, p. 187). However, he also famously emphasized the limitations of the fossil record in lacking the intermediate forms predicted by his model, for example, stating, "Why then is not every geological formation and every stratum full of such intermediate links? Geology assuredly does not reveal any such finely graduated organic chain; and this, perhaps, is the most obvious and gravest objection which can be urged against my theory. The explanation lies, as I believe, in the extreme imperfection of the geological record" (Darwin 1859, p. 280). Indeed, the recognition of the incompleteness of the biological record continues today, even in the light of rich fossil finds in both marine and terrestrial habitats (e.g., Kalmar and Currie 2010).

Given the well-documented limitations of the fossil record due to "missing data," it may seem surprising that there are numerous examples of extinct plants and animals inferred to possess signatures of web-of-life processes. However, as concluded in a study of fossil crinoids (Ausich and Meyer 1994), not only is it possible to identify hybrid forms in the biological record, but the collection of data concerning fossil hybrids allows the testing of hypotheses concerning the long-term evolutionary significance of genetic exchange. In the following sections I will discuss a series of studies from groups as diverse as flowering plants and gastropods that reflect the utility of fossil remains in testing for genetic exchange in eukaryotes. Some of the best examples of "fossil hybridization" involve mammals, particularly primates, and come from the research of Ackermann and her colleagues (Ackermann et al. 2006, 2010, 2014; Ackermann 2007, 2010; Ackermann and Bishop 2010; Fuzessy et al. 2014). Because this work is often designed to help understand the evolutionary history of *H. sapiens* and related lineages, I will defer a detailed discussion of their findings until Chapter 8. Suffice it to say; this body of work demonstrates the power of utilizing morphological surveys from contemporary natural and experimental hybrids to formulate expectations of hybrid morphology in mammalian fossil remains.

3.3.1 Genetic exchange detected in the fossil record: Paleopolyploidy in plants

"Recent genomic investigations not only indicate that polyploidy is ubiquitous among angiosperms, but also suggest several ancient genome-doubling events. These include ancient whole genome duplication (WGD) events in basal angiosperm lineages, as well as a proposed paleohexaploid event that may have occurred close to the eudicot divergence." This quote from Soltis et al. (2009) reflects not only the effect of polyploidy on plant evolution, but specifically allopolyploidy. However, genomic studies, as important as they are for testing for whole genome duplications, were preceded by studies that arrived at the same inference based on other types of data (e.g., chromosome counts; Raven 1975), including those from the fossil record.

Eckenwalder (1984), utilizing fossil leaf material and data from contemporary hybrid zones

between species of *Populus*, concluded that hybridization had been occurring for at least 12 million years among these tree lineages. Furthermore, the putative hybrid fossil leaf material (i.e., from Miocene and Pliocene sediments) was not assignable to present-day hybrid taxa and thus was designated as the extinct species, *Populus × parcedentata* (Eckenwalder 1984).

Recently, Masterson (1994) and Lomax et al. (2014) have tested the hypothesis of a polyploid origin of angiosperm lineages and clades by analyzing fossilized guard cells from extant and extinct taxa. Both of these studies used guard cell size to estimate the nuclear DNA content and thus the chromosome number of angiosperm species. Though both studies led to the conclusion of pervasive genome duplications throughout the angiosperm clade, the differing sampling schemes revealed different aspects of polyploid evolution. For example, Masterson (1994) concluded that the primitive haploid chromosome number for all angiosperms was likely seven to nine, leading to the inference that most (c. 70%) extant flowering plants had polyploid progenitors. Given that most, if not all, plant polyploid lineages derive from hybridization (Clausen et al. 1945; Stebbins 1947; Soltis and Soltis 2009), Masterson's findings support the hypothesis that a majority of angiosperms are derived from hybrid ancestors. Consistent with Masterson's conclusions, Lomax et al. (2014) also inferred genome duplication events through guard cell sizes. However, their study included a fine-scale taxonomic survey of both gymnosperms and angiosperms. This extensive survey made possible the following additional observations: (1) a large proportion of the earliest land plant taxa possessed extremely large genome sizes; and (2) though the hypothesized trend of increasing genome size through geological time was falsified, maximum genome size (often associated with polyploidy) did increase across approximately 360 million years (Lomax et al. 2014).

3.3.2 Genetic exchange detected in the fossil record: Corals

Cases of contemporary reticulate evolution in the marine organisms known as corals are numerous, and involve examples of natural hybridization, introgression, and hybrid speciation (Willis et al. 2006; Arnold and Fogarty 2009; Ladner and Palumbi 2012; Prada and Hellberg 2014). For example, Budd and Pandolfi (2010) concluded, "In reef corals, hybridization has been found at the periphery of species ranges in both the Caribbean and Indo-Pacific regions . . . Although rare on ecological time scales, hybridization has thus been hypothesized to play an important role in reef corals in range expansion and adaptation to changing environments." However, they recognized that "much of the evidence for this hypothesis is from genetic and reproductive data on living corals, which are limited to ecological time scales."

In order to test whether or not the hypothesized role of genetic exchange among coral lineages was also reflected throughout a geological timescale, these authors analyzed morphological data from fossils belonging to the *Montastraea annularis* species complex (Budd and Pandolfi 2004, 2010). The *M. annularis* species complex is ideal for such an analysis because it has been the dominant ecological form, at least on Caribbean reefs, since the Plio-Pleistocene faunal shift (*c.* 2 mya), has retained its geographic range (i.e., Caribbean, western Atlantic, and Gulf of Mexico) throughout its history, has a well-known fossil record that extends back for greater than six million years and, very importantly for testing for reticulate evolution in fossil remains, there is a strongly supported correlation between genetic and morphological data (Budd and Pandolfi 2004, 2010).

The earlier study of Budd and Pandolfi (2004) involved a comparison of the degree to which fossil forms from the Bahamas overlapped in ecological and morphological characteristics relative to patterns observed among related, extant forms of *Montastraea*. Specifically, morphological and ecological data collected from fossil colonies that existed *c.* 125,000 years ago were compared to extant species for which introgression has or has not been inferred. The comparison of the fossil and extant forms revealed a greater degree of overlap between various classes of Bahamian fossil reef corals than are found in present-day species, leading Budd and Pandolfi (2004) to infer an extended duration of introgressive hybridization among these lineages.

Budd and Pandolfi (2010) confirmed the inference of the evolutionarily significant role of introgression throughout the evolutionary history of *Montastraea*. By extending the analyses of the fossil record of this complex to samples that ranged from 0.08–0.86 million years in age, they documented a high degree of evolutionary (i.e., morphological) novelty associated with putative introgression events (Figure 3.2). Specifically, they argued for the centrality of processes such as introgressive hybridization in affecting evolutionary trajectory at the edges of species distributions, and during climatic fluctuations (Budd and Pandolfi 2010; Figure 3.2).

3.3.3 Genetic exchange detected in the fossil record: Land snails

Cerion

Land snails belonging to the genus *Cerion* are well known for demonstrating the effects of extensive introgressive hybridization among extant forms (Hearty and Schellenberg 2008; Hearty 2010).

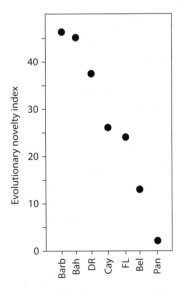

Figure 3.2 Depiction of the estimate of patterns of "evolutionary novelty" in samples of fossil and extant species of the coral genus *Montastraea*. Fossil forms were from Barbados ("Barb"), Bahamas ("Bah"), Dominican Republic ("DR"), Caymans ("Cay"), and Florida ("FL") while extant species were collected from Belize ("Bel") and Panama ("Pan"). The significantly elevated evolutionary novelty index values for Barbados and Bahamas were inferred to be at least partially due to introgressive hybridization (Budd and Pandolfi 2010).

For example, Woodruff and Gould (1987) described a 50-year-long experiment begun by the transplantation of 55 *Cerion casablancae* individuals from the Bahamas to the Florida Keys. Beginning in 1927, introgressive hybridization occurred between the introduced colony and the native *C. incanum*. The results of the genetic admixture between the two species were determined from morphological and genetic data collected in 1977 and included the disappearance of any non-introgressed *C. casablancae*, no evidence of selection against hybrid genotypes, and the presence of both intermediate and unique hybrid morphologies.

In addition to the numerous cases of natural hybridization between present-day lineages of *Cerion*, this genus is also a model system for testing the role of reticulate evolution across horizons of the biological record. The classic paper by Goodfriend and Gould (1996) illustrated the effect of introgressive hybridization involving now extinct and still existing species located on the island of Great Inagua. Specifically, they detected the origin and evolution of one population inferred to have originated from introgression between a fossil species that went extinct *c.*13,000 ybp and an extant species located near the ancient hybrid zone. In a second case, a hybrid zone persisted for thousands of years subsequent to the extinction of one of the parental forms, reflecting the apparently high fitness of some hybrid genotypes (Goodfriend and Gould 1996).

As with the previous analyses, two recent studies by Hearty and Schellenberg (2008) and Hearty (2010) have recorded potential effects of introgressive hybridization on both extinct and extant *Cerion* species. In the earlier study, morphological data were collected from 150 individuals ranging in age from 140,000 ybp to the present. Morphological variability across the various time horizons indicated periods of heightened variability, reflective of the interaction of environmental transitions and genetic admixture (Hearty and Schellenberg 2008). Likewise, Hearty (2010) described the morphological evolution of *Cerion* across a similar chronology (from *c.* 130,000 ybp to modern samples) and defined patterns of change characteristic of introgressive hybridization. In particular, he inferred a role for human transport of members of this genus throughout the Bahamian islands, resulting in

introgression between introduced and native species (Hearty 2010). It would thus appear that the ancient and recent evolutionary history of *Cerion* has been impacted by genetic admixture between divergent lineages.

Melanopsis

Heller, Sivan, and their colleagues have produced a series of studies of extinct and extant lineages from which they have inferred ancient (and recent) hybridization among species of the land snail genus, *Melanopsis*, in the Jordan Rift Valley, north of the Dead Sea (Heller and Sivan 2002; Grossowicz et al. 2003; Heller et al. 2005; Heller 2007). From their analyses, these workers have identified the signature of introgressive hybridization between the species *Melanopsis buccinoidea* and *M. costata*.

Morphologically intermediate forms of *Melanopsis* have been identified throughout the fossil chronologies from multiple sites, beginning from 1.5 mya and extending continuously through present day samples from areas of overlap (Heller and Sivan 2002; Grossowicz et al. 2003; Heller et al. 2005; Heller 2007). These results reflected "the earliest direct evidence of hybridization among molluscs that is still going on today in the same region and aquatic system, among the same species" (Grossowicz et al. 2003). Significantly, introgression has been documented between other extant *Melanopsis* species that are found throughout the fossil record as well. It is thus probable that genetic admixture is a chronologically and geographically widespread phenomenon impacting this entire gastropod clade.

3.3.4 Genetic exchange detected in the fossil record: Mammoths

It has been argued (e.g., Lister et al. 2005; Lister and Stuart 2010) that the now-extinct clade commonly known as "mammoths" (genus *Mammuthus*), rather than being typified by either anagenetic transition from one species to the next or simple bifurcating cladogenesis, instead was characterized by a more complex evolutionary history. Specifically, these authors inferred extensive geographical variation across northern Eurasia beginning with the appearance of the earliest lineage, *Mammuthus rumanus* (*c.* 3.5 mya), followed by the best known taxon, *M.*

meridionalis, during the Early Pleistocene (*c*. 2 mya). Fossil material from this latter taxon and the more recent, *M. trogontherii*, reflect signatures (e.g., bimodality of morphological traits as often found in extant hybrid populations) suggestive of introgressive hybridization between these two taxa *c*. 1.2–0.6 mya (Lister and Sher 2001). In this regard, Lister et al. (2005) inferred the presence of a hybrid zone between *M. meridionalis* and *M. trogontherii*, generating a third intermediate/mosaic morphological category.

Fossil evidence has indicated that, like the evolution of the other mammoth lineages, that of the so-called "woolly mammoth," *M. primigenius*, likewise involved a complexity indicative of web-of-life processes. Though hypothesized as being derived from *M. trogontherii*, samples of this species nonetheless displayed morphological variation best explained by a reticulate model. Lister et al. (2005) thus concluded, "There is again evidence of a complex interplay between populations in the Late Pleistocene of Europe, and the retention of both *trogontherii* and immigrant *primigenius* genes within them. Overall, in both the *meridionalis–trogontherii* and *trogontherii–primigenius* transitions, the advanced form originated in a peripheral area in a way corresponding to the first stage of an 'allopatric speciation' model, but its subsequent spread more closely approximates a multi-population, gene flow model."

Though not based on fossil evidence per se, but rather from mtDNA sequences obtained from fossil material, a study of the Late Pleistocene, North American mammoth species *M. primigenius* and *M. columbi* (i.e., Columbian mammoth) also resolved phylogenetic patterns that falsify a simple bifurcating model of evolutionary diversification (Enk et al. 2011). In particular, the *M. columbi* mtDNA genome fell within a *M. primigenius* subclade. Like the inferences based on fossil remains, this finding suggested introgressive hybridization between different mammoth species, in this case posited to have occurred in the ecotone between the glacial and the subglacial habitats characteristic of the woolly and Columbian mammoth species, respectively (Enk et al. 2011). Regardless of the specific scenario leading to the introgression between these two forms, as in many extant plant and animal taxa (see Arnold 1997, 2006 for reviews) it apparently resulted in the

replacement of the *M. columbi* mitogenome by that found in *M. primigenius* (Enk et al. 2011).

3.4 Testing the hypothesis: Contemporary hybrid zones

In Chapter 1, I discussed the historical and theoretical underpinnings of hybrid zone analyses, and illustrated how various conceptual frameworks have affected the interpretation of data, particularly with inferences concerning the forms of selection acting on hybrid genotypes (see Sections 1.2–1.4). However, in the context of the current chapter, it is necessary to illustrate how such studies have been used to test multiple hypotheses related to genetic exchange.

Inferences drawn from the analysis of contemporary hybrid zones, concerning the factors affecting ongoing genetic exchange, are best constructed through examinations of trans-generational, genetic, or phenotypic variation and habitat × genotype associations (Arnold and Martin 2010). For example, data collected from populations within and outside of putative hybrid zones allow tests for selective effects on the distribution of genotypes and phenotypes (e.g., Luttikhuizen et al. 2012; Hamilton et al. 2013; Alcaide et al. 2014; Patel et al. 2015). Though the role of genetic exchange-generated patterns will be emphasized, these types of observations also provide a robust framework for discerning the effects of factors such as incomplete lineage sorting, mutation, and genetic drift. In the following sections I will use complexes of plants and animals to exemplify the efficacy of hybrid zone analyses for a wide array of evolutionary inferences; the plant clades include irises, *Eucalyptus,* and oak species, while the animal examples come from studies of grasshoppers, house mice, and chickadees.

3.4.1 Genetic exchange and hybrid zones: Louisiana irises

In 1931, a paper appeared in the *Contributions of the New York Botanical Gardens*. Entitled "Botanical interpretation of the Iridaceous plants of the Gulf States," and authored by Small and Alexander, this paper was the stimulus for a scientific cottage-industry generating research now spanning more than eight decades (e.g., Viosca 1935; Foster 1937; Riley 1938; Randolph et al. 1967; Bennett and Grace 1990; Arnold et al. 1990a; Cruzan and Arnold 1993; Carney et al. 1994; Burke et al. 1998; Johnston et al. 2001; Cornman et al. 2004; Bouck et al. 2007; Meerow et al. 2007, 2011; Martin et al. 2008; Taylor et al. 2009; Tang et al. 2010; Dobson et al. 2011; Ballerini et al. 2013; Brothers et al. 2013; Hamlin and Arnold 2014).

The focus of Small and Alexander's (1931) study was the plant group commonly known as the Louisiana irises. Ironically, Small and Alexander's main conclusion—that there existed over 80 species of Louisiana iris within the confines of Louisiana—was demonstrably wrong. Instead, Small and Alexander had discovered, and named in some detail, products of introgressive hybridization between three morphologically, developmentally, reproductively, and ecologically distinct species, *Iris fulva*, *I. hexagona*, and *I. brevicaulis* (Viosca 1935; Foster 1937; Riley 1938). Although Small and Alexander's taxonomic work was flawed, their studies resulted in recognition of the Louisiana irises as an excellent example of the evolutionary effects arising from web-of-life processes. Specifically, these species occupy a fundamentally important scientific and historical niche as the exemplar of Anderson for the process of introgressive hybridization (Anderson 1949).

In rapid succession, Viosca (1935), Foster (1937), and Riley (1938, 1939, 1942, 1943a, b) responded to Small and Alexander's (1931) taxonomic treatment with evidence falsifying the taxonomic recognition of a majority of the "species" named by them. Instead, these taxa were shown to be recombinant (i.e., hybrid) genotypes (e.g., Figure 3.3; Riley 1938).

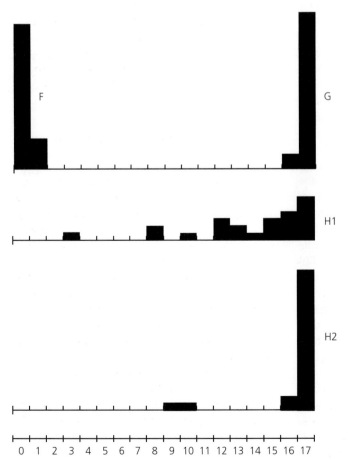

Figure 3.3 Character index distribution for populations of Louisiana irises. Each population (i.e., F, G, H1, H2) demonstrates admixtures of characters from *Iris fulva* and *I. hexagona*. A character index score of 0 indicates *I. fulva* individuals, a score of 17 typifies *I. hexagona* plants and intermediate scores reflect hybrid genotypes. The hybrid plants possess morphologies typical of different "species" recognized by Small and Alexander (1931) (from Riley 1938).

The data generated by Viosca, Foster, and Riley demonstrated that three species, *Iris fulva, I. brevicaulis*, and *I. hexagona*, had hybridized naturally to form the 70 or more remaining forms described by Small and Alexander (1931). Though comparisons of these species showed differences in pollinators, ecological setting, and flowering seasons, when in sympatry they formed hybrid zones (Viosca 1935; Riley 1938). This then was the explanation for the many forms recognized by Small and Alexander (Figure 3.3).

Anderson published his classic *Introgressive Hybridization* (Anderson 1949) subsequent to the initial flurry of research regarding the number of Louisiana iris species. In the first chapter of his book, Anderson chose the morphological characters described by Riley (1938; Figure 3.3) to illustrate the process of introgressive hybridization. In an earlier paper Anderson (1948) had used these same species to exemplify the effect of human disturbance on introgression. In both of these treatments, Anderson proposed a significant role for hybridization in the evolutionary history of Louisiana iris species. In particular, he concluded that natural hybridization had resulted in introgression. Anderson held that the general significance of introgression was to greatly increase the "variation in the participating species" and thus "far outweigh the immediate effects of gene mutation" (Anderson 1949, pp. 61–62). Anderson's viewpoint concerning the evolutionary importance of natural hybridization in the Louisiana irises (and other species; Anderson 1949) rested on his conclusion that introgression had impacted the gene pools of the hybridizing species.

A series of more recent studies, employing the discrete genetic markers not available to the earlier workers, have confirmed the presence of numerous hybrid zones between *I. fulva, I. brevicaulis,* and *I. hexagona* (e.g., Arnold et al. 1990a, b, 1991; Arnold 1993; Cruzan and Arnold 1993; Johnston et al. 2001; Hamlin and Arnold 2014). Surprisingly, while each study detected Louisiana iris hybrid zones, replete with later generation hybrid genotypes, there was a lack of F_1 hybrids. This finding suggested the presence of selection against F_1 hybrids, but selection favoring some post-F_1 hybrid genotypes, with the latter being established at significantly higher frequencies relative to the F_1 generation (Arnold 1994,

2000). The discovery of a high frequency of adult plants in natural hybrid zones demonstrating later generation hybrid genotypes is consistent with this hypothesis (Cruzan and Arnold 1993, 1994; Johnston et al. 2001). Furthermore, a study involving the introduction of experimental F_1 hybrids into a mixed population of *I. fulva* and *I. hexagona* found results consistent with this hypothesis as well. After natural pollinations, the frequencies of F_1 offspring present in fruits on *I. hexagona* and *I. fulva* plants were 0.74 and 0.03%, respectively. In contrast, the percentages of BC_1 progeny formed from pollen transfer from the introduced F_1 plants, onto *I. hexagona* and *I. fulva* flowers were 6.9 and 1.7%, respectively (Hodges et al. 1996). Taken together, the rarity of F_1 hybrid seeds and the order of magnitude greater frequency of BC_1 seeds were consistent with the hypothesis that the formation and establishment of F_1 hybrids between the Louisiana iris species was the bottleneck for introgressive hybridization. However, once this rare event occurred, the F_1-generation plants functioned as a substantial bridge for further genetic exchange and evolutionary diversification through the production of hybrid zones (Figure 3.4; Arnold 1993; Cruzan and Arnold 1993; Burke et al. 2000a; Taylor et al. 2011).

3.4.2 Genetic exchange and hybrid zones: Eucalyptus

The Australian tree genus *Eucalyptus*, known by locals as "gum trees," has been recognized as a paradigm of reticulate evolution for decades (e.g., Clifford 1954; Potts and Reid 1988, 1990; Jackson et al. 1999; McKinnon et al. 2004; Steane et al. 2011). This diverse species complex has thus been used to illustrate a diverse array of outcomes from genetic exchange, including introgressive hybridization as an important evolutionary process per se, conservation of rare forms in the face of gene flow, and the impact on phylogenetic reconstruction (Potts and Reid 1988; Butcher et al. 2005; Steane et al. 2011).

A recent series of studies by Field et al. (2008, 2009, 2011a, b) highlight well the utility of hybrid zones to test hypotheses concerning the evolutionary and ecological significance of divergence-with-gene-flow. Each of these analyses involved the assessment of the effect of introgression on the rare

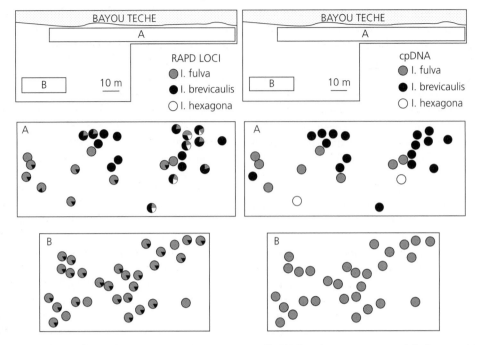

Figure 3.4 Distribution of species-specific nuclear ("RAPD") and cytoplasmic ("cpDNA") markers in a Louisiana iris hybrid zone involving *I. fulva*, *I. brevicaulis*, and *I. hexagona*. Each pie diagram represents a single plant. No F$_1$ hybrid genotypes are present; instead, there is extensive admixture of both nuclear and cytoplasmic markers indicative of advanced generation hybrid genotypes. This reflects apparent selection against F$_1$s, but the selective favoring of some later generation hybrid genotypes (from Arnold 1993).

species *Eucalyptus aggregata*. For example, in the first of the studies, the effect of population size of *E. aggregata*—both absolute size and relative size with regard to the sympatric species *E. rubida, E. viminalis,* and *E. dalrympleana*—on the generation of hybrids, levels of genetic diversity, seed production, and seedling performance was investigated (Field et al. 2008). As might be expected, when there was a predominance of potential interspecific pollen sources, hybrid seed production increased. Also in line with expectations, genetic diversity increased and seed production decreased with smaller population sizes in *E. aggregata*, suggesting that introgression might genetically enrich smaller populations, but at the same time lead to hybrid breakdown (Field et al. 2008). Overall, the results from this study, as well as Field et al. (2009) suggested that introgression in the hybrid zones between *E. aggregata* was frequent enough to lead to genetic assimilation of the rare form by its more common congeners.

The latter studies by Field et al. (2011a, b) also defined parameters contributing to the genetic

structure of hybrid zones, reproductive isolation among parental and hybrid trees, and the relative fitness of parental and hybrid genotypes. By utilizing genealogical assignments, admixture estimates, as well as recording mating system and demographic data, some of the causal factors contributing to asymmetric introgressive hybridization, hybrid formation, and inferred fitness were identified. In particular, the timing and frequency of flowering by parental and hybrid genotypes and the different floral morphologies of *E. aggregata* and its sympatric congeners affected the extent and directionality of introgression and the degree of reproductive isolation among hybrid and parental plants (Field et al. 2011a, b). Thus, as with other well-studied systems, *Eucalyptus* demonstrates a complex set of factors affecting the origin, maintenance, and evolution of hybrid zones.

3.4.3 Genetic exchange and hybrid zones: Oaks

As with many groups of organisms that demonstrate the ability to exchange genes (e.g., many mammalian lineages; Arnold et al. 2015), the evolutionary significance of natural hybridization among species belonging to the oak genus, *Quercus*, has been demonstrated repeatedly. For example, Trelease (1917) reported > 120 hybrid derivatives from natural hybridization between North American species. Likewise, Palmer (1948) concluded that introgression had been a major contributor to the evolution of the same *Quercus* clade. In the latter study, more than 70 individual hybrid types (from a variety of interspecific crosses) were identified. Grant also reflected such a conclusion when he used white oak species to exemplify the concept of a syngameon (Grant 1981, pp. 237–238).

Numerous studies of *Quercus* have also addressed the hypothesis of reticulate evolution from a population perspective. For example, in the first issue of the journal *Evolution*, Stebbins et al. (1947) tested the hypothesis that *Quercus marilandica* and *Q. ilicifolia* hybridized naturally on Staten Island, New York (Davis 1892). Describing the population, Davis (1892) observed that there were hundreds of individuals easily identified as belonging to one or the other species. However, he also observed "a number of trees whose place is not so evident, and it becomes a question when viewing them as to whether they exhibit more of the characters of *ilicifolia* or *nigra*" [now *marilandica*]. Stebbins et al. (1947) tested this hypothesis by collecting morphological data from allopatric and sympatric populations of *Q. marilandica* and *Q. ilicifolia*. These data were used for a hybrid index analysis. As Davis' insightful description suggested, the distribution of the species-specific characters made it clear that introgression had impacted these two oak species (Stebbins et al. 1947).

As with North American clades, two recent analyses of Asian oak species (Zeng et al. 2010, 2011) have also detected the effects of genetic exchange-mediated evolution in areas of ancient and recent sympatry. Both studies involved the examination of interactions between *Q. liaotungensis* and *Q. mongolica*. The first of the investigations was undertaken in order to test whether or not the two taxa were sufficiently distinctive to warrant species recognition. Zeng et al. (2010) utilized a battery of nuclear markers to determine the degree of differentiation within and between *Q. liaotungensis* and *Q. mongolica*,

whether or not any loci demonstrated greater than expected differentiation between the two taxa (i.e., reflective of divergent selection), and the pattern of introgression in hybrid zones. The results obtained confirmed the specific status of the two lineages. In addition, a set of six loci possessed a signature of divergent selection between the *Q. liaotungensis* and *Q. mongolica* genomes, also consistent with the recognition of these two taxa at the species level. Zeng et al. (2010) did, however, detect numerous hybrid zones, all reflecting introgression between the two species.

Zeng et al. (2011) extended the analysis in order to define the pattern of introgression between *Q. liaotungensis* and *Q. mongolica* by collecting sequence data for both cpDNA and nuclear loci. Though this second analysis detected consistent asymmetric introgression from *Q. mongolica* into *Q. liaotungensis*, the genetic structure varied between different hybrid zones. Hybrid zones in northeast China were characterized by patterns indicative of ancient introgression and possibly increased reproductive isolation between the two species, while those zones located in north China reflected recent introgressive hybridization and hybrid swarms (Zeng et al. 2011). It was concluded that the alternate evolutionary outcomes observed in the different hybrid zones likely resulted from the effects of both history and natural selection (Zeng et al. 2010, 2011).

3.4.4 Genetic exchange and hybrid zones: Australian grasshoppers

The Australian/Papuan grasshopper genus *Caledia* was developed into a remarkable model system for studying evolutionary phenomena by the late David Shaw and his colleagues. The range of topics addressed included population genetics, chromosomal evolution, speciation, adaptation, and the role of climate change in organismal distributions. Remarkably, all of the topics were investigated with a genus that ostensibly contains only two species, *Caledia captiva* and *Caledia species nova 1* (Shaw 1976), with much of the work involving a series of hybrid zones and clinal variation for chromosomal and genetic markers. The presence of only two recognized species (however, see later for the discussion of a sibling species) within this genus belies the remarkable genetic, ecological, behavioral, and developmental diversity housed within various sub-specific taxa (e.g., Shaw 1976; Shaw et al. 1976, 1980, 1983; Shaw and Wilkinson 1978; Daly et al. 1981; Colgan 1985, 1986; Arnold et al. 1986; Kohlmann et al. 1988; Groeters and Shaw 1992, 1996; Groeters 1994). I am going to review, in some detail, the findings of studies addressing various topics associated with the evolution of this genus. There are two reasons for my presenting an extended discussion of this group. First, it is an excellent example of reticulate evolutionary processes. Second, and in my estimation of equal importance, I hope to motivate new research into this model system.

Some of the earliest studies of *Caledia* detected distinct chromosomal forms that occurred in populations from southern Victoria, up the east coast of Australia to northern Queensland, along the north coast to the Northern Territory and across the Torres Strait to Papua New Guinea (Shaw 1976; Shaw et al. 1976). Subsequent analyses of chromosome, allozyme, highly repeated DNA, morphological, and experimental hybridization data further defined this genus into eight differentiated lineages (Shaw 1976; Shaw et al. 1976, 1980, 1988; Moran and Shaw 1977; Daly et al. 1981; Arnold and Shaw 1985; Arnold 1986; Arnold et al. 1986, 1987a, b; Marchant et al. 1988). These "races" and species included the southeast Australian, Moreton, southern Torresian, northern Torresian, Northern Territory Torresian, Papuan Torresian, Daintree, and *C. species nova 1* taxa.

Additional resolution of *Caledia* populations into evolutionary categories was accomplished through a comparative approach, utilizing numerous data sets. Thus, although a north to south cline in centromere position for each chromosome (Figure 3.5) resulted in completely acrocentric linkage groups in Victoria and metacentric chromosome complements in the extreme northern populations (Shaw et al. 1988), the southeast Australian and Moreton taxa were recognized as belonging to the same subspecies. This inference was drawn due to similarities in chromosome banding patterns (Shaw et al. 1988), allozyme variation (Daly et al. 1981), mtDNA (Marchant 1988; Marchant and Shaw 1993; Shaw et al. 1993), and highly repeated DNA sequences (Arnold and Shaw 1985), along with reproductive compatibilities detected through experimental hybridization (Shaw et al. 1980). Taken

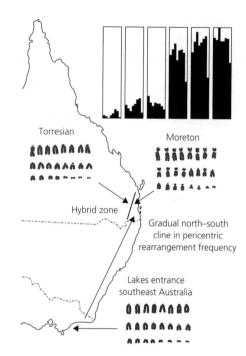

Figure 3.5 Distribution of chromosome variation within the Torresian, Moreton, and southeast Australian races of the grasshopper *Caledia captiva*. The location of the contemporary hybrid zone between the Torresian and Moreton chromosomal races is noted. The changeover in the frequency from the Moreton to the Torresian chromosomal form across the present-day hybrid zone located in southeastern Queensland is indicated by the histogram at the top. This changeover occurs over a distance of only 200 m (from Shaw and Coates 1983).

together, these data suggested that the populations located along this cline belonged to a single evolutionary clade. The same class of data also indicated that all of the Torresian forms should be placed into a second subspecies, that the Daintree form was a sibling species of *C. captiva*, and that *C. species nova 1* represented a third, morphologically-differentiated, and reproductively-isolated species (Arnold et al. 1987b).

The richness of the *Caledia* species complex for tests of evolutionary pattern and process per se is obvious from the findings. However, these organisms are likely best known from detailed, multi-year analyses elucidating hybridization, introgression, and reproductive isolation within and between taxa. As with the complexity of chromosomal rearrangement and highly repeated DNA sequence differences (reflected in vastly different chromosome banding patterns), levels of reproductive isolation also varied greatly depending upon the taxa examined (Figure 3.6; Shaw et al. 1980, 1982; Shaw and Coates 1983). For example, the Daintree sibling species produced few F_1 progeny when crossed with any of the other *C. captiva* taxa, with any hybrid progeny produced being sterile (Figure 3.6; Shaw et al. 1980). In contrast, the viability and fertility of F_1 individuals produced by matings between Moreton/southeast Australian and Torresian individuals was not significantly reduced compared to control (i.e., within-subspecies) crosses (Shaw et al. 1980). However, a significant reduction in the viability of the reciprocal BC_1 generations and the F_2 generation produced by the same Moreton/southeast Australian × Torresian crosses was detected (Figure 3.6; Shaw et al. 1980, 1993). Indeed, matings between

Torresian individuals and the various chromosomal races of the southeast Australian/Moreton taxa resulted in BC_1 viabilities ranging from *c.* 40 to 70% and F_2 viabilities of *c.* 0 to 45% (Shaw et al. 1993).

It is apparent that barriers to genetic exchange among the subspecies and species of *Caledia* are quite strong (Figure 3.6). In the context of the present discussion, we must therefore ask the question of whether multiple data sets support or reject a hypothesis of genetic exchange between the various taxa. In particular, does multigenerational sampling from inside and outside contemporary hybrid zones detect patterns of genetic variation suggesting introgressive hybridization? Furthermore, if the hypothesis of genetic exchange is supported, does the pattern of genetic variation suggest processes that differentially affect the extent of introgression

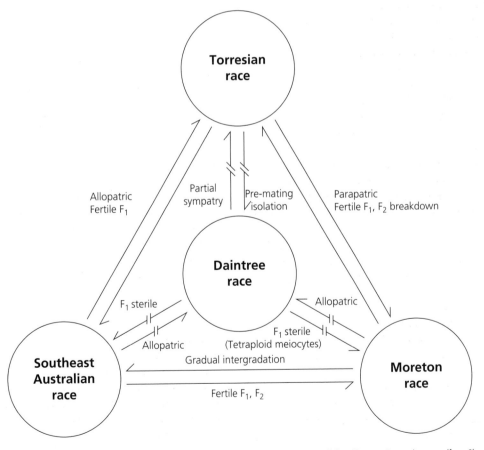

Figure 3.6 The pattern of reproductive isolation found among the various *C. captiva* forms, as defined by experimental crosses (from Shaw et al. 1980).

for different genomic (cytoplasmic and nuclear) elements?

Hybrid zone analyses within the *C. captiva* species complex focused on the parapatrically distributed Moreton and Torresian subspecies. These subspecies were found to be diagnostically different for chromosome rearrangements, chromosome banding, cytological distribution of repetitive DNA families, allozymes, as well as DNA sequence variation for repetitive DNA families, rDNA, and mtDNA (Shaw 1976; Shaw et al. 1976; Daly et al. 1981; Arnold and Shaw 1985; Arnold et al. 1986, 1987a, b; Marchant 1988; Marchant et al. 1988; Marchant and Shaw 1993). The first data used to test for genetic exchange involved investigations of the chromosomal rearrangement differences between these two subspecies near and within the region of parapatry (Moran and Shaw 1977; Moran 1979). Populations of the two subspecies, near the area of parapatry, differ cytologically in that Moreton individuals have a preponderance of submetacentric or metacentric chromosomes while Torresian individuals possess acrocentric or telocentric chromosomes (Figure 3.5). By sampling from one side of the zone of parapatry to the other, Moran and Shaw (1977) detected individuals polymorphic for the "diagnostic" chromosome rearrangements and the presence of Torresian-specific chromosome variants on the Moreton side of the zone of parapatry, but no Moreton-specific variants on the Torresian side of the zone. These findings led to the conclusion that introgressive hybridization was indeed occurring and that it was asymmetric; that is, only from the Torresian into the Moreton race (Moran and Shaw 1977).

A second analysis of the diagnostic rearrangement differences was also made during a subsequent generation. This study involved the assaying of chromosomal variation for both the original transect across the zone of parapatry assayed by Moran and Shaw and also for a second transect (Shaw et al. 1979). In addition, allozyme variation was surveyed across the two transects. In the case of the chromosomes, asymmetric introgression from Torresian into Moreton populations was once again supported, this time reflected by two transects. Like the chromosome markers, the allozyme variation also suggested introgression between the subspecies. However, unlike the

cytological markers, introgression was estimated to be symmetrical, with the allelic variants typical for each subspecies found in populations on both sides of the zone of overlap (Shaw et al. 1979; Moran et al. 1980). The asymmetry of the chromosome introgression was hypothesized to reflect the selective advantage for the Torresian chromosomes (Shaw et al. 1979, 1993). It was concluded that the advantage was environmentally-mediated. In this case, the ecological determinant was inferred to have been a drought that led to environment × genotype interactions that favored introgression from the xeric-adapted Torresian individuals into the mesic-adapted Moreton grasshoppers (Kohlmann et al. 1988). Consistent with this conclusion was the observation that the asymmetric introgression was reversed in generations of grasshoppers that experienced more rainfall (Shaw et al. 1985, 1993). This reversal suggested that greater rainfall favored components of the Moreton (mesic-adapted) genome and thus caused their introgression into populations on the Torresian side of the hybrid zone (Shaw et al. 1993).

Given the support for the hypothesis of genetic exchange between the Moreton and Torresian subspecies, it was then possible to address whether or not molecular markers demonstrated symmetrical or asymmetrical patterns of introgression. Like the chromosomal markers, an asymmetry was also detected when the Moreton and Torresian sides of the hybrid zone were compared for both nuclear and cytoplasmic markers (i.e., rDNA and mtDNA; Arnold et al. 1987a; Marchant 1988; Marchant et al. 1988). Surprisingly, when these same individuals and populations were surveyed for the diagnostic allozymes, the same asymmetrical pattern was found for these markers as well (Shaw et al. 1990). This result was unexpected because the earlier studies had resolved allozyme variation suggesting symmetrical introgression (e.g., Moran et al. 1980). The finding of asymmetry in the later studies likely reflected the more intensive sampling regime of these analyses.

The asymmetric introgression detected with the rDNA, allozyme, and mtDNA markers was consistent in that it largely reflected the transfer of Moreton markers into Torresian populations (Shaw et al. 1990, 1993). However, the extent of introgression (i.e., the distance of introgressed markers from the current hybrid zone) of molecular markers compared to that of chromosome structural polymorphisms was significantly different (Arnold et al. 1987a; Marchant et al. 1988). Near the zone of contact the chromosome rearrangement and C-band differences changed from 100% Moreton to 100% Torresian within 1 km of the hybrid zone. In contrast, the Moreton molecular markers were found up to 450 km from the present-day hybrid zone (Shaw et al. 1990). The most parsimonious explanation for the coincidental occurrence of Moreton allozyme, rDNA, and mtDNA markers up to 450 km north of the present-day zone of overlap was that the zone had moved from north to south, leaving the Moreton markers in its wake (Marchant et al. 1988). Given the well-established pattern of environmental change that has caused a switch from mesic to xeric habitats north of the present-day hybrid zone, it would be expected that the xeric-adapted Torresian form would have displaced Moreton populations.

In addition to the distance of introgression detected north of the contemporary zone of overlap, the frequency of the introgressed allozyme, rDNA, and mtDNA markers is also of interest. The Moreton allozymes were not in high frequency in the more northerly populations, suggesting a model of neutral diffusion (Marchant et al. 1988). In contrast, the Moreton mtDNA and rDNA markers were found to be the most frequent haplotypes/genotypes in the introgressed populations—indeed they were found to be fixed in some of these chromosomally Torresian populations (Shaw et al. 1990). Various scenarios can be proposed to explain this pattern. However, comparisons with other systems and experimental analyses have suggested that the introgressing Moreton mtDNA may have been selected for in the Torresian nuclear background (Arnold 2006), while the Moreton rDNA was inferred to have acted as the template for gene-conversion events, leading to its preponderance in the introgressed Torresian populations (Arnold et al. 1988).

The trans-generational surveys of the *Caledia* species complex reflect a robust method for testing for genetic exchange. In addition, the multi-year analyses allowed numerous inferences concerning the pattern and underlying causes of introgression between the hybridizing taxa. As stated at the outset, not only was this species complex developed into a model for testing numerous evolutionary hypotheses, it represents an unrealized goldmine for additional studies that apply the newest approaches such as next-generation sequencing and evolutionary development experiments.

3.4.5 Genetic exchange and hybrid zones: Chickadees

One of the predictions concerning the relationship between climate change and hybrid zones is that with alterations in the environment and thus the distribution of the hybridizing taxa, areas of overlap will change in their geographical position. Since environmentally-mediated perturbations of the distributions of taxa will have been (and will continue to be) a reoccurring process, it is expected that their effects on genetic exchange will have been pervasive throughout the biological record (Anderson and Stebbins 1954; Swenson and Howard 2005). For example, long-term studies of the cricket genus *Allonemobius* have resulted in the detection of hybrid zone evolution consistent with a causal role for environmental changes. Britch et al. (2001), using the cumulative 14 years' worth of data for transects across "east coast" and "mountain" hybrid zones, detected differential responses. In particular, the mountain transect showed an increase in the frequency of *Allonemobius socius* genotypes in many of the populations. Britch et al. (2001) inferred that "these increases suggest a northward movement of this southern cricket and may reflect response to warming of the climate in the Appalachian Mountains." In contrast, *A. socius* did not increase in frequency along the east coast transect. In answer to the question of why there were not similar increases in frequencies along both transects, Britch et al. (2001) suggested that "the Atlantic Ocean buffers temperature and moisture fluctuations brought about by climate warming."

As with the *Caledia* and *Allonemobius* complexes, morphological and genetic analyses within the chickadee genus, *Poecile*, have detected genomic characteristics consistent with hybrid zone movement due to ancient and recent climatic fluctuations. In terms of ancient interactions, Spellman et al. (2007) defined patterns of mtDNA variation in the mountain chickadee (*Poecile gambeli*) that implicated glacial cycling in both causing divergence and introgression during the evolution of this species.

They thus detected two strongly differentiated clades (an eastern and western form), which shared mtDNA haplotypes indicating post-glacial genetic exchange (Spellman et al. 2007). Furthermore, there was also evidence of at least limited introgression during the divergence of lineages within both the eastern and western clade (Spellman et al. 2007).

Consistent with the ancient expansions and contractions among the *P. gambeli* lineages, the occurrence and possible causal factors of historical hybrid zone movement have been documented between the black-capped (*P. atricapillus*) and Carolina chickadees (*P. carolinensis*). For example, transgenerational population genomic analyses of a hybrid zone in southeastern Pennsylvania detected an *c.* 11.5 km northward shift of the hybrid zone over a single decade (Taylor et al. 2014b). The major factor argued to be causal in the distributions of the northern, *P. atricapillus*, and the southern, *P. carolinensis* (and thereby the position of the hybrid zone between them), was winter temperatures. Specifically, Taylor et al. (2014b) concluded that minimum winter temperatures limited the northern range of the Carolina chickadee (possibly due to differential metabolic performances of the two species at colder temperatures; Olson et al. 2010). Other factors identified as important for species' distributions and the location of hybrid zones included mating preferences (Bronson et al. 2003b; Reudink et al. 2006) and hybrid fitness (Bronson et al. 2003a; Taylor et al. 2014a), however, global climate change will likely continue to affect greatly the pattern of genetic exchange between the black-capped and Carolina chickadee (Taylor et al. 2014b).

3.4.6 Genetic exchange and hybrid zones: House mice

One of the truly classic mammalian systems typifying divergence-with-introgression involves the house mice taxa *Mus musculus musculus* and *Mus musculus domesticus* (sometimes referred to as *M. musculus* and *M. domesticus*). The vast majority of studies of hybridization, hybrid zones, and/or experimental manipulations between these lineages have addressed the genomic architecture of reproductive isolation. For example, Janoušek et al. (2012) examined *c.* 1400 loci distributed throughout the genome in individual mice from two natural hybrid zones. By comparing the patterns of variation they were able to identify specific loci that contributed differentially to reproductive isolation, some of which were associated with hybrid male sterility. Likewise, Turner et al. (2014) utilized an experimental F_2 hybrid population to infer genomic components that cause hybrid male sterility. Their comparison of mapped gene expression traits (i.e., associated with testes function), sterility phenotypes, and associated QTL allowed the identification of several thousand loci (acting either in cis or trans) that affected expression variation for male fertility (Turner et al. 2014). These analyses also allowed an increased understanding of how genes that contribute to reproductive isolation in *Mus* may actually interact with many other loci to produce a sterility phenotype (Turner et al. 2014).

Even though an emphasis has been placed on identifying the genomic regions that provide various levels of reproductive isolation between house mouse lineages (e.g., Baird et al. 2012; Dzur-Gejdošová et al. 2012; Giménez et al. 2013; Phifer-Rixey et al. 2014; Turner and Harr 2014; Dumas et al. 2015), a number of authors have discussed the significance of introgression. Thus, patterns of introgression have alternatively been utilized to infer such disparate factors as colonization histories and adaptive trait transfer (Payseur et al. 2004; Gompert and Buerkle 2009; Jones et al. 2010; Staubach et al. 2012). Jones et al. (2010) analyzed a variety of genomic regions (Y-chromosome, autosomal, and mitochondrial DNA) as well as morphological variation to construct the history of the human-mediated dispersal of *M. m. musculus* and *M. m. domesticus* into Norway. Involved in this dispersal was the establishment of ancient and recent hybrid zones when their human carriers brought the two subspecies into contact. The colonization scenario favored by Jones et al. (2010) involved *M. m. domesticus* being brought to Norway during the Viking era and encountering the previously-imported *M. m. musculus*. Regardless of when these forms arrived, their dispersal into what is now Norway was accompanied by extensive introgressive hybridization. In a more recent hybrid zone analysis, Staubach et al. (2012) not only detected introgression between *M. m. musculus* and *M. m. domesticus*, but

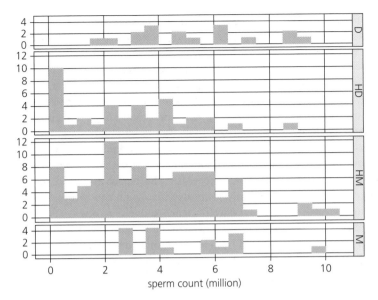

Figure 3.7 Sperm counts for experimentally produced *Mus musculus domesticus* ("D"), *M. m. musculus* ("M"), hybrids most similar to *M. m. domesticus* ("HD"), and hybrids most similar to *M. m. musculus* ("HM"). The largest range of variation in counts was found in the hybrid genotypes, with both hybrid classes demonstrating significantly lower counts than either subspecies. However, both hybrid classes contained genotypes demonstrating equivalent or higher amounts of sperm relative to individuals of the two subspecies (from Turner et al. 2012).

also were able to estimate the fraction of the genome involved. Surprisingly, these analyses detected that a minimum of 10% of the genomes possessed evidence of introgression.

Though a portion of the introgression between the house mouse taxa is consistent with "neutral diffusion," some instances of genetic exchange reflect apparent adaptive trait transfers (also detected in interspecific crosses as well; see Song et al. 2011; Liu et al. 2015). Gompert and Buerkle (2009), in an analysis of variation across X-chromosome loci in a *M. m. musculus*/*M. m. domesticus* hybrid zone (data reported by Payseur et al. 2004), inferred directional selection favoring introgression mainly from *M. m. musculus* into *M. m. domesticus*. Such directional introgression supports the hypothesis that alleles from one lineage are adaptive in the alternate lineage as well. Indeed, Staubach et al. (2012) inferred just such adaptive trait introgression from an analysis of high-density SNP typing arrays. These authors concluded that "natural genomes are subject to complex adaptive processes, including the introgression of haplotypes from other differentiated populations or species at a larger scale than previously assumed for animals" (Staubach et al. 2012).

Each of the studies discussed indicates the importance of reticulate evolution in the house mouse. As with all cases of hybridization (Barton and Hewitt 1985; Arnold 1997), each of the *Mus* hybrid zones

reflects selection against some hybrid genotypes (Figure 3.7; Turner et al. 2012). Thus, some regions of the genomes of these two subspecies are resistant to introgression (Janoušek et al. 2012). However, also as detected in other cases of genetic exchange that proceed past the initial F_1 generation (Arnold 2006), some hybrid genotypes demonstrate elevated fitness relative to their parental taxa (Figure 3.7; Turner et al. 2012). This increased fitness indicates the presence of regions of the genomes of the *Mus* subspecies in which recombination with the alternate taxon is promoted.

3.5 Testing the hypothesis: Intragenomic divergence

As discussed throughout this book, the emphasis of tests for genetic exchange is normally placed on detecting similarities between divergent evolutionary lineages. For example, when complete genomes are sequenced they are searched for similarities with other genomes. What can be considered the other side of the methodological coin are searches that examine within-gene/-genome sequences in an effort to find intraspecific divergences that are much higher than expected by chance. This class of hypothesis testing is conceptually similar to analyses such as ABBA/BABA, however, intragenomic divergence analyses do not necessarily involve population-level comparisons. To illustrate this approach, I will discuss findings from studies of various bacterial species belonging to the genera *Escherichia, Shigella,* and *Salmonella*. Each of these studies included tests for islands of intraspecific sequence divergence that would indicate horizontal gene transfer events between divergent organisms.

The first analysis involved pathogenic (enteropathogenic and enterohemorrhagic) *Escherichia coli* strains and the gene *eae* encoding the outer membrane protein intimin. McGraw et al. (1999) defined the evolutionary genetics of this putative virulence factor by sequencing the *eae* genes from three enteropathogenic and four enterohemorrhagic *E. coli* strains, respectively. The *eae* locus belongs to the LEE pathogenicity island, a chromosomal region thought to be transferred horizontally between different bacterial strains (see McGraw et al. 1999 for a discussion and additional references). Consistent with this hypothesis, McGraw et al. (1999) detected large regions within the LEE island that possessed sequence characteristics indicative of genetic transfer. The focus of the study by McGraw et al. (1999)— the *eae* genes—also demonstrated the effects of horizontal transfer/recombination between *E. coli* strains and between these strains and "sources outside of the *E. coli* population." Thus genetic transfer, particularly within the 3´ end of the various intimin genes led to mosaic genes that possessed high interstrain divergence in the 3´ portions of the *eae* gene, while retaining relatively low levels of divergence in other portions of the gene (McGraw et al. 1999).

The conclusion that intragenic mosaicism was likely caused by HGT was also reached by Feil et al. (2000) in a study of *Neisseria meningitidis* and *Streptococcus pneumonidae*. They described the mosaicism in the following manner: "For example, most of a gene may be identical in sequence for two isolates of a species, whereas a 500-bp region in the middle may differ at 5% of nucleotide sites. Significant mosaic structure is indicative of recombinational exchanges" (Feil et al. 2000). By examining multilocus sequence-typing data sets, Feil et al. (2000) estimated the number of alleles that had arisen through genetic exchange compared with point mutations. The admixed nature of the two genomes was reflected in the ratios of 10:1 and 4:1 for *S. pneumonidae* and *N. meningitidis*, respectively. This result reflected the

widespread effects of horizontal gene transfer in producing numerous mosaic alleles in the genomes of these two pathogens (Feil et al. 2000).

Another common approach to test for mosaicism in prokaryotic genes involves the estimation of the relative proportions of synonymous substitutions per site (pS) and non-synonymous substitutions per site (pN). Two studies by Hughes and his colleagues exemplify this methodology. In a comparison of two completely sequenced *Mycobacterium tuberculosis* genomes, Hughes et al. (2002) placed genes into two categories (high or low pS) using a probabilistic (Bayesian) model. The observation that a proportion of gene pairs from the two genomes showed high levels of divergence was seen as evidence for the possible role of several different processes, including HGT. Indeed, Hughes et al. (2002) concluded that a majority of the examples of highly divergent gene regions was due to genetic transfer (with the remainder explained by "differential deletion" of genes). In the second study, Hughes and Friedman (2004) compared sequence divergence in 5´ intergenic regions with genes linked to these nongenic sequences. The species *Streptococcus agalactiae* was found to possess numerous genes and 5´ spacers that were divergent from surrounding regions, indicative of their being received as a unit from a distantly related genome. In contrast, *Chlamydophila pneumoniae* also possessed numerous spacers and gene units that had been transferred from divergent genomes, but few of these spacer/gene combinations appeared to have been donated by the same distant genome (Hughes and Friedman 2004). These two divergent patterns, however, reflect the pervasive effects of genetic exchange on prokaryotic evolution and also indicate the utility of tests of intragenomic/intraspecific sequence divergence for detecting such exchange.

Two final examples of genetic exchange in prokaryotes detected, at least partially, through intragenomic comparisons involved analyses of pathogenic and commensal lineages of *E. coli, Shigella,* and *Salmonella*. In the first of these studies, Touchon et al. (2009) utilized genomic sequences from 20 commensal and pathogenic lineages of *E. coli* and *Shigella* to test for intragenomic divergences indicative of genetic exchange events. Though their results agreed with previous findings of the *E. coli* (sensu lato) genome being greatly affected by HGT, they also detected stretches of high intragenomic divergence, reflective of "recombination hotspots." Certain genes belonging to the core genome (i.e., genes found in a majority of *E. coli* strains) thus marked insertion points for horizontally acquired genes (Touchon et al. 2009).

Karberg et al. (2011) also documented the labile nature of prokaryotic genomes due to horizontal acquisition of new genes. However, rather than gene sequence data per se, information concerning codon usage was analyzed. From these data, it was concluded that the genomes of both *E. coli* and *Salmonella enterica* had evolved through the acquisition of genes from a similar pangenome, reflected by greater intra- than intergenomic variability for acquired versus nonacquired genomic elements (Karberg et al. 2011). As with all of the other examples discussed, these findings reflect the fundamental effect of genetic exchange on genomic and organismic evolution of prokaryotic species.

3.6 Conclusions

The purpose of this chapter was to demonstrate several approaches commonly used to test the hypothesis of divergence-with-gene-flow. Of primary concern has been discerning the relative roles of deep coalescence (i.e., incomplete lineage sorting) and genetic exchange in producing patterns of phylogenetic discordance. Additionally, ancient population structure must also be accounted for in models that are affected by this parameter. Significantly, a number of recent analytical approaches provide estimates of the effects of genetic exchange, coalescence, and population subdivision thus allowing the inference of genetic exchange events. A number of studies of plant and animal complexes were reviewed that utilized explicit tests for incomplete lineage sorting and genetic exchange, indicating the ability to distinguish between these alternate processes.

Data sets from the fossil record were also reviewed illustrating the efficacy of tests for ancient admixture, involving both extinct and extant lineages. Significantly, patterns of morphological variability in the fossil record, suggestive of past introgressive hybridization, were often mirrored in present-day populations of the same taxa. Such observations not only indicate the strength of inferences of genetic exchange based upon the fossil material, but as well emphasize that genetic exchange events often have an extended rather than ephemeral effect on lineages through the formation of hybrid zones. Indeed, the utility of analyses of hybrid zones (in fossil and extant populations), both for detecting reticulate evolution and for defining the multitude of processes associated with genetic admixture was illustrated in groups as divergent as irises and mammoths.

In the final section, the power of combining genomic sequencing with tests for intragenic and intragenomic divergence in prokaryotic lineages was demonstrated. The focus on examples from bacterial systems is not meant to suggest that this type of approach cannot be applied to eukaryotic systems. Indeed, it was pointed out that these types of studies are conceptually similar to methodologies developed to test for genetic exchange in eukaryotes as well (e.g., "ABBA/BABA" analyses). It is important to note that even with the increased availability and importance of analytical techniques for inferring genetic exchange from genomic information, comparative approaches will continue to provide the clearest insight into the evolutionary patterns and processes associated with reticulate evolution. Where possible, the combining of findings from explicit tests of deep coalescence versus introgression/HGT, transgenerational hybrid zone analyses, fossil data, and patterns of intragenomic divergence will always allow the greatest resolution for detecting various evolutionary processes, including genetic exchange.

The studies reviewed illustrate the broad phylogenetic distribution of horizontal gene transfer and introgressive hybridization and the many approaches for testing the hypotheses proceeding from the web-of-life metaphor. In addition, the examples discussed in this chapter should also make it clear that it is not necessary to limit oneself to organisms with sequenced genomes to study the evolutionary processes and effects associated with genetic exchange. Furthermore, experimentally difficult systems (e.g., higher organisms difficult to maintain in captivity or microorganisms that are difficult to culture) are no longer intractable. There would thus seem to be a method available to test for the evolutionary role of genetic exchange in any organismic clade.

CHAPTER 4

Genetic exchange, reproductive barriers, and the mosaic genome

"If you bring on the same brush a plant's own pollen and pollen from another species, the former will have such a prepotent effect that it will invariably and completely destroy . . . any influence from the foreign pollen."
(Darwin 1859, pp. 98)

"The fundamental importance of isolation in the evolutionary process has been recognized for a long time . . . The only way to preserve the differences between organisms is to prevent their interbreeding, to introduce isolation."
(Dobzhansky 1937, pp. 228)

"We found that YBF30, the only fully sequenced example of HIV-1 group N . . . is a recombinant of divergent viral lineages . . . This mosaic genome structure of YBF30 implies previous co-infection and recombination of divergent SIVcpz strains in a P. t. troglodytes host."
(Gao et al. 1999)

"Most speciation genomic studies thus aim to identify exceptionally divergent loci between populations, but divergence will be affected by many processes other than reproductive isolation (RI) and speciation."
(Lindtke et al. 2012)

"We find evidence for multiple incompatibilities in most crosses, including failure to store sperm after mating, failure of sperm to reach the site of fertilization, failure of sperm to fertilize eggs, and failure of embryos to develop."
(Rose et al. 2014)

4.1 Genetic exchange and reproductive isolation: Two sides of the same coin

I often use the analogy of genetic exchange and reproductive isolation as occupying either side of the same coin. In seminars on this topic, I also often show a photo of two North American Rocky Mountain bull elk with their antlers locked in a sparring match to illustrate this analogy. I believe that both the analogy and the photo illustrate the conclusions to be drawn from the accumulated evidence of the past several decades regarding both the observation of the semipermeability of reproductive barriers between divergent lineages (Key 1968) and the

processes that cause this semipermeability. As already defined in Chapter 1 and discussed in Chapters 2 and 3, the transfer of some loci, but not others, between different evolutionary units leads to "mosaic genomes" (Harrison 1986, 1990; Wu 2001).

A premise of this book is that genetic exchange occurs throughout (and, sometimes, between) viral, prokaryotic, and eukaryotic clades and that these exchange events can have major evolutionary effects through the production of novel genotypes that are more fit than their progenitors in certain environments. Mechanisms that limit or promote genetic exchange are thus of primary importance in determining the evolutionary trajectory of the products of the admixture events. The quotations at the beginning of this chapter indicate that processes limiting genetic exchange have been recognized for well over 100 years. These statements also reflect the two major concepts that I wish to emphasize regarding barriers to genetic exchange. First, reproductive isolation consists of not a single barrier, but a series of processes each contributing only partially to whatever degree of isolation is present (e.g., Scopece et al. 2013; Bi et al. 2015). Second, unless genetic exchange is blocked completely it can occur regardless of the strength of the isolating barriers. To paraphrase Dr. Ian Malcolm from the movie *Jurassic Park*, life can and often does find a way around reproductive barriers.

The idea that reproductive barriers are multistaged is not new. This was, for example, a major tenet of Dobzhansky's (1937) classic, *Genetics and the Origin of Species*. It is also now recognized that in spite of multiple barriers, organisms exchange genes (e.g., *Mimulus guttatus* and *Mimulus nasutus*: Kiang and Hamrick 1978; Diaz and Macnair 1999; Sweigart and Willis 2003; Martin and Willis 2007; Brandvain et al. 2014; Oneal et al. 2014). Because of this, the most informative studies will simultaneously examine the creative potential of genetic exchange *and* the genomic architecture of the adaptive phenotypes held together in spite of hybridization/horizontal gene transfer/viral recombination. Furthermore, because reproductive isolation is a major factor that can affect evolutionary diversification (e.g., Noutsos et al. 2014), it is most instructive that it be examined as a multistage process.

In addition to considering barriers to genetic exchange as multistaged, such limitations should be recognized as varying episodically in intensity. For example, in an analysis of reproductive isolation between the plant species, *Mimulus cardinalis* and *M. lewisii*, Ramsey et al. (2003) stated, "In aggregate, the studied reproductive barriers prevent, *on average* [my emphasis], 99.87% of gene flow, with most reproductive isolation occurring prior to hybrid formation." The key, I believe, is the phrase "on average." As I will emphasize throughout this text, rare events can be extremely important. If evolutionary biologists did not believe this, they would not point to the vanishingly infrequent mutations

(or recombination events) that result in increased fitness, as the foundation for evolutionary change. Thus, "average effects" should be considered a null hypothesis, and not necessarily predictive of evolutionary pattern or process. It is likely that, as with many other similar plant examples, *M. cardinalis* and *M. lewisii* go through rare, episodic bouts of introgressive hybridization, regardless of what the "average" strength of barriers might be.

Another key aspect of the two-sided coin metaphor is the recognition that the relative difficulty in forming certain hybrid generations (e.g., F_1 hybrid plants, first-generation recombinant viruses, etc.) can determine where hybridization is most likely to occur (i.e., ecotones or disturbed habitats) and the types of later generation hybrid/recombinant genotypes and phenotypes that can be produced. In this chapter, I will illustrate some of the processes that can limit genetic exchange. There are two reasons for my covering this topic. First, it is important to illustrate the lack of correlation between the often severe restrictions on hybrid formation and the extensive occurrence of introgressive hybridization, horizontal gene transfer, and viral recombination. This paradox, I will argue, is resolved by the hypothesis that once "hybrids" are synthesized, it is likely that positive selection favors some of the recombinant genotypes in certain environments. Exogenous selection may indeed play a major role in the stabilization of hybrid types by favoring hybrids over parents in certain environments (see Chapter 5). However, it would also appear necessary to invoke the simultaneous importance of reproductive isolation for the "stabilization" of hybrid lineages (Grant 1981). The barriers to genetic exchange between divergent taxa discussed in this chapter may thus subsequently assume a significant role in isolating newly formed recombinant viruses, prokaryotes, plants, or animals from their progenitors.

In the following sections I will describe studies that have demarcated mechanisms affecting the probability that certain recombinant genotypes will be produced. As stated, I will use these analyses to illustrate the difficulty of forming admixed genomes. Furthermore, I wish to highlight the fascinating processes that underlie the barriers to genetic exchange. The structure of this chapter will follow various "stages" that precede the establishment of recombinant progeny. I do not intend to suggest that these stages are necessarily discrete. However, as reflected by the likes of Darwin (1859) and Dobzhansky (1937), using this approach to illustrate limitations to hybrid formation does permit the identification of at least general categories. Though the processes discussed will not be exhaustive, I hope to illustrate why reproductive isolation should be defined as multistaged. Furthermore, though some of the mechanisms discussed will not have easily definable counterparts in all taxonomic groups, each of the examples given can be broadly categorized as either "pre-exchange" or "postexchange" which are roughly analogous to "prezygotic" or "postzygotic" or, alternatively, the "premating" and "postmating" classes of Dobzhansky (1937).

I will first exemplify the multistage character of reproductive barriers by once again considering work on species of Louisiana irises. I will then emphasize the similarity of effects from various processes on genetic exchange within viral, prokaryotic, and additional eukaryotic assemblages. In keeping with the thesis of this book, I will review examples that not only exemplify the multistage nature of genetic isolation, but also, in the face of these barriers, reflect genetic exchange.

4.2 Genetic exchange and reproductive isolation in Louisiana irises: Prezygotic processes

Lexer and Widmer (2008) reviewed the research findings for five "model" plant systems, one being the Louisiana iris species complex. As in the present text, these authors emphasized the genic speciation model to illustrate the fact that evolutionary divergence often reflected both reproductive isolation and permeability of different portions of the same genome. Thus, as we consider the role of various isolating mechanisms, it is important to always place them in the context of genomic rather than whole-organism barriers. This necessity should already be evident from the discussions in the preceding chapters. However, given that this chapter will provide a plethora of examples of mechanisms leading to limitations

for divergence-with-gene-flow, it is of particular importance to once again emphasize that if there is "gene flow" during divergence then, by definition, at least some of the genomic regions of the organisms involved are open to admixture. For example, speaking of hybrid fertility of something greater than zero, regardless of how low the fitness might be, indicates that recombination has occurred leading to introgression, horizontal gene transfer, or viral recombination. I realize that I will appear to belabor this point, but it seems that humans, including those interested in evolutionary biology, have very short memories concerning well-substantiated conclusions.

Lexer and Widmer's (2008) review of the Louisiana iris work emphasized findings from genetic mapping studies. In particular, they discussed the observations suggesting the role of differential hybrid fitness (i.e., postzygotic barriers) on the distribution of hybrid genotypes in both artificial and natural environmental settings. However, the differential environmental settings of the various species also reflect prezygotic barriers.

4.2.1 Niche differentiation and reproductive isolation

The Louisiana irises demonstrate distributions among habitats that contribute to limitations to interspecific gene flow. Because most pollinator visits occur between nearest neighbors (Figure 4.1; Wesselingh and Arnold 2000) and plants belonging to the different species occupy separate niches, interspecific pollen transfer and thus hybrid formation will be limited by habitat associations. For example, while *I. fulva* and *I. brevicaulis* occur in shaded, bayou, hardwood forest, and swamp habitats, *I. hexagona* is found in open, freshwater marshes (Viosca 1935; Bennett and Grace 1990; Cruzan and Arnold 1993; Johnston et al. 2001). In addition, consistent with its occurrence near coastal environments and thus environment-dependent selective constraints, *I. hexagona* demonstrates a higher tolerance of salinity stress, relative to at least *I. fulva* (Arnold and Bennett 1993; Van Zandt and Mopper 2002, 2004; Van Zandt et al. 2003). Likewise, the homoploid hybrid species, *I. nelsonii*, appears to be limited by a combination of environmental factors, including light and water availability, leading to its occupation of a distinctive habitat relative to

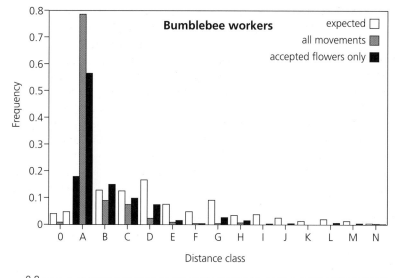

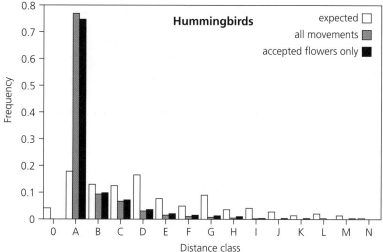

Figure 4.1 Results from pollinator observations in an experimental array consisting of individuals of *I. fulva*, *I. brevicaulis*, F$_1$, and the first backcross generations toward each parent (genotypes not indicated). The distribution of pollinator transitions between plants over various distance classes is shown for both bumblebee workers and hummingbirds. Expected values are based on the frequency of occurrence of each distance class in the array and data are presented for all flowers approached and for "accepted" flowers only (from Wesselingh and Arnold 2000).

its progenitors (Randolph 1966; Taylor et al. 2011). All four of the Louisiana iris species thus sort into different habitats defined by their ability to grow in different levels of soil moisture and light (Bennett and Grace 1990; Arnold and Bennett 1993; Cruzan and Arnold 1993; Emms and Arnold 1997; Johnston et al. 2001; Martin et al. 2006; Taylor et al. 2011), leading to limitations for interspecific pollen transfer. However, the mosaic nature of microhabitat distributions allows the different species to grow in close proximity and hybridize (Viosca 1935; Riley 1938; Arnold et al. 1990a, b; Arnold 1993; Cruzan and Arnold 1993; Johnston et al. 2001; Hamlin and Arnold 2014).

Another component of the prezygotic barriers to hybridization presented by the ecological setting of Louisiana irises involves the response of pollen vectors to different floral syndromes. Viosca (1935) suggested that hummingbirds preferred the pollination syndrome possessed by *I. fulva* (i.e., flowers of this species are solid red with protruding anthers and highly reflexed sepals, and lack nectar guides and a strong scent—characteristics normally associated with hummingbird pollination). Likewise, the description of the homoploid hybrid species, *I. nelsonii*, as having a similar floral coloration to *I. fulva* (Randolph 1966) suggested a preference by hummingbirds for its flowers as well (Taylor et al. 2012). Bumblebees were thought to be the major pollinators for *I. brevicaulis* and *I. hexagona*, as reflected by their floral characteristics (i.e., blue flowers marked with prominent white and yellow nectar guides, stiff upright sepals, and strong scent—characteristics normally associated with insect pollinators; Viosca 1935). In contrast, the nectar rewards (a major benefit gained by the pollinators) present in each of the three species do not possess characteristics that clearly place them into classical nectar-pollen vector categories (Burke et al. 2000b; Emms and Arnold 2000; Wesselingh and Arnold 2000). For example, although *I. fulva* possesses lower nectar sugar concentrations than *I. brevicaulis*, as expected for a hummingbird- versus a bumblebee-pollinated floral syndrome, the nectar concentration of *I. brevicaulis* (i.e., possessing a bumblebee-syndrome-type floral pattern) falls well within concentrations found in numerous hummingbird-pollinated species (Burke et al. 2000b; Wesselingh and Arnold 2000). Furthermore, no significant differences were found in the nectar concentrations of *I. fulva* and the other bumblebee-type floral syndrome species, *I. hexagona* (Emms and Arnold 2000).

To determine the effect of divergent floral syndromes on reproductive isolation and genetic exchange, our group and that of Martin et al. performed a series of experiments involving *I. fulva*, *I. hexagona*, *I. brevicaulis*, *I. nelsonii*, and experimental hybrids (Emms and Arnold 2000; Wesselingh and Arnold 2000; Bouck et al. 2007; Martin et al. 2008; Ballerini et al. 2012; Taylor et al. 2012; Brothers et al. 2013). These studies allowed a test for pollinator preference as a pre-pollination barrier to genetic exchange. The findings from pollinator observations by Emms and Arnold (2000), Wesselingh and Arnold (2000), and Taylor et al. (2012) led to the conclusion that different pollinator classes had preferences for certain floral syndromes, and that this preference would lead to assortative mating and reproductive isolation. However, the pollinators (particularly bumblebees) were also revealed to be a bridge for the initiation of interspecific hybridization (e.g., Figure 4.2; Wesselingh and Arnold 2000).

The genetic architecture of prezygotic reproductive barriers has now been studied in a variety

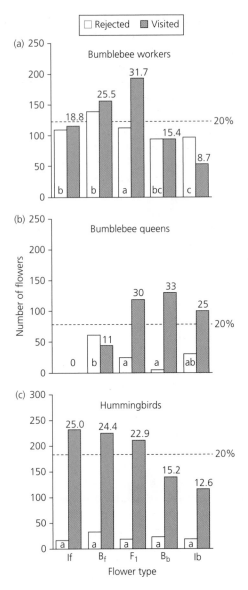

Figure 4.2 Results from pollinator observations in an experimental array consisting of individuals of *I. fulva*, *I. brevicaulis*, F₁, and the first backcross generations toward each parent. Distribution of pollinator approaches to each of the flower/genotypic classes for (a) bumblebee workers, (b) bumblebee queens, and (c) hummingbirds. The approaches are divided into flowers rejected (white bars) and flowers visited (shaded bars). The percentage of the total number of flowers visited is given above the shaded bars. The 20% line indicates the expected number of flowers visited when visitation rates would be equal for all flower types. At the bottom of the white bars, letters indicate significant differences in the fraction of flowers rejected between flower types: If = *I. fulva*; Ib = *I. brevicaulis*; Bf = first backcross generation toward *I. fulva*; Bb = first backcross generation toward *I. brevicaulis* (from Wesselingh and Arnold 2000).

of plant and animal systems (e.g., Bradshaw and Schemske 2003; Ortíz-Barrientos and Noor 2005; Fishman et al. 2008; Sweigart 2010; Hermann et al. 2013). Likewise, experiments designed to discern the number, locations, and strength of effect of loci affecting such barriers among the Louisiana iris species proceeded from the earlier behavioral experiments discussed. A common finding from analyses of the genetic architecture of floral traits thought to affect pollinator preferences was the detection of (1) both large- and small-effect QTL that contributed to floral traits and (2) the presence of transgressive phenotypes affected by some of the QTL (Bouck et al. 2007; Martin et al. 2008; Brothers et al. 2013; Taylor et al. 2013). Each of these analyses thus reflected similar genetic architectures underlying the floral morphologies of *I. fulva, I. hexagona, I. brevicaulis*, and *I. nelsonii*.

As indicated by the studies of Emms and Arnold (2000) and Wesselingh and Arnold (2000), the differential pollinator preferences for divergent, interspecific floral forms led to partial reproductive isolation. Martin et al. (2008) tested whether the genetic architecture of the pollinator behavior affected by floral phenotypes could be defined. They did this by utilizing experimental arrays containing *I. fulva, I. brevicaulis*, and experimental F_1 and backcross hybrids (i.e., backcross, genetic mapping populations). By observing pollinator behavior in these arrays they described regions of the plant genomes that affected pollinator preferences for the two species and their artificial hybrids. As expected from the previous behavioral experiments, hummingbirds preferred *I. fulva* and under-visited both *I. brevicaulis* and backcrosses toward *I. brevicaulis*. Likewise, lepidopterans preferred *I. fulva* and backcrosses toward *I. fulva*, but also under-visited *I. brevicaulis* and *I. brevicaulis* backcrosses. Bumblebees preferred *I. brevicaulis* and F_1 hybrids, but avoided *I. fulva*. Although each of the pollen vector classes preferred one or the other iris species, these preferences did not prevent visitation to other hybrid/parental classes. Furthermore, QTL mapping with the reciprocal backcross populations detected multiple loci that differentially affected the acceptance and rejection of floral phenotypes by the different pollinator classes. These findings led Martin et al. (2008) to infer that quantitative genetic factors affected the reproductive isolation, pattern

of pollinator-mediated introgressive hybridization, and thus natural hybrid zone evolution between *I. fulva* and *I. brevicaulis*.

4.2.2 Phenology, pollen competition, and reproductive isolation

Flowering time (i.e., phenology) differences act as a significant prezygotic reproductive barrier between some of the Louisiana iris species. For example, Figure 4.3 illustrates the phenological differences between the early flowering *I. fulva* and the later flowering *I. brevicaulis* in an area of natural sympatry. Because *I. fulva* and *I. hexagona* demonstrate largely overlapping flowering seasons (Viosca 1935; Arnold et al. 1993), genetic exchange between *I. hexagona* and *I. brevicaulis* will also be greatly limited by phenological differences.

Martin et al. (2007) tested for the processes leading to the genotypic and phenotypic evolution of flowering time in *I. fulva* and *I. brevicaulis*. They utilized experimental backcross populations, in both natural and greenhouse settings, across two flowering seasons. As expected, *I. fulva* initiated and terminated flowering significantly earlier than *I. brevicaulis*. Examination of line crosses of the reciprocal F_1 and BC_1 hybrids indicated that flowering time was polygenic in nature. Martin et al. (2007) also defined QTL that affected the initiation of flowering in each of the parental species. For the BC_1 population toward *I. fulva*, 14 of 17 detected QTL caused flowering to occur later in the season when *I. brevicaulis* alleles were introgressed, while the remaining three caused flowering to occur earlier. In the BC_1 hybrid population toward *I. brevicaulis*, 11 of 15 detected QTL caused flowering to occur earlier in the season when introgressed *I. fulva* alleles were present, while the remaining four caused flowering to occur later. Martin et al. (2007) concluded, "These ratios are consistent with expectations of selection (as opposed to drift) promoting flowering divergence in the evolutionary history of these species. Furthermore, epistatic interactions among the QTL also reflected the same trends." Likewise, Ballerini et al. (2012), in a separate QTL analysis of prezygotic isolating barriers between these two iris species, detected clustering of QTL for different traits also suggestive of natural selection,

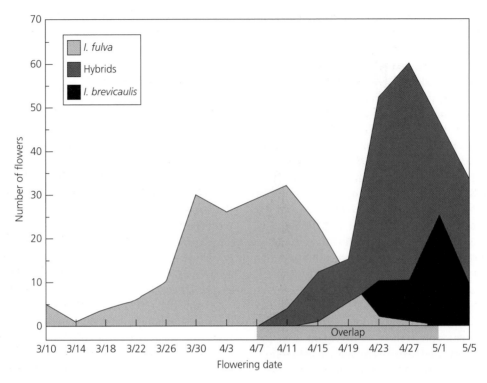

Figure 4.3 The number of flowers produced each day by *I. fulva*, *I. brevicaulis*, and hybrid genotypes over a single flowering season in a natural hybrid zone. The time period when both parental species were flowering is indicated as "overlap" (from Cruzan and Arnold 1994).

in this case limiting the breaking up of coadapted gene complexes.

4.2.3 Pollen competition and reproductive isolation

Within plant species, a series of processes must be successful for fertilization to occur following pollen deposition (Knox 1984; Williams et al. 1999). In interspecific crosses the same processes may limit the production of hybrid progeny. For example, it has been recognized for many decades that the length of the style of the pollen parent is often correlated with the speed at which pollen tubes develop and/or how far they will grow in a foreign style (Table 4.1; Buchholz et al. 1935; Blakeslee 1945). Pollen transfer between taxa with widely differing style lengths can thus lead to aberrant pollen tube development and concomitant barriers to gene exchange (Buchholz et al. 1935; Blakeslee 1945; Williams et al. 1986; Williams and Rouse 1988).

Table 4.1 Pollen tube lengths as related to style length for ten species of *Datura* (Buchholz et al. 1935). The values given in the column marked "Pollen tubes" are the sum of the percentages of pollen tubes that grew at least 50% as far as the farthest growing tube in conspecific and heterospecific crosses (Buchholz et al. 1935). The larger the value, the more tubes that grew a greater distance. The average style length (in mm) for each species (Buchholz et al. 1935) is given in the column marked "Style."

Species	Style	Pollen tubes
D. stramonium	70	382
D. quercifolia	40	289
D. ferox	38	290
D. pruinosa	40	346
D. leichhardtii	30	318
D. discolor	140	639
D. ceratocaula	135	519
D. meteloides	190	547
D. metel	120	529
D. inoxia	150	559

In the Louisiana iris complex, there are large differences in floral structures (including style length), suggesting the possible role of post-pollination processes in limiting hybrid formation. Indeed, a set of observations suggested the effect of interspecific pollen competition on the fertilization success of conspecific and heterospecific gametes. The observations included: (1) the presence of numerous hybrid zones between various combinations of *I. fulva, I. brevicaulis, I. hexagona*, and *I. nelsonii* in southern Louisiana (Viosca 1935; Riley 1938; Anderson 1949; Randolph 1966; Randolph et al. 1967; Arnold et al. 1990a, b; Arnold 1993; Johnston et al. 2001; Cornman et al. 2004; Hamlin and Arnold 2014); (2) hybrid populations that often consisted mainly of hybrid individuals possessing genotypes that identified them as advanced generation (e.g., B_2 or later) recombinants (Nason et al. 1992; Cruzan and Arnold 1993, 1994); (3) few, if any, naturally occurring adult F_1 individuals were ever identified although hundreds of experimental F_1 individuals had been produced (Arnold 1994; Johnston et al. 2001); and (4) natural pollinations in a population consisting of *I. fulva* and *I. hexagona* individuals resulted in less than 1% F_1 hybrid seeds (Arnold et al. 1993; Hodges et al. 1996). Taken together, these data suggested that the formation and establishment of the initial (i.e., F_1) hybrid generation was relatively difficult. In addition, the almost total lack of hybrid seeds in the *I. fulva* and *I. hexagona* mixed population (Arnold et al. 1993; Hodges et al. 1996) suggested that some of the processes responsible for the absence of F_1 hybrids might occur post-pollination, but pre-fertilization. Subsequent analyses supported this latter inference, specifically indicating a causal role for interspecific pollen competition.

The initial study designed to test for the effect of post-pollination processes in limiting F_1 hybrid formation also allowed an assessment of the effect of phenology. Arnold et al. (1993) recorded the flowering times for each *I. hexagona* and *I. fulva* plant from a mixed population (Figure 4.4). This population was established by introducing 200 rhizomes of *I. hexagona* into a natural population of *I. fulva* (Arnold et al. 1993). Unlike the phenological differences found between *I. brevicaulis* and *I. fulva* (Cruzan and Arnold 1994), the latter species and *I. hexagona*

demonstrated broadly overlapping flowering times (Figure 4.4; Arnold et al. 1993). Yet, when seeds from this population were genotyped, less than 1% of them were F_1 hybrids. Consistent with an important role for prezygotic reproductive barriers, the only fruits containing hybrid seeds were produced when there was a vast excess of one species' flowers relative to the alternate species (Figure 4.4; Arnold et al. 1993).

As discussed, the floral displays of *I. fulva* and *I. hexagona* should lead to assortative pollinations due to pollinator preferences. However, an additional factor known to affect reproductive isolation in other plant groups was that of pollen competition (e.g., Heiser et al. 1962, 1969; Smith 1968, 1970). Arnold et al. (1993) tested for the role of interspecific pollen competition in limiting hybrid formation by examining the genotypes of seeds produced from hand pollinations of the two species using a 50%:50% mixture of *I. fulva* and *I. hexagona* pollen. These crosses did not result in any F_1 hybrid individuals, suggesting that pollen competition might be affecting the potential for seed siring by heterospecific gametes. A series of subsequent experiments confirmed a role for pollen competition in heterospecific crosses between these two species, and between *I. fulva* and *I. brevicaulis* as well.

Carney et al. (1994) examined the patterns of seed siring using a replacement series of different pollen mixtures. The results from this analysis included a significant reduction, from expected values, in the frequency of hybrid seed formation at each of the pollination treatments. The pattern of hybrid seed formation indicated that post-pollination (but pre-fertilization) phenomena were limiting the formation of F_1 hybrid seeds. As the proportion of heterospecific pollen in the mixtures increased, so did the proportion of hybrid seeds formed, albeit at a significantly lesser frequency than expected (Carney et al. 1994).

An additional set of experiments involving *I. fulva* and *I. hexagona* also demonstrated the effect of pollen competition leading to preferential fertilization by conspecific, male gametes. This subsequent study included an examination of the impact of various time intervals between the application of heterospecific and conspecific pollen on (1) seed siring ability of the two classes of pollen and (2) the

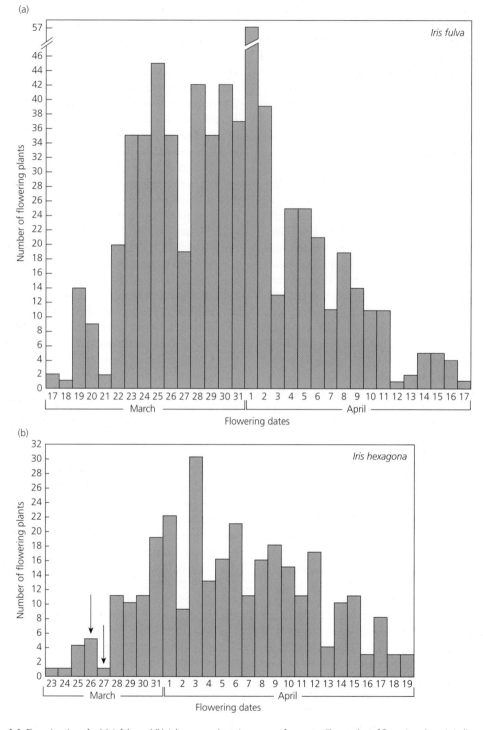

Figure 4.4 Flowering times for (a) *I. fulva* and (b) *I. hexagona* plants in an area of sympatry. The number of flowering plants is indicated along the vertical axis, and the days of the month are given along the horizontal axis. Arrows indicate the flowering dates of the only two flowers from *I. hexagona* that contained hybrid seeds (from Arnold et al. 1993).

relative position of hybrid and conspecific progeny in the fruits of the two species (Carney et al. 1996). Results from this analysis again suggested the role of interspecific pollen competition as a factor in reproductive isolation between *I. fulva* and *I. hexagona*. Both species produced significantly more hybrid progeny following longer delays between the application of heterospecific and conspecific pollen. However, *I. fulva* and *I. hexagona* differed significantly from one another in the strength of the barrier to hybrid progeny formation. Thus, delays of 0–6 hr between the application of *I. hexagona* pollen onto *I. fulva* flowers, followed by the application of *I. fulva* pollen, did not produce significantly different numbers of hybrid seeds; only after a delay of 24 hr was there a significant increase in F$_1$ hybrids in *I. fulva* fruits (Carney et al. 1996). In contrast, there was a significant increase in hybrid formation by *I. hexagona* plants between the "0" hr delay (i.e., the application of a 50%: 50% mixture of hetero- and conspecific pollen) and those fruits that developed from flowers that experienced 1–24 hr pollination delays (Carney et al. 1996).

These results are consistent with the hypothesis that pollen competition affected F$_1$ formation by *I. fulva* and *I. hexagona* plants. However, the reproductive systems of these two species showed differences in their siring ability on the alternate species. Although both species sired fewer seeds on the opposite species' flowers than expected, *I. fulva* demonstrated a propensity to sire significantly greater numbers of hybrid seeds in *I. hexagona* fruits than did the latter species in *I. fulva* fruits. Thus, not only was *I. fulva* a relatively better heterospecific sire, but its pollen was also a relatively better sire on its own flowers when in competition with *I. hexagona*. Consistent with this result were findings from a naturally pollinated mixed population, in which *I. fulva* acted as the maternal parent for significantly fewer (0.03%) hybrid seeds than did *I. hexagona* (0.74%; Hodges et al. 1996).

Emms et al. (1996) also detected siring successes in experimental crosses consistent with interspecific pollen competition having a role in limiting hybrid formation, but in this case involving *I. fulva* and *I. brevicaulis*. Though higher frequencies of F$_1$ formation were found relative to the outcomes of the *I. fulva* and *I. hexagona* crossing trials, there was still a significant reduction of hybrid progeny formation. Specifically, the application of 50%: 50% mixtures of *I. fulva* and *I. brevicaulis* pollen to the stigmas of both species resulted in 24% and 39% F$_1$ hybrid seeds in *I. fulva* and *I. brevicaulis* fruits, respectively. These frequencies were significantly different from one another and from the expected 50%: 50% ratio (Emms et al. 1996).

The post-pollination processes defined by these analyses lead to great restrictions on hybrid formation. However, the analyses also suggest a set of conditions that could lead to the formation of natural hybrid zones (Arnold 1994). First, an excess of heterospecific relative to conspecific pollen would increase the likelihood of hybrid progeny being formed. Second, a time lag between the deposition of the heterospecific pollen and pollen from the same species would promote the formation of hybrid offspring. Both an excess of interspecific pollen and a time lag between the arrivals of the two pollen types, on a particular stigma, would be likely when there was an excess of flowers of one of the species (Figure 4.4; Arnold et al. 1993).

4.3 Genetic exchange and reproductive isolation in Louisiana irises: Postzygotic processes

Numerous studies have documented the presence of an increased frequency of postzygotic selection against some hybrid genotypes (see Arnold and Hodges 1995; Arnold 1997, 2006 for examples). Furthermore, elevated levels of inviability and infertility have been assumed by some to be common characteristics for all hybrid individuals, regardless of genotype (Mayr 1963). Data to test this assumption will be discussed in Chapter 5; however, findings from the Louisiana iris complex can be used to demonstrate the variable nature of postzygotic reproductive isolation. That postzygotic reproductive isolation is present among the iris species is hardly surprising given the prevalence of such barriers throughout plant and animal assemblages. It is thus likely that the low fitness of hybrid genotypes (either in terms of viability or fertility) has contributed to the occurrences of introgression and hybrid speciation documented in organisms as diverse as sunflowers, humans, monkey flowers, fruit flies, and birds (Heiser et al. 1969; Rieseberg 1991; Martin and Willis 2010; Currat and Excoffier 2011; Rabosky and Matute 2013; Turelli et al. 2014).

4.3.1 Hybrid viability and reproductive isolation

As discussed, various prezygotic isolating barriers combine to limit greatly the formation of F_1 hybrids between the Louisiana iris species. Likewise, processes leading to selection against hybrid genotypes (i.e., postzygotic barriers) also contribute to reproductive isolation among these taxa. In particular, Louisiana irises illustrate well genotype specific hybrid inviability and sterility. For example, several experimental and field analyses have detected hybrid inviability in crosses between *I. fulva* and *I. brevicaulis*. The first indication of genotype specific inviability among advanced generation hybrids between these two species came from a study of genetic variation in a natural hybrid zone (Cruzan and Arnold 1994). In this analysis, Cruzan and Arnold (1994) examined the genotypic classes present in adult, flowering plants and in the seed arrays produced by these plants (Figure 4.5). These authors detected significant differences in the estimates of linkage disequilibrium for the parental plants and their seed cohorts, with the adults demonstrating significantly higher levels of disequilibrium (Cruzan and Arnold 1994). This pattern suggested that postzygotic selection had acted against certain recombinant genotypes, between the seed and adult life history stages, thus leading to more conspecific allele associations in the adults than in the seed population.

To test whether such selection could have acted during the life history stage at which the progeny were collected, Cruzan and Arnold (1994) compared the genotypes of viable and inviable seeds. Consistent with selective constraints at this life history stage, hybrid genotypic classes most dissimilar to their parents (i.e., *I. fulva* or *I. brevicaulis*) had the highest levels of inviability (Figure 4.5). Likewise, results from a recent analysis of herbivory and fungal infection rates among parental and hybrid genotypes (Dobson et al. 2011) detected the lowest resistance to attack in hybrid genotypes similar to those found to possess the lowest fitness in the

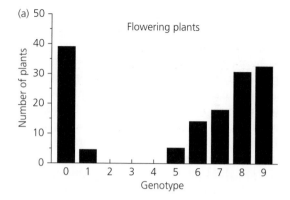

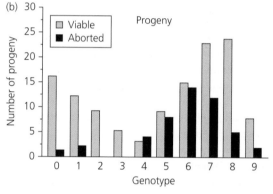

Figure 4.5 Distribution of genetic markers among (a) adult, flowering plants and (b) seeds—both viable and aborted—collected from the individuals in (a). 0 and 9 are the genotypic scores for *I. fulva* and *I. brevicaulis* plants, respectively. Scores of 1–8 are various hybrid classes (from Cruzan and Arnold 1994).

natural hybrid zone (Cruzan and Arnold 1994). That viability selection was a major isolating factor for these taxa was also indicated by a series of experimental crosses, including one study that examined reciprocal F_2 populations. Burke et al. (1998) detected a significant deficit of intermediate hybrid genotypes, but an excess of *I. fulva*-like and *I. brevicaulis*-like hybrid genotypes, in the *I. fulva* × *I. brevicaulis* F_2 progeny. The under- and over-representation of certain genotypes was apparently due to both maladaptive nuclear–nuclear and cytonuclear interactions, resulting in postzygotic barriers, including zygotic inviability (Burke et al. 1998). Indeed, one hybrid genotype was absent from both F_2 cohorts, indicative of the synthesis of a "recombinant lethal" genotype (Dobzhansky 1946).

More recent analyses, utilizing genetic mapping of *transmission ratio distortion* (i.e., "TRD"; Huang et al. 2013) and QTL associated with hybrid fitness, have allowed tests for the genetic architecture of postzygotic barriers between *I. fulva* and *I. brevicaulis*. Bouck et al. (2005) tested for segregation distortion as an indicator of postzygotic (i.e., viability) selection on reciprocal, first-generation backcross hybrids. These workers detected genomic regions that reflected significantly lower than expected rates of introgression; these regions reflected portions of the genome that were reproductively isolated between the two species. Martin et al. (2005) and Tang et al. (2010), examining long-term survivorship in these same BC_1 hybrid populations, also detected the effect of interactions between introgressed and non-introgressed regions that lowered hybrid viability under greenhouse conditions, limiting recombination and introgression. Both of these studies, though separated by five growth seasons, detected similar levels and patterns of TRD among the hybrid progeny, indicating the contribution of postzygotic reproductive isolation for portions of the genomes of *I. fulva* and *I. brevicaulis* (Martin et al. 2005; Tang et al. 2010). Likewise, two studies that assayed short-term and long-term survivorship under natural conditions, once again for the same BC_1 hybrid genotypes, also detected lower viability for some recombinant genotypes relative to parental individuals (Martin et al. 2006; Taylor et al. 2009). QTL mapping revealed both additive and epistatic interactions leading to the postzygotic reproductive barriers (Martin et al. 2006; Taylor et al. 2009). Furthermore, the contributions of both intrinsic and extrinsic selection, as well as year-to-year fluctuations in the patterns of fitness, indicated the complex nature of postzygotic reproductive barriers for certain portions of the genomes of *I. fulva* and *I. brevicaulis* (Taylor et al. 2009).

As discussed in Section 4.2, floral traits and interspecific pollen competition contribute to prezygotic reproductive isolation between *I. fulva* and *I. hexagona*. As with *I. fulva* and *I. brevicaulis*, reduced viability of some hybrid genotypes contributes to reproductive isolation between *I. fulva* and *I. hexagona* as well (Arnold and Bennett 1993). For example, Cornman et al. (2004) used a paternity analysis to determine the contribution of pollen flow (both long distance and short distance) and postzygotic selection in the structuring of a population containing "*I.*

fulva-like" and "*I. hexagona*-like" flowering plants. The comparison of adult and progeny genotypes led Cornman et al. (2004) to the conclusion "that partial reproductive isolation can occur over small spatial scales relative to pollen flow in Louisiana iris hybrids. Thus, in this system, life history characteristics such as selfing and asexual propagation do appear to contribute to the stabilization of hybrid lineages as predicted by theory . . . but postzygotic selection against subsequent recombination is also significant."

4.3.2 Hybrid fertility and reproductive isolation

Though more limited, the data on hybrid fertility in Louisiana irises reflect the expectation that this stage can also act as a barrier for gene flow among the various species (Riley 1938, 1939; Anderson 1949; Randolph et al. 1967; Bouck 2004; Ballerini et al. 2012). For example, Anderson (1949, p. 3) reported pollen fertilities for *I. fulva*, *I. hexagona*, and naturally occurring hybrids. Pollen fertilities for individuals of the two parental forms ranged from 89 to 99% (mean, 95.7%), while the hybrid plants possessed fertilities of 52–98% (mean, 85%). Randolph et al. (1967) collected pollen fertility data from numerous natural populations, including (1) individuals from allopatric populations of *I. fulva*, *I. brevicaulis*, and *I. hexagona*, (2) individuals of *I. fulva*, *I. brevicaulis*, and *I. hexagona* from sympatric populations, and (3) hybrid plants from natural hybrid zones. These investigators found that the allopatric, sympatric, and hybrid samples had pollen fertilities of 45–94% (mean, 93.5%), 15–95% (mean, 93%), and 15–95% (mean, 89%), respectively.

Consistent with the results from the earlier studies, the two reciprocal *I. fulva* × *I. brevicaulis* BC_1 mapping populations discussed also demonstrated a range of pollen fertilities among the hybrid genotypes (Bouck 2004). However, there was an asymmetry in the reduction in pollen fertility with the backcross population toward *I. brevicaulis* demonstrating a much lower (mean, 65%) value than the backcross population toward *I. fulva* (mean, 91%). Furthermore, this asymmetry was associated with differential introgression of different alleles, as detected in a QTL analysis (Ballerini et al. 2012). This supports the hypothesis that the genetic architecture of hybrid fertility likely affects the potential for recombination and thus introgression between *I. fulva* and *I. brevicaulis* in natural hybrid zones.

In the sequence of reproductive isolating barriers, both viability and fertility selection apparently act as narrow sieves for the transfer of genetic material among Louisiana irises. However, as described in Chapter 3, these species were used as *the* paradigm for the process of introgressive hybridization (Anderson 1949). Recent analyses have confirmed the evolutionarily significant role played by genetic exchange among these species (Arnold et al. 2012). Notwithstanding the restriction imposed by the lower viability of some hybrids—and indeed imposed by numerous reproductive barriers—genetic exchange has apparently played a significant role in structuring the population genetics, ecological amplitude, and taxonomic diversity of this plant species complex.

4.4 Genetic exchange and reproductive isolation in viral clades

The Louisiana irises illustrate well the sequential and multistaged nature of reproductive isolation. Yet, they are but one of countless such examples. In the following sections, I will highlight the multiple stages that can act as reproductive barriers using a variety of microorganisms, plants, and animals. However, these examples once again will reflect the two sides of the reproductive isolation/genetic exchange "coin"; each could thus be used to illustrate the creative role of genetic exchange in the evolution of biological life. For example, in Chapter 6, I will discuss a number of cases in which genetic exchange among microorganisms led to evolutionary novelty. Indeed, the role of reassortment and horizontal gene transfer between members of divergent lineages is now known to be of primary importance for the evolution of viruses and prokaryotes, respectively (e.g., O'Keefe et al. 2010; Vijaykrishna et al. 2010; Li et al. 2013; Noda-García et al. 2013). In the current section, I will consider the reproductive barriers that may limit the earliest stages of genetic exchange in both HIV and influenza.

4.4.1 Pre-exchange isolating barriers and viral reassortment: HIV

Early studies of retroviruses confirmed the mechanisms by which reassortment could take place (e.g., Coffin 1979). In spite of this information, Castro-Nallar et al. (2012) pointed out that recombination between divergent lineages of retroviruses "was regarded as almost nonexistent mainly because it was thought that multiple infections within the same individual were rather unlikely. This led to the general thought that recombination could not contribute to HIV-1 evolution." In contrast to this viewpoint is the now well-understood fact that "HIV adapts not only by positive selection through mutation but also by recombination of segments of its genome in individuals who become multiply infected" (Heeney et al. 2006). Yet, the earlier workers were correct in that multiple infections should be unlikely thus leading to "assortative mating"—i.e., if reassortment does occur it should be between viruses from the same evolutionary lineage that propagated in a single host. Furthermore, I have argued elsewhere (Arnold 2006) that the recombinant evolution of HIV could be considered analogous to the prezygotic, ecological, isolating barrier between *I. fulva* and *I. hexagona* that occur in different niches, in that it reflects the overcoming of ecological barriers.

Castro-Nallar et al. (2012) also reviewed the history of the controversy surrounding the origin of the two HIV "species," HIV-1 and HIV-2. Both species belong to the genus *Lentivirus* and have been defined using "genome organization, phylogenetic relationships, clinical characteristics, virulence, infectivity and geographic distribution" (Castro-Nallar et al. 2012). The definition of which primate species acted as the donor(s) for the human retrovirus uncovered the breaching of multiple, ecological barriers preventing genetic exchange among viral lineages. By definition, crossing the first ecological barrier involved zoonotic (i.e., cross-species) infections from other primates into humans (Hutchinson 2001; Castro-Nallar et al. 2012). Such zoonotic infections did not, however, occur at random. Specifically, simian immunodeficiency viruses (SIVs; the ancestral lineages of HIV) were apparently most likely to be transmitted among closely related primates (Moya et al. 2004). Given the phylogenetic distance between humans and some of the potential primate sources of SIV/HIV, the zoonotic event(s) must be considered a major barrier for the initial transfer of SIV/HIV and thus for the subsequent genetic exchange among HIV lineages. In particular, it has been concluded that HIV-1 (the major variant found in the human population) derived from a SIV from chimpanzees (specifically *Pan troglodytes troglodytes*; Gao et al. 1999) and also *Gorilla gorilla gorilla* (Plantier et al. 2009), while HIV-2 (mainly found in West Africa and India) was transferred from sooty mangabeys (Van Heuverswyn and Peeters 2007).

The opportunity for the viral reassortment that was the basis for so much of the evolutionary diversification of HIV resulted from a significant host/ecological shift from one primate species to another. However, the host shifts necessary for the genetic exchange among SIV/HIV types must also have involved transfers within primate lineages and within the same individual. First, for viral recombination between divergent types to occur, the types must be cocirculating in the same individual. Such cocirculation requires that either the divergent lineages arose independently in the same individual, or there was a dispersal event by one or both into the same habitat (i.e., an individual organism). The co-infection of an individual host is therefore similar to the dispersal by different species of animals or plants across ecological barriers into a geographic zone of overlap. Second, in the case of the diploid HIVs (containing two RNA molecules per virion), hybrid RNA molecules can only arise when virions from two or more divergent lineages infect the same cell and RNA molecules from divergent virions are packaged into a single virion (Hu and Temin 1990; Chin et al. 2005).

The failure of ecological barriers to isolate the various HIV forms was itself a multiphase process. First, the crossing of the between-species/within-individual ecological barriers to form hybrid SIV/HIV lineages predates the transfer into *H. sapiens*. For example, viral exchange—likely occurring within chimpanzees—resulted in a mosaic, ancestral HIV-1 (Gao et al. 1999; Paraskevis et al. 2003). Second, subsequent to the chimpanzee → human transfer event, recombination was rife, resulting in hybrid HIV-1 lineages accounting for 10–40% of the infections in Africa and 10–30% of the infections in Asia. This has

resulted in 49 variants and such a pervasive pattern of recombination that it has been questioned whether HIV actually should be considered as possessing discrete subtypes (Castro-Nallar et al. 2012). Finally, within individual humans, intertype genetic exchanges are extremely frequent, leading to the production of new genotypes at frequencies that equal or exceed mutation rates (Shriner et al. 2004).

The ecological barriers to genetic exchange among HIV lineages have not prevented recombination. Instead, reassortment between divergent HIVs has provided the basis for a large proportion of their diversity and evolution. Furthermore, recombination may result in viruses with increased fitness reflected in their resistance to antiviral medications and host-immune responses (Carvajal-Rodríguez et al. 2007; Castro-Nallar et al. 2012).

4.4.2 Pre-exchange isolating barriers and viral reassortment: Influenza

The definition of lineages of influenza viruses is based upon the DNA sequence of the genes coding for the hemagglutinin (HA) and neuraminidase (NA) as well as the antigenic properties of these glycoproteins (Figure 4.6). The HA and NA glycoproteins play important roles in determining host specificity, propagation of progeny viruses, and thereby the spread and severity of the disease (Medina and García-Sastre 2011). Influenza kills an estimated 250,000–300,000 people annually (World Health Organization 2003), with pandemics such as the "Spanish flu" accounting for ten times that number (Gibbs et al. 2002). Of the numerous lineages (i.e., subtypes; Figure 4.6) of influenza that have arisen and circulated in animal populations, only three have resulted in pandemics: H1N1 during 1918 and 2009, H2N2 during 1957, and H3N2 during 1968 (Garten et al. 2009; Medina and García-Sastre 2011). However, some subtypes that have not progressed to the category of pandemic are still of extreme concern due to the high frequency of lethality they cause to humans (e.g., H7N9 and H5N1: Kandun et al. 2006; Wang et al. 2008; Belser et al. 2013; Watanabe et al. 2013).

In their review article, Medina and García-Sastre (2011) reflected that "it is well known that simultaneous infection of a single cell by two distinct

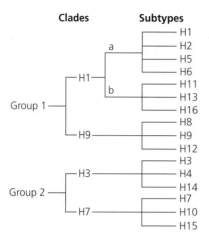

Figure 4.6 The definition of lineages of influenza viruses based upon the DNA sequence of the genes coding for hemagglutinin and neuraminidase along with the antigenic properties of these glycoproteins. This results in the placement of variants into two major groups (1 and 2), five clades, and 16 subtypes (from Medina and García-Sastre 2011).

influenza A viruses can lead to gene mixing, or reassortment, which can result in the generation of a novel influenza virus strain, and it is believed that most human pandemic viruses arose in this manner." That reassortant viruses are the basis for the majority of influenza pandemics would seem surprising given the many barriers that would prevent the genetic exchange events thus causing inter-subtype recombination to be rare. The genesis and spread of the pandemic-causing H1N1 subtype illustrates the sequence of rare events that must occur to produce novel influenza variants. In the context of this book, reviewing the origin and epidemiology of this viral lineage is also important because they reflect the evolutionary significance of rare events in the generation of "hybrid" lineages—in this case reassortant influenza.

The two pandemics caused by H1N1 variants occurred in 1918 and 2009 (Frost 1919; Taubenberger and Morens 2006; Fraser et al. 2009; Smith et al. 2009). It has been estimated that the 1918 pandemic, known as the "Spanish flu," killed approximately 50 million people and was unusual in that it had relatively mild effects on infants and the elderly, but caused extensive mortality in adults aged 20–40 (Frost 1919; Taubenberger and Morens 2006). The sequence and serology of the Spanish flu virus

led Worobey et al. (2014) to propose the following model to explain the evolution of this subtype: (1) the H1 gene variant arose in humans in *c.* 1907; (2) the remaining seven genes in the influenza genome were introduced into a human host directly from an avian source (possibly involving an H7N1 virus); (3) the avian source virus reassorted with the human virus containing the H1 gene that had arisen in *c.* 1907 thus giving rise to the pandemic-causing H1N1 viral lineage; and (4) the H1N1 found in swine evolved from the human, Spanish flu virus, rather than the human variant deriving from swine viruses. Each of the steps in this model reflects preexchange, reproductive barriers that would have limited the generation of the new H1N1 influenza lineage responsible for the 1918 pandemic. They included the generation of point mutations in the H1 gene found in humans, a minimum of one interspecies transfer event (i.e., avian → human) and a minimum of one reassortment event between the avian and human virus that resulted in an asymmetric viral hybrid (i.e., containing one "human" and seven "avian" genes; Worobey et al. 2014).

It can be expected that during the evolution of influenza very few reassortant, relative to nonreassortant, lineages were formed. Given that the majority of human pandemic influenza will arise via reassortment, the rarity of such events would seem fortuitous for our species. Yet, as evidenced by the observation of extensive reassortment within and between influenza lineages (Medina and García-Sastre 2011; Dudas et al. 2015; Pu et al. 2015), the barriers present are not sufficient to prevent evolutionarily effective genetic exchange.

4.5 Genetic exchange and reproductive isolation in bacterial clades

Polz et al. (2013) recently observed that "many bacterial and archaeal lineages have a history of extensive and ongoing horizontal gene transfer and loss, as evidenced by the large differences in genome content even among otherwise closely related isolates." In some cases, whole metabolic pathways have been transferred between divergent bacterial clades (e.g., the "whole-pathway tryptophan operon" present in *Corynebacterium glutamicum* and *C. diphtheriae*; Noda-García et al. 2013). Furthermore, it has been inferred that entire clades originated through the transfer of genes underlying metabolic functions (Nelson-Sathi et al. 2015). However, in spite of the widespread HGT-mediated evolution in prokaryotes, Polz et al. (2013) also presented evidence that habitat and niche associations among bacterial species limited genetic exchange among different lineages. In Section 4.5.1, I will illustrate both the reproductive barriers that limit HGT and the profound evolutionary significance of such transfers for prokaryotic lineages. In particular, I will illustrate both the pre-HGT barriers and the significance of horizontal exchanges on the evolutionary history of the bacterial genus *Yersinia*.

4.5.1 Pre-exchange isolating barriers and HGT: Yersinia

Plague (caused by the bacterial species *Yersinia pestis*) is estimated to have killed 200 million people during the period of human recorded history (Perry and Fetherston 1997). Alexandre Yersin isolated the causative agent for plague in 1894 (Perry and Fetherston 1997). At the time of the discovery, the third pandemic of plague was underway (Achtman et al. 1999). A large proportion of plague fatalities occurred during the three pandemics that swept through different portions of the known world: (1) the Justinian plague from 541 to 544 AD in the Mediterranean basin, Mediterranean Europe, and the Middle East; (2) the European Black Death from 1347 to 1351 AD; and (3) the pandemic begun in the Yunnan province of China in *c.* 1855 that spread around the globe via steamship routes (Perry and Fetherston 1997).

Yersinia pestis is now understood to be a derivative of the rarely lethal, enteric bacterial species *Y. pseudotuberculosis* (Achtman et al. 1999, 2004; Parkhill et al. 2001; Chain et al. 2004). For the thesis of the present text, it is significant to note the central role played by genetic transfer in the evolution of this pathogen. Indeed, several pathogenic species in this genus reflect parallel evolutionary steps involving horizontal gene transfer resulting in the acquisition of virulence determinants (Reuter et al. 2014). In regard to all of the pathogenic lineages, rare genetic transfers resulted in the origin of adaptations necessary for the ecological shift from one to another niche. Specifically for *Y. pestis*, this involved the unlikely series of evolutionary steps leading from a gastrointestinal bacterial species, transferred through contaminated food and water, into a "systemic invasive infectious" disease-causing pathogen (Parkhill et al. 2001) transferred either subcutaneously by an insect intermediate or through the air by infected humans. The details of the complex *Y. pestis* evolutionary history have been defined using data from studies of both the population genetic structure and genomic constitution of *Y. pestis* and its congeners (Achtman et al. 1999, 2004; Parkhill et al. 2001; Chain et al. 2004). As mentioned, the series of events leading to the acquisition and loss of numerous genetic pathways resulting in the pathogenicity seen in *Y. pestis* has been repeated in the evolutionary history of unrelated congeners (Reuter et al. 2014).

Achtman and his colleagues carried out two genetic surveys (Achtman et al. 1999, 2004) to define microevolutionary patterns within *Y. pestis*. In the first of these they estimated the population genetic structure of three *Yersinia* species (i.e., *Y. pestis*, *Y. pseudotuberculosis*, and *Y. enterocolitica*) by collecting partial sequence data from six genes (*dmsA, glnA, manB, thrA, tmk, trpE*) and RFLP data using the insertion element IS100 as a probe. The gene sequence information was collected for worldwide samples of *Y. pestis* (36 strains), *Y. pseudotuberculosis* (12 strains), and *Y. enterocolitica* (13 strains). The RFLP data were collected from 49 strains, representing three biotypes (i.e., biovars) of *Y. pestis*. Both data sets allowed an assessment of microevolutionary patterns, independent of phenotype, within *Y. pestis*. The conclusions from the first analysis were summarized by the title of the paper containing these data, "*Yersinia pestis*, the cause of plague, is a recently emerged clone of *Yersinia pseudotuberculosis*" (Achtman et al. 1999). The plague bacteria thus differed little in sequence variation from either of its congeners and, in particular, was highly similar to *Y. pseudotuberculosis*. This study did, however, define the three biovars as reciprocally monophyletic from one another. The second study of *Y. pestis* did not find reciprocal monophyly for the various biovars. Instead, geographic origin of the samples was a better predictor of phylogenetic patterning. This latter study also provided an estimated time of origin for the *Y. pestis* lineage of *c.*10,000–13,000 ybp. A second divergence event, occurring *c.* 6500 ybp, gave rise to lineages more commonly associated with human populations (Achtman et al. 2004).

Three analyses that applied genomic approaches have also helped define the diverse array of genetic processes contributing to the evolution of *Y. pestis*. Conclusions from a study by Parkhill et al. (2001) emphasized both the role of HGT and gene loss in the evolutionary trajectory resulting in the plague bacteria. First, these authors detected sequence footprints (i.e., insertion sequence element perfect repeats and anomalous GC base-composition biases) in the genome of *Y. pestis* indicating numerous recombination events (Parkhill et al. 2001). Second, they discovered evidence for widespread gene inactivation in the *Y. pestis* genome, relative to its sister taxon *Y. pseudotuberculosis*. The large-scale gene inactivation was indicated by the presence of *c.* 150 pseudogenes. Specifically, they detected numerous genes thought to be associated with the ancestral, enteric bacterial habitat that had been preferentially silenced (Parkhill et al. 2001). Third, these authors argued that *Y. pestis* acted as the recipient of DNA from multiple donors, including bacteria and viruses (Parkhill et al. 2001). The data assembled in this analysis suggested that *Y. pestis* had developed adaptations as a result of both horizontal exchange and gene silencing.

As with Parkhill et al. (2001), Chain et al. (2004) also detected a role of both gene acquisition and gene loss in the evolution of the plague bacteria. However, these latter authors particularly emphasized the role of gene loss and modification in the evolutionary pathway leading to the plague bacterium. Yet, even with this emphasis, they identified horizontal gene transfer events that yielded 32 chromosomal genes as well as two plasmids. One of the plasmids carried a gene that encoded phospholipase D (Hinnebusch et al. 2002; Hinchliffe et al. 2003). This gene product has been shown to be necessary for the viability of the plague bacteria in the midgut of its vector, the rat flea *Xenopsylla cheopis* (Hinnebusch et al. 2002).

The most extensive, genomic analysis of *Yersinia* to date involved the sequencing of > 200 genomes, encompassing all the species from this genus. Unlike the previous genomic studies, Reuter et al. (2014) found that "contrary to hypotheses that all pathogenic *Yersinia* species share a recent common pathogenic ancestor, they have evolved independently but followed parallel evolutionary paths." The evolutionary relationships defined from the whole-genome sequence information thus separated the various pathogenic species from each other, placing them closest to independent, non-pathogenic forms. Notwithstanding the difference between the findings of this recent analysis and previous studies, the whole-genome sequencing of all the species from *Yersinia* defined the major and, surprisingly, repeated role played by the improbable acquisition of certain genes from divergent evolutionary donors (Reuter et al. 2014).

In *Yersinia*, pre-exchange barriers predicted to act as a formidable sieve for the transfer of genes between species have not stopped the acquisition of both the genes and the adaptations they provide. As already discussed, this same conclusion has also been drawn for numerous plant and animal examples. In the remaining sections of this chapter, I will highlight findings from the mosquito genus *Anopheles* and the sunflower genus *Helianthus* that reflect the role that *post*-exchange (i.e., postzygotic) reproductive barriers play in determining the type of evolutionary products produced by natural hybridization.

4.6 Genetic exchange and postzygotic reproductive isolation in eukaryotes: Hybrid viability and fertility

In *Genetics and the Origin of Species*, Dobzhansky discussed hybrid inviability as one of several isolating mechanisms that "engender such disturbances in the development that no hybrids reach reproductive stage" (Dobzhansky 1937, p 231). Though a useful general descriptor, it has been repeatedly demonstrated that all hybrids are not created equal (i.e., they are not always unfit relative to their parents). Instead, certain hybrid genotypes are expected to develop normally and demonstrate relatively high viability (see Arnold and Hodges 1995; Arnold 1997; Arnold 2006 for examples). However, the presence of some frequency of hybrid inviability is a universal observation when matings occur between genetically divergent lineages (e.g., Charron et al. 2014). This breakdown in the zygotic development of some hybrid genotypes, but not in others, is one of the factors leading to the observed differential introgression resulting in mosaic/recombinant genomes.

4.6.1 Hybrid inviability and introgressive hybridization: Anopheles

The mosquito genus *Anopheles* provides an excellent example of how differential hybrid viability may affect organismic evolution. As the vectors of malaria, species of *Anopheles* are of major interest in terms of genetic variation and evolutionary differentiation. Of particular importance is an understanding of the number of differentiated forms present in nature. This knowledge is needed to design control methods, including those that depend upon gene flow between introduced, genetically modified mosquitoes and naturally occurring individuals (Cohuet et al. 2005). Furthermore, whether there is a single, or many, species that must be controlled has obvious implications for the types of control measure—for example, the types of pesticide—that can be used effectively.

Three of the species of most interest as malaria vectors belong to the African assemblage known as the *Anopheles gambiae* species complex. Until recently, this complex consisted of seven morphologically indistinguishable species (della Torre et al. 1997; Stump et al. 2005b). Recently, two new species were recognized (Coetzee et al. 2013). One of the newly named taxa, *Anopheles coluzzii*, was previously designated the "M form" of *An. gambiae*. The "S form" of this species retained the name, *An. gambiae* (Coetzee et al. 2013). *Anopheles gambiae* (encompassing the recently named *An. coluzzii*) and *An. arabiensis* have been considered the two most important malarial vectors in sub-Saharan Africa (White 1971). It is now well established that *Anopheles* species complexes in general demonstrate the capacity for some frequency of intertaxonomic gene flow (e.g., Walton et al. 2000, 2001; Stump et al. 2005b; Morgan et al. 2011). This is particularly true for the *An. gambiae* complex, including *An. gambiae*, *An. arabiensis*, and *An. coluzzii* (Figure 4.7; Lanzaro et al. 1998; Onyabe and Conn 2001; Coluzzi et al. 2002; Gentile

et al. 2002; Tripet et al. 2005; Yawson et al. 2007; Parmakelis et al. 2008; Mendes et al. 2010; Neafsey et al. 2010; Weetman et al. 2012; Lee et al. 2013). Furthermore, a proportion of this genetic exchange apparently involves adaptive trait transfers. For example, both Fontaine et al. (2015) and Norris et al. (2015) inferred adaptive trait transfers among these species. Norris et al. (2015) detected the transfer of insecticide resistance genes from *An. gambiae* into *An. coluzzii*. This introgression was coincident with the increased use of insecticide-treated bed nets in Mali, suggesting the action of selection favoring resistant mosquitoes (Norris et al. 2015). Consistent with this inference, a phylogenomic analysis by Fontaine et al. (2015) led to the conclusion that "traits enhancing vectorial capacity may be gained through interspecific gene flow, including between nonsister species."

Though introgression among lineages has been recognized (Figure 4.7), it is also known that postzygotic reproductive barriers likely contribute to limitations for genetic exchange both within and between *An. gambiae* complex species (White 1971; Slotman et al. 2005; Turner et al. 2005; Lee et al. 2013; O'Loughlin et al. 2014). For example, significant inviability has been detected in the first backcross generation between *An. gambiae* and *An. arabiensis*, as reflected by lower than expected frequencies of certain genotypes. Slotman et al. (2004) found that the *An. gambiae* X chromosome (X_G) occurred at a frequency of *c.* 10% (the expected frequency was 50%) in the backcross toward *An. arabiensis*. Thus, the X_G chromosome causes inviability when placed onto a largely *An. arabiensis* genetic background. By examining autosomal loci, Slotman et al. (2004) were able to demonstrate that this large-scale inviability was due to an incompatibility between X_G loci and one or more loci on each of the autosomes. It is also possible that interactions between the *An. gambiae* X chromosome and the *An. arabiensis* Y chromosome resulted in some level of inviability in the backcross populations (Slotman et al. 2004). Regardless of the identity of the interacting genomic factors, introgression between *An. gambiae* and *An. arabiensis* is greatly restricted by hybrid inviability.

As with *An. gambiae* and *An. arabiensis*, both pre- and postzygotic hybrid viability appear to limit genetic exchange between *An. gambiae* and *An. coluzzii*

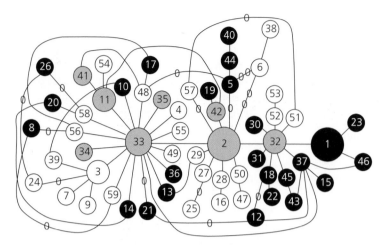

Figure 4.7 Parsimony network constructed using mtDNA haplotypes detected in *Anopheles* populations. Numbers in the circles indicate the haplotype designation, and the size of the circles is roughly proportional to the haplotype's frequency. Open and black circles indicate haplotypes unique to *An. gambiae* and *An. arabiensis*, respectively. Shaded circles reflect haplotypes shared between the species (from Besansky et al. 1997).

(Figure 4.8; Diabaté et al. 2005, 2008, 2009; Weetman et al. 2012; Lee et al. 2013; Cassone et al. 2014). Reidenbach et al. (2012) thus stated, "No intrinsic postmating barriers have been detected between M and S [i.e., *An. gambiae* and *An. coluzzii*, respectively] . . . However, extrinsic, environment-based postmating barriers are presumed to be strong and to act against maladapted hybrids in the alternate M versus S larval habitats." Their statement concerning a lack of "intrinsic barriers" reflected laboratory-crossing experiments that yielded no evidence of lower viability in hybrid, relative to nonhybrid, genotypes (e.g., Hahn et al. 2012). In contrast, reciprocal transplant experiments of *An. gambiae* and *An. coluzzii* into natural environments (Diabaté et al. 2005, 2008) as well as numerous, genomic analyses (Onyabe and Conn 2001; Stump et al. 2005a, b; Reidenbach et al. 2012; Weetman et al. 2012; Lee et al. 2013) detected evidence for strong, environment-dependent selection against at least some hybrid genotypes.

In spite of the genetic discontinuity between *An. gambiae*, *An. coluzzii*, and *An. arabiensis*, caused at least partially by viability selection against some hybrid genotypes, introgressive hybridization has impacted greatly the population genetics of these three species. This conclusion has been reached on the basis of data indicating the sharing of chromosomal inversions, mtDNA, autosomal, and X chromosomal loci in areas of sympatry (e.g., Figure 4.8; Besansky et al. 1997, 2003; Lanzaro et al. 1998; Gentile et al. 2002;

Donnelly et al. 2004; Stump et al. 2005a, b; Yawson et al. 2007; Weetman et al. 2012; Lee et al. 2013; Nwakanma et al. 2013).

Past and ongoing introgression has resulted in mosaic genomes in *Anopheles*, similar to those found in influenza A, Darwin's finches, Louisiana irises, etc. Genetic exchange has thus structured the genomic constitution of *An. gambiae*, *An. coluzzii*, and *An. arabiensis*. The patterns of introgression also reflect the role that selection plays in constraining and facilitating genetic exchange. I discussed above the selective sieve that prevents the expected frequency of introgression between *An. arabiensis* and *An. gambiae* for the X_G chromosome (Slotman et al. 2004). Similarly, pericentromeric "islands" have also been identified in which introgression is greatly impeded between *An. gambiae* and *An. coluzzii* (Figure 4.8; Turner et al. 2005; Reidenbach et al. 2012; Lee et al. 2013; but see Cruickshank and Hahn 2014). However, selection may also promote the introgression of certain loci. For example, population genetic variation and experimental hybridization supported the inference of adaptive introgression between *An. arabiensis* and *An. gambiae* (della Torre et al. 1997). Indeed, it has been argued that selective introgression from *An. arabiensis* (a xeric-adapted species) into *An. gambiae* (a mesic-adapted species) may have led to the ecological expansion of *An. gambiae*, thus allowing it to become the predominant vector of malaria in sub-Saharan Africa (Besansky et al. 2003).

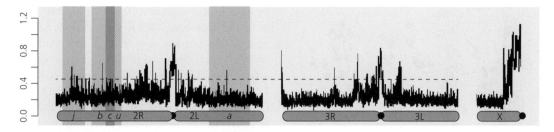

Figure 4.8 Genome scan of divergence along the chromosomes of *An. gambiae* and *An. coluzzii* (i.e., "M" and "S" forms, respectively) found in sympatric Cameroon populations. Horizontal dashed line is the threshold for significant divergence between the *An. gambiae* and *An. coluzzii* loci (Reidenbach et al. 2012).

4.6.2 Hybrid sterility and introgressive hybridization: Helianthus

Another widely estimated component of hybrid fitness is fertility. The reason for the frequent use of this factor to estimate hybrid fitness likely relates to its relative ease of measurement (especially for hybrid males). For plants and animals, numerous methodologies have been applied to define the fertility of hybrid genotypes (e.g., Arnold and Jackson 1978). Often such methods allow inferences not only of the relative fitness (in terms of fertility) of hybrids and their parents, but also facilitate hypotheses concerning causality; for example, the presence of chromosomal structural heterozygosity leading to unbalanced meiotic products and eventually to nonviable gametes (Heiser 1947). Genic components can also contribute to lower fertility in early- and later-generation hybrids. As discussed in Section 4.3.2, *I. fulva* × *I. brevicaulis* BC$_1$ hybrids possessed a range of pollen fertilities (i.e., from 0 to 100%; Bouck 2004). This effect on male fitness was apparently not due to chromosomal structural differences in the parental species since F$_1$ hybrids demonstrated *c.* 100% fertile pollen. As with hybrid viability, and regardless of the factors that cause the variation in fertility, it is again a truism that different hybrid genotypes can vary in terms of their estimated fitness.

The annual sunflower species complex is probably best known as an evolutionary model system developed through the research of Heiser, Rieseberg, and their colleagues, in particular with regard to the definition of introgressive hybridization and hybrid speciation (e.g., Heiser 1947, 1951a, b; Stebbins and

Daly 1961; Rieseberg et al. 1988, 2003; Rieseberg 1991; Ungerer et al. 1998; Carney et al. 2000; Lexer et al. 2003; Gross et al. 2004; Ludwig et al. 2004; Rosenthal et al. 2005; Whitney et al. 2006, 2010; Donovan et al. 2009; Renaut et al. 2014). Yet, these same workers have shown that hybridization between various sunflower taxa is greatly restrained by fertility barriers. This has been particularly well demonstrated for the species *H. annuus* and *H. petiolaris*. For example, Heiser demonstrated that F$_1$ hybrids possessed pollen fertilities of 0–30% (Heiser 1947). In addition, crosses designed to form the F$_2$ and first backcross generations resulted in a maximum seed set of 1% and 2%, respectively (Heiser et al. 1969). This extremely low fertility is likely due to the numerous chromosomal rearrangement differences present between these species (Figure 4.9; Heiser 1947; Heiser et al. 1969), and apparently has accumulated much faster in annual, relative to perennial, species from this genus (Owens and Rieseberg 2014). Indeed, the great limitations placed upon genetic exchange between *H. annuus* and *H. petiolaris* by such strong fertility selection is a partial cause of the repeated occurrence of similar hybrid genotypes from independent crossing experiments (Rieseberg et al. 1996). Furthermore, the observation that hybrid species tend to have the same genomic organization as found in the experimental hybrids suggests that the intense fertility selection against many hybrid genotypes has significantly affected natural genetic exchange as well (Rieseberg et al. 1996, 2003).

Taking into account all barriers to gene flow—including the strong fertility selection against hybrids—the overall reproductive isolation

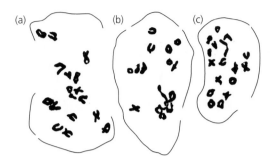

Figure 4.9 Drawings of metaphase I chromosomal configurations for (a) *H. annuus*, (c) *H. petiolaris*, and (b) F$_1$ hybrids formed between these two species. Note that only bivalents occur in (a) and (c), but multivalents are present in (b), indicating chromosome-rearrangement differences between the two species (from Heiser 1947).

between *H. annuus* and *H. petiolaris* has been estimated to be >0.999 (Sambatti et al. 2012). Such an extremely high level of isolation could be used to argue against the possibility of evolutionarily important genetic exchange. Contrary to this prediction, natural hybridization between *H. annuus* and *H. petiolaris* has given rise to "rampant" introgression between these two species (Yatabe et al. 2007; Kane et al. 2009); this genetic exchange has extended over the entire period since they diverged from a common ancestor, *c.* 1 mya (Strasburg and Rieseberg 2008). This longstanding introgressive hybridization has resulted in numerous natural hybrid zones and at least three hybrid species (*Helianthus anomalus*, *H. deserticola*, and *H. paradoxus*; Heiser et al. 1969; Rieseberg et al. 1990, Rieseberg 1991). In addition, Rieseberg et al. (2003), Whitney et al. (2006, 2010), and Donovan et al. (2009) documented hybridization among various annual *Helianthus* taxa, resulting in recombinant genotypes capable of invading novel habitats. Therefore, genetic exchange has not only played a major role in producing the mosaic genomes possessed by taxa in this complex (Yatabe et al. 2007; Strasburg and Rieseberg 2008; Kane et al. 2009; Sambatti et al. 2012), but has also resulted in evolutionary and ecological novelty reflected in species with new adaptations (Lexer et al. 2003; Rieseberg et al. 2003; Ludwig et al. 2004; Rosenthal et al. 2005; Donovan et al. 2009). Again, it is apparent that an incredibly strong selective sieve that allowed the production of only a few hybrid genotypes did not prevent significant evolutionary effects from genetic exchange.

4.7 Conclusions

The purpose of this chapter, like reproductive isolation, is multipartite. I wanted to emphasize that reproductive isolation should be assumed to consist of (until proven otherwise) a sequential, multitiered series of processes that limit genetic exchange. To illustrate that this conclusion is not organism-specific, I have used the [taxonomically] broadest possible set of examples. Notwithstanding the significant reproductive barriers present in all the organismic groups, evolutionary diversification—whether in terms of new lineages, novel gene combinations, and/or new adaptations—was detected. As I have stated in the previous chapters, this observation is one factor that undermines the dismissal of introgressive hybridization, hybrid speciation, or horizontal gene transfer as evolutionarily unimportant due to the presence of strong isolating barriers. In conclusion, reproductive isolation should be considered from the standpoint that barriers to genetic exchange are often permeable. Given that evolutionary diversification is best described by a web-of-life metaphor, such an approach should offer the best opportunity to realize a coherent understanding of evolutionary process and pattern.

CHAPTER 5

Genetic exchange and fitness

"Pure species have of course their organs of reproduction in a perfect condition, yet when inter-crossed they produce either few or no offspring. Hybrids, on the other hand, have their reproductive organs functionally impotent"
(Darwin 1859, p. 246)

"In the light of the above results it would seem likely that the different habitat pattern of the Appalachians more readily provides intermediate zones for hybrids between these two species."
(Hubricht and Anderson 1941)

"The overall lack of evidence for reduced hybrid fitness is inconsistent with either the dynamic-equilibrium or reinforcement models."
(Moore and Koenig 1986)

"Strong selection occurred against the parental genotypes in the middle hybrid zone garden in middle hybrid zone soil; F_1 hybrids had the highest fitness under these conditions"
(Miglia et al. 2007)

"Fitness of hybrid treatments showed declines relative to midparent values followed by rapid recovery, with two hybrid replicates ultimately showing higher fitness than parentals"
(Hwang et al. 2011)

"This suggests that when multiple traits must function together, novel combinations of traits in hybrids might reduce performance below that expected for an intermediate phenotype."
(Arnegard et al. 2014)

5.1 Genetic exchange and fitness: Recombinant genotypes are just like everybody else

It has been argued since the time of Darwin's *Origin* that natural hybridization between divergent populations is an evolutionary dead end. A common rationale for this conclusion is the view that the outcome of hybridization episodes has been governed by selection against hybrids, independent of the environment. Historically, plant biologists (e.g., Anderson 1949; Stebbins 1959) have been more open to environment-dependent models of hybrid fitness, but until the later decades of the twentieth century, process-oriented studies of plant hybrid populations were far fewer than those from animal systems (Harrison 1990; Arnold 1992, 1997). Because animals are often much more difficult to manipulate experimentally and to monitor in nature (but see e.g., Martin and Wainwright 2013; Grant and Grant 2014a), this resulted in a lack of data to rigorously test for patterns of hybrid fitness and the importance of environment-dependent selection on the outcome of episodes of genetic exchange (Harrison 1990; Arnold and Hodges 1995; Arnold 1997).

Conceptual assumptions and empirical observations concerning "hybrid" fitness have been discussed in the preceding chapters in the context of research on hybrid zones, species concepts, and adaptive evolution. It could be argued that the subject has been well illustrated by these previous discussions. However, the fitness of hybrids has been used as the fulcrum for those who wished to either discount or emphasize the evolutionary significance of genetic exchange (e.g., Mayr 1942, 1963; Arnold 1992, 1997, 2006; Dowling and DeMarais 1993; Rieseberg 1997; Coyne and Orr 2004; Seehausen 2004). Likewise, it still seems unclear to many that there are inherent problems associated with how the term "hybrid" is applied to discuss the fitness of recombinants. This term is quite appropriate for indicating an individual of mixed ancestry. Yet, inclusion of all recombinant classes under this term has led to incorrect conclusions concerning the fitness of hybrids relative to their progenitors. For example, it is often stated that "hybrids" are uniformly unfit. However, what is generally being referred to by the term "hybrids" is a genotypically heterogeneous group of individuals. In contrast, when investigators divide admixed individuals into well-defined genotypic classes, fitness estimates for these classes can range from greater, equivalent, intermediate, or less than the parental forms (Arnold and Hodges 1995; Stelkens et al. 2014). In the following sections I will thus illustrate the two observations that need to be recognized for understanding the potential evolutionary impact from genetic exchange: (1) admixed genotypes can vary in fitness, often in an environment-dependent manner and (2) recombinant genotypes that demonstrate lower inferred fitness, relative to non-recombinant and parental genotypes, can still be the basis for evolutionary innovations (Arnold and Hodges 1995; Arnold 1997, 2006; Arnold et al. 1999).

5.2 Genetic exchange and fitness: Environment-independent and -dependent selection on animal hybrids

If I may digress into informality for a moment in order to reemphasize a common theme of this book, I think nowadays when I hear someone repeat the mantra that "hybridization is unimportant in animal taxa" I less often bristle with frustration and more often sigh with weariness—at least, I

hope that is what I do. Thankfully, this is a much less common occurrence now than in earlier times when arguments for a heightened recognition of the importance of reticulate evolutionary processes in animals were made (e.g., Lewontin and Birch 1966; Arnold 1992, 1997; Dowling and DeMarais 1993). Yet, it is still correct that some evolutionary zoologists struggle with recognizing the influence of natural hybridization and genetic exchange *sensu lato* (e.g., horizontal transfer of retroviral sequences into mammalian genomes thus affecting gene function; Weiss 2006; Cordaux and Batzer 2009; Kuhn et al. 2014; Section 1.7.1) in the evolutionary diversification of animal lineages and clades. However, it is also correct that the recent emphasis on testing for introgression accompanying and shaping evolutionary divergence in many animal groups reflects a new appreciation of reticulate evolution.

The appreciation, along with the development of techniques for the collection and analysis of genomic sequences, is also making straight the path for tests of the fitness of hybrid genotypes relative to their parents. In Section 3.4.6 I discussed a model for such analyses involving the genus *Mus*. In the following sections I will provide further illustrations of the fact that animal hybrids vary in fitness, sometimes reflective of intrinsic (i.e., environment-independent) and sometimes extrinsic (i.e., environment-dependent) natural selection. Regardless of the particular form of selection, a range of hybrid genotypes from a cross between divergent animal lineages almost always results in a range of fitness estimates, including those genotypes that reflect positive selection. For example, it is important to point out that the discussion of adaptive trait introgression in Chapter 1 illustrates examples of elevated hybrid fitness. However, the cases discussed in Sections 5.2.1 to 5.2.3 will also illustrate the observation that even hybrids with lower relative viability and/or fertility can be the harbingers of novel adaptations and lineages.

5.2.1 Genetic exchange and hybrid fitness in animals: Drosophila

A major drawback to studies of genetic variation in natural populations is the relative weakness of such analyses for estimating the fitnesses of different genotypes (Lewontin 1974). In the context of this book, it thus remains extremely difficult to falsify the null hypothesis of genetic exchange due to selective neutrality (Clark 1985). Indeed, only by experimentation, long-term assays across generations, and/or meta-analyses can this hypothesis be tested adequately. Fortunately, such analyses have been undertaken in a number of animal taxa, including the genus *Drosophila* (e.g., Gomes and Civetta 2014). Though much of the research on hybrid fitness in this genus has been designed to demonstrate pre- and postzygotic reproductive barriers in the context of "biological speciation" (Koopman 1950; MacRae and Anderson 1988; Coyne and Orr 1989; Zeng and Singh 1993; Kilpatrick and Rand 1995; Jenkins et al. 1996; Coyne et al. 2002; Tao and Hartl 2003; Bayes and Malik 2009; Barbash 2010; Dickman and Moehring 2014; Matute et al. 2014; Turelli et al. 2014), these estimates often provide the necessary data for inferring the relative fitness of hybrids. These inferences come not only from the experiments themselves, but also by comparing the experimental findings to genetic variability in natural populations (e.g., Solignac and Monnerot 1986; Llopart et al. 2005a, b, 2014; Nunes et al. 2010; Garrigan et al. 2012; Herrig et al. 2014).

Figure 5.1 reflects the results from a meta-analysis utilizing 173 pairwise crosses, allowing a test for a correlation between genetic divergence and postzygotic reproductive isolation ("RI"), among *Drosophila* species (Rabosky and Matute 2013). From these data (and those in a similar data set from birds), Rabosky and Matute (2013) concluded that RI (i.e., degree of hybrid inviability and infertility) was not correlated with speciation rate. For the purpose of the present discussion, the results provided in Figure 5.1 also illustrate well the observation of variation in hybrid fitness estimates. Specifically, values of RI obtained from the interspecific crosses ranged from 0 to 1, with 0 indicating the recovery of viable and fertile hybrids relative to control crosses and 1 indicating completely inviable or sterile hybrids (Rabosky and Matute 2013). Thus, some crosses between recognized species of *Drosophila* gave rise to hybrid genotypes that were as fit as intraspecific progeny. Furthermore, those crosses that produced hybrid progeny arrays with a *mean* value between 0 and 1 also likely included genotypes that possessed fitnesses equivalent to non-hybrid progeny (Arnold and Hodges 1995; Arnold 1997).

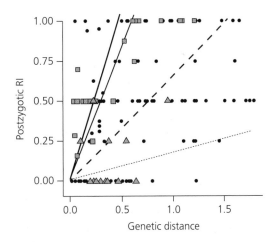

Figure 5.1 Pairwise postzygotic reproductive isolation between species of *Drosophila*. The genetic distance and postzygotic reproduction isolation (i.e., RI) is shown for 173 pairwise crosses. Squares indicate species complexes with a "fast" evolution of RI evolution and triangles reflect species that demonstrate a "slow" evolution of RI. The solid and dashed lines indicate the trajectories of RI evolution from various fitted linear models (i.e., thick and thin lines indicate the fit for sympatric and allopatric species pairs, respectively.) Values of RI range from 0 (hybrids were found to be fertile and viable) to 1 (all hybrid offspring were sterile or inviable) (from Rabosky and Matute 2013).

Drosophila simulans, D. mauritiana, and D. sechellia

Historically, hybridization in *Drosophila* was considered to be rare. One explanation for this assumption was the observation of reduced, mean hybrid fitness due to the sterility of F_1 males. However, hybridization was not unknown in this genus, and its effects were sometimes suggested to be of evolutionary importance (Kaneshiro 1990). Several species groups have now become models of reticulate evolutionary processes, one of which contains *D. simulans*, *D. sechellia*, and *D. mauritiana* (Solignac and Monnerot 1986; Aubert and Solignac 1990). Though these three species are homosequential (i.e., possess identical polytene chromosome banding patterns; Lemeunier and Ashburner 1984), experimental crosses between them reveal both pre- and postzygotic reproductive isolation. For example, there were extreme difficulties in crossing female *D. mauritiana* with male *D. simulans* (David et al. 1974; Robertson 1983), but no significant restrictions with the reciprocal cross (Aubert and Solignac 1990). All

of the reciprocal crosses among these three species result in sterile or highly infertile F_1 males, but
fertile females (David et al. 1974; Robertson 1983;
Zeng and Singh 1993; Tao and Hartl 2003). In addition to the variation in F_1 male fertility, Dickman
and Moehring (2014) demonstrated that the relative
fitness, in terms of longevity, likewise fluctuated
significantly among male and female hybrid genotypes. In this latter analysis, male hybrids often displayed no fitness reduction relative to non-hybrid
genotypes, while female hybrid genotypes demonstrated significantly reduced fitnesses (Dickman
and Moehring 2014).

In spite of the restrictions placed on initial and
advanced generation hybrid formation among *D.
simulans*, *D. sechellia*, and *D. mauritiana*, past and
contemporaneous introgression has impacted patterns of divergence and adaptation in this complex.
Figure 5.2 illustrates the results from an analysis
by Solignac and Monnerot (1986) of mtDNA variation in these three species. This figure illustrates a
discordant pattern between the mtDNA haplotype
and the species assignment of some of the populations sampled. Specifically, the mtDNA haplotypes
present in some of the *D. mauritiana* populations
were identical to some identified in *D. simulans*
samples (Figure 5.2). Solignac and Monnerot (1986)
proposed three hypotheses to explain the mtDNA
variation detected in *D. simulans* and *D. mauritiana*
populations. Two of these hypotheses incorporated
introgression as an explanation for the mtDNA variation, with *D. mauritiana* or *D. simulans* alternately
acting as the donor species for the mtDNA found in
contemporary populations (Solignac and Monnerot
1986). These investigators concluded that the most
parsimonious explanation involved introgression
from *D. simulans* into *D. mauritiana*. Support for this
hypothesis came from the experimental crosses (discussed earlier) in which matings between female *D.
mauritiana* with male *D. simulans* (David et al. 1974;
Robertson 1983; Zeng and Singh 1993) were largely
unsuccessful, but with the reciprocal cross being
viable (Aubert and Solignac 1990). Finally, Aubert
and Solignac (1990) argued for the rapid incorporation of mtDNA from *D. simulans* into *D. mauritiana*,
supporting the hypothesis that introgression was
promoted by a selective advantage of some hybrid
genotypes (Aubert and Solignac 1990).

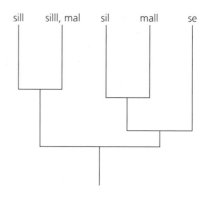

Figure 5.2 Mitochondrial haplotype network for *Drosophila
simulans* (si), *D. mauritiana* (ma), and *D. sechellia* (se) (from Solignac
and Monnerot 1986).

More recent analyses of *D. simulans*, *D. sechellia*,
and *D. mauritiana* (Ballard 2000; Nunes et al. 2010;
Garrigan et al. 2012; Brand et al. 2013; Matute and
Ayroles 2014) have both confirmed and extended
an understanding of the role of genetic exchange in
the origin, diversification, and adaptive evolution
within this group. By analyzing both nuclear and
mitochondrial loci, Nunes et al. (2010) inferred that
introgression observed between *D. simulans* and *D.
mauritiana* was the result of at least three bouts of natural hybridization. Intriguingly, the results from this
study detected bidirectional introgression of mtDNA
(Nunes et al. 2010) rather than the asymmetric transfer inferred from earlier studies and predicted by
experimental crosses (see earlier in this section).
Though the addition of nuclear loci suggested some
introgression of these genomic components as well,
Nunes et al. (2010) pointed to the need for genomic
sequencing data to address questions concerning
the process and pattern of evolution in the species
group. Garrigan et al.'s (2012) paper entitled "Genome sequencing reveals complex speciation in the
Drosophila simulans clade" described just such an
analysis of whole genome alignments for *D. simulans*,
D. sechellia, and *D. mauritiana*. The "complexity" associated with diversification in this group included
the observations that "up to 4.6% of autosomal and
2.2% of X-linked regions have evolutionary histories
consistent with recent gene flow between the mainland species (*D. simulans*) and the two island endemic
species (*D. mauritiana* and *D. sechellia*). Our findings
thus show that gene flow has occurred throughout

the genomes of the *D. simulans* clade species despite considerable geographic, ecological, and intrinsic reproductive isolation" (Garrigan et al. 2012). Furthermore, analyses of these data detected not only frequent genetic exchange among the three species, but apparent adaptive trait introgression as well (Garrigan et al. 2012; Brand et al. 2013).

One of the hypotheses tested by these analyses was the duration of introgression—i.e., whether introgression was ancient, recent, or both. In this regard, mtDNA and nuclear data sets led to the inference of divergence-with-gene-flow that spanned the *c.* 250,000 years since *D. simulans*, *D. sechellia*, and *D. mauritiana* diverged from a common ancestor. Once again quoting from the conclusions of Garrigan et al. (2012), and consistent with the findings of Ballard (2000) and Nunes et al. (2010), "the distributions of the times to exchangeability (τ) suggest long histories of modest gene flow with strong modes of $\tau \to 0$, suggesting very recent introgression within the last ~2400 yr . . . It seems possible that these recent introgression times could be the consequence of human activity, moving flies between islands during very recent history." With regard to the possible role of ongoing natural hybridization, Matute and Ayroles (2014) detected *D. simulans* × *D. sechellia* hybrid individuals from three Seychelles archipelago islands where the two species overlap.

Drosophila pseudoobscura and D. persimilis

Evolutionary biologists have utilized the clade containing *Drosophila pseudoobscura pseudoobscura, D. p. bogotana*, and *D. persimilis* to illustrate processes as divergent as chromosomal evolution, *gene conversion*, and sexual selection (Schaeffer et al. 2003; Schaeffer and Anderson 2005; Kim et al. 2012). However, this species complex is probably best known from the work of Dobzhansky et al. on the evolution of pre- and postzygotic reproductive barriers (Dobzhansky 1936, 1946; Tan 1946; Koopman 1950; Jenkins et al. 1996; Noor et al. 2001; Chang and Noor 2007; McDermott and Noor 2011, 2012; Phadnis 2011; Kim et al. 2012) and from studies by Noor and others on speciation via reinforcement (Noor 1995; Kelly and Noor 1996; Ortíz-Barrientos et al. 2004; Lorch and Servedio 2005; Ortíz-Barrientos and Noor 2005). All of these studies have contributed to

the development of the *D. pseudoobscura / D. persimilis* complex as a model system for understanding the evolution of highly stringent barriers to genetic exchange, reflecting some hybrid genotypes with low relative fitnesses. Indeed, Dobzhansky (1973) identified only one naturally occurring *D. pseudoobscura* × *D. persimilis* F_1 individual from *c.* 35 years of collections, and thus concluded that hybrid formation could not be greater than 1 in 10,000 interspecific matings. In contrast, though also detecting extremely low frequencies of natural F_1 hybrid formation between these species, Powell (1983) initially concluded that significant introgression had occurred, as reflected by cytoplasmic (i.e., mtDNA) introgression in areas of sympatry. However, he later suggested that the sharing of the mtDNA haplotypes in sympatry might have instead resulted from incomplete lineage sorting (Powell 1991). This latter conclusion enforced the idea of limited, or no, gene exchange between these species (Dobzhansky 1973) and, as an aside, illustrated the difficulty (as discussed in Chapter 3) in separating the effects of genetic exchange from retention of ancestral polymorphisms.

Given these findings, it may seem surprising that *D. p. pseudoobscura, D. p. bogotana*, and *D. persimilis* are now also paradigms for divergence-with-gene-flow. For example, while pointing to the cumulative effect of [hybrid] "male sterility, hybrid inviability, sexual isolation, and a hybrid male courtship dysfunction" in the reproductive isolation of *D. pseudoobscura* and *D. persimilis*, Noor et al. (2001) likewise concluded "that gene flow between these species via hybrid males may be possible at loci spread across much of the autosomes." In this regard, studies by Hey and his colleagues have indeed documented the signature of reticulate evolution between these two species (Wang and Hey 1996; Wang et al. 1997; Machado et al. 2002, 2007; Machado and Hey 2003; Hey and Nielsen 2004). The results from the analyses of both nuclear and mitochondrial genomes supported the conclusions of Key (1968), Harrison (1986), and Wu (2001) in that the genomes of these species recombined during their divergence from a common ancestor and were thus semi-permeable.

The paradigm of mosaic genomes made up of portions from multiple species was first supported by an examination of variation at the period (i.e., *per*) locus (Wang and Hey 1996); in *Drosophila* this locus is known to affect both circadian rhythm and male courtship song (Konopka and Benzer 1971; Kyriacou and Hall 1980; Vanin et al. 2012; Pegoraro et al. 2014). In their analysis of *per* alleles from *D. p. pseudoobscura*, *D. p. bogotana*, *D. miranda*, and *D. persimilis*, Wang and Hey (1996) detected a putative "hybrid" sequence in the latter species (i.e., "*persimilis*-40"). The presence of the *persimilis*-40 variant in the gene genealogy that included all *per* alleles caused paraphyly among the *D. persimilis* samples. Divergence-with-gene-flow, rather than incomplete lineage sorting, was considered the most likely explanation because the calculated time of the introgression event was well after the initial divergence (*c.* 1 million ybp) of *D. pseudoobscura* and *D. persimilis* from a common ancestor (Wang and Hey 1996). The introgression episode was, however, inferred to be relatively ancient, occurring >100,000 ybp (Wang and Hey 1996).

In a subsequent analysis of *D. pseudoobscura* and *D. persimilis*, the data from Wang and Hey (1996) were augmented with sequence information from *Adh* (Schaeffer and Miller 1993) and the heat-shock protein gene *Hsp82* (Wang et al. 1997). In this second study, the methodology of Wakeley and Hey (1997) was used to fit allopatric models of speciation to the sequence data. This model assumed that two lineages diverged from an ancestral form at a single time point and that there had been no introgression since divergence (Wakeley and Hey 1997). This exercise indicated that all of the models assuming divergence in allopatry were incompatible with at least some of the data from the three loci, suggesting that the divergence of *D. pseudoobscura* and *D. persimilis* was accompanied by gene flow between the two lineages (Wang et al. 1997). However, as expected from the concept of a semipermeable genome/species barrier, the three loci differed with regard to the amount of estimated introgression. There was large, limited, and no inferred genetic exchange since divergence, from the *Adh*, *per* and *Hsp82* sequence data, respectively (Wang et al. 1997). In general then, a model that assumed divergence and gene flow (the isolation with migration or IM model) fitted the multilocus data for these taxa (and for other species complexes as

well: e.g., cichlids: Hey et al. 2004, Won et al. 2005; chimpanzees: Won and Hey 2005). The application of the separate "IM" model of Nielsen and Wakeley (2001) also gave results consistent with the hypothesis that the two *Drosophila* species had undergone introgressive hybridization (Hey and Nielsen 2004). Furthermore, Machado et al. (2002) and Machado and Hey (2003) inferred introgression between *D. pseudoobscura* and *D. persimilis* by applying estimates of linkage disequilibrium and a phylogenetic approach. These studies demonstrated that genes closely linked to genomic regions containing hybrid sterility loci were not exchanged, but regions distal to such loci were able to recombine in hybrids.

Like these studies, an analysis of sequence variation associated with loci inside and outside of chromosomal inversions (focusing particularly on the rearrangements found on the second chromosome) also detected a pattern indicative of variable levels of introgression since divergence (Figure 5.3). Machado et al. (2007) detected a large variance in sequence divergence across all regions, but with greater divergence occurring between *D. pseudoobscura* and *D. persimilis* loci located within inverted regions relative to those outside the inversions. Estimates of both divergence and Nm (i.e., migration; Figure 5.3) were found to be "consistent with a model of species divergence in which gene flow between *D. pseudoobscura* and *D. persimilis* has occurred at some loci outside the inverted region but has not occurred inside the inversion since species divergence" (Machado et al. 2007). Thus, the location of regions relative to both "sterility loci" and inversions (Figure 5.3) likely contribute to the semipermeable nature of the *D. pseudoobscura* and *D. persimilis* genomes through the sieve of hybrid fitness.

5.2.2 Genetic exchange and hybrid fitness in animals: Tigriopus

Burton, Edmands, and their colleagues have produced an unparalleled set of experimental data addressing the genomic and ecological underpinnings of reproductive isolation in animals—as reflected by hybrid fitness—using the intertidal copepod, *Tigriopus californicus*. Numerous studies have thus documented the role of genic interactions contributing to reduced fitness among a range of hybrid genotypes

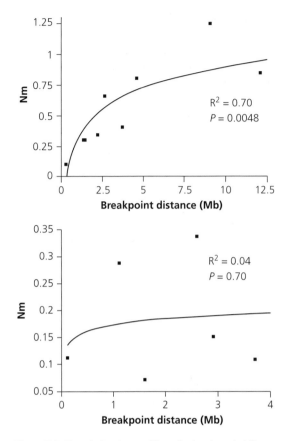

Figure 5.3 Plots of migration rate (*Nm*; reflecting the probability of introgression events between *D. pseudoobscura* and *D. persimilis* since divergence from a common ancestor) against distance to the inversion on the second chromosome for loci outside (upper plot) or inside (lower plot) the inverted region (from Machado et al. 2007). Reproduced with permission from the Genetics Society of America.

(i.e., "hybrid breakdown") under both laboratory and natural environments (Burton 1990; Harrison and Edmands 2006; Willett and Berkowitz 2007; Hwang et al. 2012; Barreto et al. 2015). For example, in an analysis of long-term viability, admixed genotypes demonstrated consistently lower fitness relative to parental classes (Hwang et al. 2012). Likewise, an earlier study designed to test the fitness effects of two malic enzyme loci in F_2 hybrid populations produced patterns indicative of selection against certain hybrid genotypes (Willett and Berkowitz 2007).

One of the hallmarks of the *Tigriopus* studies has been the elucidation of the role of cytonuclear interactions in the evolution of reproductive isolation. Specifically, extensive and elegant experimental

analyses have provided details of how nuclear-mtDNA gene interactions can greatly limit the potential for genetic exchange between divergent lineages (Burton et al. 1999; Edmands and Burton 1999; Ellison and Burton 2006, 2008a, b; Harrison and Burton 2006; Barreto and Burton 2013; Foley et al. 2013). In their review, "Cytonuclear genomic interactions and hybrid breakdown," Burton et al. (2013) used findings from *Tigriopus* to highlight how mismatching of protein products from the nucleus and mitochondria of alternate lineages could lead to lower fitness among hybrid genotypes. Figure 5.4 illustrates the expected fitnesses of hybrid and parental genotypes given such cytonuclear interactions along with the observed fitnesses of various hybrid and parental classes from an analysis of *Tigriopus*. With regard to hybrid breakdown, at least the F_3 and paternal backcross generations demonstrate significantly lower inferred fitness (Figure 5.4; Ellison and Burton 2008b).

The studies discussed illustrate conclusively the development of postzygotic reproductive barriers, in the form of reduced hybrid fitness, among evolutionary lineages of *Tigriopus*. Yet, studies by these same research groups have also detected a range of hybrid fitness estimates, including genotypes that exceed the fitness of parental forms in some environments (Willett 2008; Edmands 2008; Edmands et al. 2009; Hwang et al. 2011; Pritchard and Edmands 2013; Pritchard et al. 2013; Pereira et al. 2014). Thus, Edmands (2008) detected the "creation of favorable gene combinations by disrupting parental linkage groups" as reflected in faster development by some first generation backcross hybrid (i.e., BC_1) genotypes. Likewise, a long-term assay of experimental hybrid swarms found reduced relative fitness in the earlier stages of hybrid swarm development, but with the final census (c. 20 generations) detecting two hybrid populations that exceeded parental fitness (Hwang et al. 2011).

A recent analysis of thermal tolerances displayed by parental F_1 and advanced generation hybrid genotypes detected not only a range of inferred hybrid (and parental) fitness estimates, but also the occurrence of novel adaptations of hybrids relative to parental lineages (Pereira et al. 2014). Figure 5.5 presents the results of thermal tolerances found among the parental "BR" and "SD" lineages

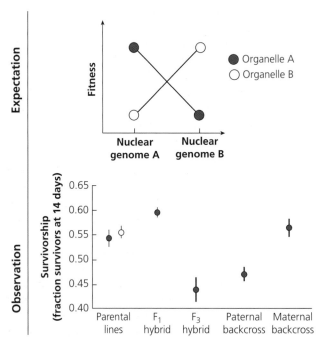

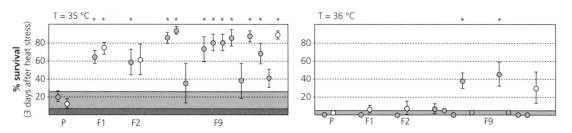

Figure 5.4 The upper graph illustrates the expectation that intrinsic selection will lead to the higher fitness of individuals possessing the nuclear and organellar (i.e., in this example, mitochondrial) genes from the same evolutionary lineage. The lower graph reflects the findings from experimental crosses between divergent lineages of the intertidal copepod, *Tigriopus californicus*. Hybrid breakdown, due to the mismatch of mitochondrial and nuclear genomes, is evident in both the F_3 and Paternal backcross classes (from Burton et al. 2013).

Figure 5.5 Survivorship (mean ± 1 SE), at both a sublethal (T = 35°) and lethal (T = 36°) temperature treatment, of two *Tigriopus* parental lineages ("P") and F_1, F_2, and F_9 hybrid genotypes derived from crosses between the two parents. Each genotype is colored to reflect which of the two parental mitochondrial haplotypes is present. The left panel illustrates the survivorship at temperatures near the physiological limit of the parental taxa. The right panel indicates the survivorship of parental and hybrid lineages at a lethal temperature to one or both parents. Asterisks (*) above the panels indicate hybrid genotypes that demonstrate significantly increased fitness relative to their parents (Mann-Whitney *U* test; all $P < 0.016$) (from Pereira et al. 2014).

and F_1, F_2, and F_9 hybrids from BR × SD crosses. All hybrids demonstrated either equivalent or higher tolerance (i.e., fitness) relative to their parents at temperatures that were either sublethal or lethal for parental genotypes (Figure 5.5; Pereira et al. 2014). Furthermore, some of the hybrid genotypes possessed significantly higher fitnesses relative to their parents, even at the "lethal" temperature (Figure 5.5), indicating the generation of novel ecological adaptations among the admixed lineages (Pereira et al. 2014).

5.2.3 Genetic exchange and hybrid fitness in animals: Rana

Studies of members of the frog genus, *Rana*, have produced a wealth of data to test hypotheses concerning hybridogenetic reproduction, adaptation, biogeography, reproductive isolation, speciation, and reticulate evolution (Moore 1947; Littlejohn and Oldham 1968; Sage and Selander 1979; Green et al. 1996; Christiansen and Reyer 2011; Zhou et al. 2012). In regard to the development of pre- and postzygotic barriers to reproduction and divergence-with-introgression, the New World clade of the leopard frog genus has been examined in great detail for over 70 years. For example, Moore reported a series of analyses that not only defined morphological and physiological variability in the *Rana pipiens* complex (Moore 1939, 1944), but also investigated patterns of hybrid fitness (in the form of viability) through an extensive array of crossing experiments involving samples from Canada to Mexico (Moore 1944, 1946a, b, 1947). Likewise, a number of more recent studies also identified the presence of reproductive barriers—both pre- and postzygotic—between different geographic samples from across the range of the species complex (Ruibal 1955; Littlejohn and Oldham 1968; Mecham 1968; Cuellar 1971; Frost and Platz 1983). In terms of postzygotic barriers, Table 5.1 reflects typical results obtained from the earlier and later analyses. Thus, when different "races" and/or species within this complex were crossed experimentally, lower viability of hybrids was routinely observed (Moore 1944, 1946a, b, 1947; Ruibal 1955; Cuellar 1971; Frost and Platz 1983). However, in spite of the great restrictions on hybrid formation from both behavioral

(i.e., mating calls; Littlejohn and Oldham 1968) and developmental processes (Moore 1939, 1944, 1946a, b, 1947; Ruibal 1955; Cuellar 1971; Frost and Platz 1983), numerous hybrid zones were reported (Kruse and Dunlap 1976; Sage and Selander 1979; Kocher and Sage 1986; Di Candia and Routman 2007). Given the presence of these admixed populations, it would seem likely that hybrid fitness varies between genotypes and across environments.

By far, the most extensive analyses of parental and hybrid fitness within the New World leopard frog species complex have been those of Parris and his colleagues. Parris et al. (Parris 1999, 2000, 2001a, b, c, 2004; Parris et al. 1999, 2001; Semlitsch et al. 1999) estimated numerous components of fitness, across a number of environments, for the species *R. blairi* and *R. sphenocephala* and various synthetic hybrids in both experimental and natural settings. The components analyzed across these studies included patterns of development, degree of predation, and resistance to pathogens. These analyses detected a complex set of genotype × environment fitness effects in the parental and hybrid classes. For example, the rearing of *R. blairi*, *R. sphenocephala*, F_1, BC_1, and BC_2 larvae in experimental

Table 5.1 Relative fitness of F_1 male, hybrid individuals synthesized from experimental crosses between species of leopard frogs (genus *Rana*). Male fitness estimates were derived from the number of sperm from the testicular tissue of sexually mature control (i.e., "Lowland form controls") and hybrid frogs (from Frost and Platz 1983).

Lowland form controls		Rana chiricahuensis × Rana pipiens		Rana chiricahuensis × Rana blairi		Rana chiricahuensis × lowland form	
Specimen	Sperm #	Specimen	Sperm #	Specimen	Sperm #	Specimen	Sperm #
431	464	271	64	346	0	432	0
393	624	345	17	368	0	424	0
373	540	367	31	369	43	421	2
419	356			370	13	418	8
382	438			371	52	412	15
420	368			376	53	411	20
425	688			377	31	402	0
423	260					401	0
484	528					344	10
						409	41
						433	27
						341	2

ponds, in both single-genotype and two-way (i.e., larvae from both parents or from one parent and one hybrid class) mixtures, allowed developmental assays. From his observations Parris (1999) concluded, "On average, primary-generation (F_1) hybrids experienced either increases or decreases in larval fitness component values relative to parental species, whereas advanced generation (BC_1 and BC_2) hybrids had equivalent or higher values for most fitness components."

A second study by Parris (2001a) inferred the fitness of parental species and F_1 genotypes under natural conditions. In this analysis, three classes of genotypes (both species and F_1 hybrids) were placed into enclosures within ponds in three different habitats that were representative of the ecological amplitudes of the species and hybrids—prairie, woodland, and river floodplain. Parris (2001a) did not detect reduced hybrid fitness in any of the environments. Instead, for one of the fitness components (body mass at metamorphosis), the F_1 hybrid demonstrated higher and equivalent fitness to *R. sphenocephala* and *R. blairi*, respectively, in the river floodplain habitat (Parris 2001a). Significantly, the river floodplain is a primary site in which the species overlap and hybridize (Parris 2001a), leading to the inference that the initial hybrid generation could act as an effective bridge for the evolution of hybrid zones. However, a subsequent analysis of hybrid and parental fitness in the absence or presence of an emergent fungal pathogen of amphibians (i.e., *Batrachochytrium dendrobatidis*) detected a lower fitness of F_1 hybrids relative to the parental genotypes (Parris 2004). Thus, though reticulate evolutionary processes have been, and continue to be, of significance for lineages within the leopard frog genus, the evolutionary role of introgression is likely affected by differential, often environmentally-dependent, selection.

5.3 Genetic exchange and fitness: Environment-independent and -dependent selection on plant hybrids

A main focus for our research group has been the estimation of hybrid and parental fitness in various organisms in both experimental and natural habitats (e.g., Shoemaker et al. 1996; Williams et al. 1999; Promislow et al. 2001). In particular, we have drawn attention to the fact that hybrid genotypes can display a range of fitness estimates, sometimes due to environmental setting. Whether in the context of adaptive trait transfer (Section 1.5.2), surveys of natural hybrid zones (Section 3.4.1), or barriers to introgression (Sections 4.2 and 4.3), the majority of the data used to illustrate this conclusion have come from studies of the Louisiana iris species *I. fulva, I. hexagona,* and *I. brevicaulis*. The studies of our group, and more recently those of Martin et al., have included experimental manipulations in both greenhouse and natural settings as well as from natural hybrid zones. Each of the studies has supported the hypothesis that hybrids vary in fitness and thus vary in the likelihood of their contributing to long-term evolutionary effects.

Plants, in general, have been tractable in terms of defining the causes of reproductive barriers and hybrid fitness (e.g., Ruhsam et al. 2013; Alcázar et al. 2014; Lindtke et al. 2014; Weller et al. 2014). A classic example of this was the work of Clausen, Keck, and Hiesey (1939) and their insistence that species be defined on the basis of experimental analyses. Their conceptual framework was based on data from experimental crosses, reciprocal transplants into the habitats of the various taxa, and morphological analyses. The work of Clausen, Keck, and Hiesey (e.g., 1939, 1945) thus illustrated the diagnostic power of experimental crosses coupled with manipulations of the environmental setting in which different genotypes were placed, in testing various hypotheses concerning hybrid fitness and the process of speciation.

Recent analyses have continued to validate the power of the experimental approach emphasized by Clausen, Keck, and Hiesey. For example, an important role for epigenetic remodeling, in both heterosis and lethality in F_1 hybrids, has been inferred from experimental crossing studies combined with analyses of hybrid and parental methylomes (Shen et al. 2012; Greaves et al. 2014; Ng et al. 2014; Burkart-Waco et al. 2015). Likewise, an analysis of the viability and fertility of experimental hybrid genotypes constructed by crossing *Silene latifolia* and *S. diclinis*—two plant species that possess sex chromosomes—provided a definition of the causes of the phenomena of Haldane's rule (Demuth et al. 2014). Finally, Brennan et al. (2014) used experimental crosses and transmission ratio distortion mapping to define regions of the genomes of two *Senecio* species that were more or less likely to introgress across natural hybrid zones. In the following two sections I will provide illustrations of how experimental analyses of the genera *Ipomopsis* and *Mimulus* have led to insights into the causes of variable hybrid fitness and evolutionary diversification.

5.3.1 Genetic exchange and hybrid fitness in plants: Ipomopsis

For over 15 years, the research group of Diane Campbell has been investigating the evolution of floral traits in the genus *Ipomopsis* (Campbell and Waser 1989; Campbell et al. 1991, 1994; Campbell and Dooley 1992; Meléndez-Ackerman et al. 1997; Bischoff et al. 2014, 2015). Along with, and informed by, the studies of floral traits has been an extensive series of analyses on the origin and evolution of (1) reproductive barriers and (2) natural hybrid zones between the species *Ipomopsis aggregata* and *Ipomopsis tenuituba* (Meléndez-Ackerman 1997; Meléndez-Ackerman and Campbell 1998; Aldridge 2005; Wu and Campbell 2005; Aldridge and Campbell 2006, 2007, 2009; Bischoff et al. 2014, 2015). Likewise, the studies of floral trait evolution and natural hybrid zones led this group into tests of the null hypothesis which arose from the modern synthesis: that hybrid genotypes are less fit than non-recombinant genotypes. Indeed, Campbell's group has generated one of the most extensive data sets on this topic (Campbell et al. 1997, 2002a, b, 2005, 2008, 2010; Campbell 2003; Wu and Campbell 2006, 2007; Campbell and Waser 2007; Campbell and Wendlandt 2013), including the holy grail for process-oriented evolutionary biologists—i.e., net/lifetime fitness estimates in natural environments (Lewontin 1974, pp. 235–239).

Both pre- and postzygotic reproductive isolation between *I. aggregata* and *I. tenuituba* have been documented, with the effects of the various barriers on the origin and evolution of hybrid zones being estimated from both experimental and natural hybrid populations (Meléndez-Ackerman and Campbell 1998; Campbell et al. 2002a, b; Wu and Campbell 2005; Aldrich and Campbell 2007, 2009).

The fact that areas of sympatry and introgression between these two species have been detected reflects the fact that in spite of strong pre- and postzygotic selection against interspecific genetic exchange, there is likely differential selection operating on hybrid genotypes. This has indeed been well documented by the numerous studies of natural hybrid zones and the transplantation of parental and [experimentally-produced] admixed genotypes into *I. aggregata, I. tenuituba*, and hybrid habitats (Campbell et al. 1997, 2002a, b, 2005, 2008, 2010; Wu and Campbell 2006, 2007; Campbell and Waser 2007; Campbell and Wendlandt 2013).

Some of the earliest analyses that estimated parental and hybrid fitness suggested that recombinant genotypes were being uniformly selected against. For example, pollinators, specifically hummingbirds, demonstrated floral preference for the *I. aggregata* floral type compared to the *I. tenuituba* form (Figure 5.6; Campbell et al. 1997, 2002b; Meléndez-Ackerman et al. 1997). This strong preference is likely due to a suite of floral characteristics possessed by *I. aggregata* (Meléndez-Ackerman et al. 1997). In contrast, *I. tenuituba* possesses floral traits that are more attractive to insect pollinators such as hawkmoths (Meléndez-Ackerman et al. 1997). Regardless

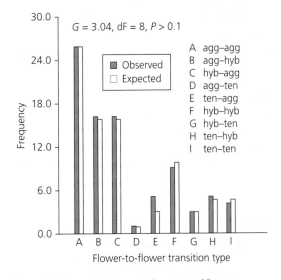

Figure 5.6 Observed and expected frequencies of flower to flower transitions, carried out by hummingbirds, between *Ipomopsis aggregata* (agg), *I. tenuituba* (ten), and natural hybrids (hyb) between these two species (from Meléndez-Ackerman et al. 1997).

as to the exact cues that are causing the differential pollinator preferences for either *I. aggregata* or *I. tenuituba*, this component of the ecological setting forms a barrier for the origin of natural hybrid zones. Furthermore, early analyses of pollinator behavior in natural hybrid zones indicated selection against hybrid genotypes as reflected by the following conclusion of Campbell et al. (1997): "visitation patterns by hummingbirds resemble an advancing wave model in which *I. aggregata* is selectively favored everywhere in the hybrid zone." In spite of the pollinator preferences that restrict introgression between *I. aggregata* and *I. tenuituba*, these species form numerous hybrid zones in areas of sympatry (Aldridge 2005). Furthermore, hybrid genotypes produced by crosses between these taxa have been inferred to have a range of fitnesses relative to their parents (Campbell and Waser 2001, 2007; Campbell et al. 2002a, 2008; Wu and Campbell 2006; Campbell and Wendlandt 2013).

Campbell and Waser (2007) measured lifetime fitness for *I. aggregata, I. tenuituba*, and F_1 hybrid genotypes using a reciprocal transplant analysis within a natural hybrid zone. Each of the parental species and their reciprocal F_1 hybrids were represented in the study using full-sib seed families synthesized from experimental, greenhouse crosses. Seeds from each of these families were planted into parental and hybrid habitats in a hybrid zone in Colorado, USA. Campbell and Waser (2007) then collected the following data for each family: survival to reproduction, mean seed production for flowering plants, and mean age at reproduction. Based on these data, the finite rate of increase (λ; calculated from age-specific survival and reproduction) was recorded in order to estimate lifetime fitness (Lande 1982; McGraw and Caswell 1996; Caswell 2001). Figure 5.7 summarizes the lifetime fitness estimates for *I. aggregata, I. tenuituba*, and F_1 full-sib families planted in either the parental or hybrid habitats. The results reflect well the complexities of fitness estimates affected by intrinsic and extrinsic selection on parental and hybrid genotypes. Thus, *I. aggregata* and *I. tenuituba* genotypes had high fitnesses in their own habitat, but low relative fitness in the alternate parent's environment (Campbell and Waser 2007). Both of the reciprocal F_1 hybrids had the highest fitness in the center of the hybrid zone

(i.e., the "hybrid" habitat; Figure 5.7), but the set of full-sib F_1 hybrid families formed with *I. tenuituba* as the maternal parent survived well only in the hybrid niche (Figure 5.7; Campbell and Waser 2007). Campbell and Waser (2007) inferred that the difference in fitnesses of the reciprocal F_1 hybrid families in different habitats was likely due to a cytonuclear × environment effect.

Combining the data on hybrid and parental fitness from previous studies and the above analysis led to the general conclusion that "both vegetative adaptation to physical environment and floral adaptation to pollinators contribute to observed patterns of phenotypic expression in this hybrid zone and

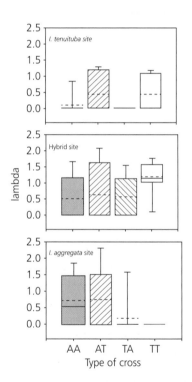

Figure 5.7 Box plots illustrating lifetime fitness based on the finite rate of increase (λ, "lambda") of full-sib families for *I. aggregata* ("AA"), *I. tenuituba* ("TT"), F_1s with *I. aggregata* maternal parents ("AT"), and F_1s with *I. tenuituba* as the maternal parent ("TA") planted as seeds into either parental or hybrid habitats. The median value is indicated with a solid line. The dimensions of the box indicate the 75th and 25th percentiles. The 95th and 5th percentiles are indicated by bars extending beyond each box. Dashed lines indicate the mean values. TA families planted in the *I. tenuituba* site and TT families planted in the *I. aggregata* site had values of 0 (from Campbell and Waser 2007).

to persistence of the hybrid zone" between *I. aggregata* and *I. tenuituba* (Campbell and Waser 2007). Consistent with this conclusion were results from a subsequent study in which not only parental and F_1, but also F_2 and first generation backcross genotypes were utilized in reciprocal transplants across natural parental and hybrid habitats (Campbell et al. 2008). Lifetime fitness estimates for all of the various recombinant and parental categories resulted in a similar conclusion as that arrived at by Campbell and Waser (2007). In particular, hybrid genotypes demonstrated a range of lifetime fitness estimates, from lower to higher, than parental genotypes, varying in an environment-dependent manner, leading the authors to conclude, "These results run counter to any model of hybrid zone dynamics that relies solely on intrinsic nuclear genetic incompatibilities" (Campbell et al. 2008).

5.3.2 Genetic exchange and hybrid fitness in plants: Mimulus

Though still small, the number of plant species complexes being developed as evolutionary and ecological models is increasing. As discussed in the previous chapter (Section 4.2), Lexer and Widmer (2008) recognized five genera to which these models belonged—*Iris, Helianthus, Silene, Populus*, and *Mimulus*. In the case of *Mimulus*, a wonderfully rich set of examples have been recently explored by many research groups, including those led by Toby Bradshaw, Lila Fishman, Doug Schemske, Andrea Sweigart, and John Willis. However, like the Louisiana iris complex, monkey flower species (i.e., *Mimulus*) have been the focus of research for many decades, with many of the studies testing hypotheses related to adaptation, reproductive isolation, divergence-with-gene-flow, and hybrid speciation (e.g., Vickery 1959, 1964; Kiang and Libby 1972; Kiang and Hamrick 1978; Waser et al. 1982; Fishman et al. 2001, 2002, 2008, 2013; Beardsley et al. 2003; Fishman and Willis 2005; Whittall et al. 2006; Sweigart et al. 2008; Barr and Fishman 2010; Lowry and Willis 2010; Modliszewski and Willis 2012; Streisfeld et al. 2013; Vallejo-Marin and Lye 2013; Yuan et al. 2013; Oneal et al. 2014).

Among the analyses within the *Mimulus* complex are many that provide inferences of parental and hybrid fitness, particularly those that address the genetic architecture of pre- and postzygotic reproductive barriers within and between species. In particular, the work by the Willis and Fishman research groups has yielded numerous tests for the strength and genetic underpinnings of various mechanisms that limit genetic exchange. For example, Fishman et al. (2008), using crosses between *Mimulus guttatus* and *M. nasutus*, were able to estimate the effect of prezygotic barriers, in the form of differential selection on different recombinant male gametes (i.e., reflecting "conspecific pollen precedence"). Observations of transmission ratio distortion (TRD) in backcross generation progeny were used to map QTL contributing to selection on hybrid pollen grains/tubes that were competing to fertilize ovules in the same ovary. Eight genomic regions were identified as having TRD caused by conspecific pollen precedence (Fishman et al. 2008). Though primarily designed to identify loci that caused limitations in the formation of the F_1 generation, this analysis also provided data indicating that different recombinant gametes possessed low to high relative fitnesses, mainly as a result of being placed into different stylar environments (Fishman et al. 2008).

The effect of *postzygotic* barriers on hybrid formation and introgression has also been inferred through multiple experimental and population level analyses. Using a series of controlled crosses between a wide array of geographical samples of *M. guttatus* and *M. nasutus*, Martin and Willis (2010) were able to document the role of intrinsic hybrid incompatibility due both to nuclear–nuclear and cytonuclear genetic interactions. However, the genetic incompatibilities detected did not always lead to lower estimated fitness for all hybrid genotypes relative to those of *M. guttatus* and *M. nasutus*. Figure 5.8 presents a portion of the data that indicates a range of inferred fitnesses for hybrids, as estimated from proportion of seeds germinated; such was also the case when male fertility (i.e., pollen viability) was estimated (Martin and Willis 2010). In particular, various F_1 and F_2 genotypes were found to possess fitnesses ranging from lower than both parents to equivalent to the parent demonstrating the highest estimated fitness (Figure 5.8). From their data, Martin and Willis (2010) drew the conclusion that "intrinsic postzygotic isolation is common in hybrids between these

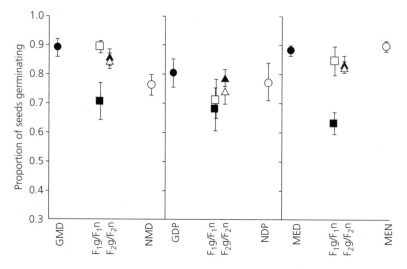

Figure 5.8 Proportion of seeds that germinated that were produced from crosses within *M. guttatus* (i.e., GMD, GDP, MED), within *M. nasutus* (i.e., NMD, NDP, MEN), and between the two species (i.e., left panel = GMD × NMD; middle panel = GDP × NDP; right panel = MED × MEN). The fill-color of hybrid designations indicates that the cytoplasmic genome corresponds to the parent having the same color. Circles represent conspecific seed cohorts, squares represent reciprocal F_1 cohorts, and triangles represent F_2 cohorts (from Martin and Willis 2010).

Mimulus species, yet the particular hybrid incompatibilities responsible for effecting this isolation differ among the populations tested." It would seem that this conclusion—that there is genetic variation distributed across the ranges of the two species that contributes to reproductive isolation—could also be extended to the estimates of hybrid fitness. Thus, genetic variability in different populations of *M. guttatus* and *M. nasutus* when combined in hybrids sometimes yields hybrids with lower, equivalent, or higher fitness relative to conspecific genotypes (Figure 5.8; Martin and Willis 2010).

Though often used to emphasize reproductive isolation and the lower fitness of some hybrid genotypes (e.g., Schemske and Bradshaw 1999; Ramsey et al. 2003; Stathos and Fishman 2014; Sweigart and Flagel 2015), the range of fitnesses estimated for many intra- and interspecific hybrids suggest that the *Mimulus* clade could likely also be a paradigm of divergence-with-gene-flow. Such was the conclusion of Brandvain et al. (2014) from their analysis of 19 whole genome sequences from *M. guttatus* (13 genomes), *M. nasutus* (5 genomes) and an outgroup species (1 genome). The availability of the whole genome data sets as well as computational methodologies necessary to analyze such large scale data provided the resolution needed to detect signatures of both divergence and reticulation. Brandvain et al. (2014) thus identified genomic signatures correlated with the ecological, genetic, and evolutionary diversification of this pair of species. For example, the

genome analyses demonstrated that the transition from an outcrossing (i.e., *M. guttatus*) to selfing (i.e., *M. nasutus*) species resulted in a significant reduction in diversity (Brandvain et al. 2014). In addition, the hypothesis that *M. guttatus* and *M. nasutus* are on separate evolutionary and ecological trajectories was supported by evidence for selection against introgression. Yet, in spite of apparent selection against some hybrid genotypes, divergence-with-introgression was also detected. Though introgression was largely from *M. nasutus* into *M. guttatus*, there was also a small amount of introgression detected from the outcrossing species into the selfing form. The possible significance of the latter process was indicated in the following quote: "Evidence of introgression from *M. guttatus* into *M. nasutus* is subtler, but is potentially critically important. Even relatively low levels of introgression into a selfer may rescue the population from a build up of deleterious alleles, and reintroduce adaptive variation, and so may lower its chances of extinction, a fate considered likely for most selfing lineages." (Brandvain et al. 2014).

5.4 Genetic exchange and fitness: Horizontal gene transfer and selection in prokaryotes

Genetic exchange is one of several processes that contribute to prokaryotic evolution. The evolutionary trajectory of bacterial species is also modified by gene duplication, deletion, inactivation, and ultimately loss (Mira et al. 2001). Indeed, to explain the relatively small DNA content and lack of nonfunctional DNA in bacterial lineages that are regularly increasing DNA content through horizontal gene transfer and gene duplication (Zhaxybayeva and Gogarten 2004; Zhaxybayeva et al. 2004), Mira et al. (2001) hypothesized a bias in genomic evolution in which DNA deletions outnumbered additions. Kurland (2005) agreed with this conclusion arguing, "Genomes that continuously expand due to the uninhibited acquisition of horizontally transferred sequences are genomes that are earmarked for extinction." However, it is undeniable that the acquisition of genetic material via horizontal gene transfer is a major facilitator of prokaryotic evolution (Ochman et al. 2000; Daubin and Ochman 2004). Horizontal transfer events are ubiquitous in bacterial clades, contributing to the fitness of genotypes and acting as agents for adaptive evolution. To support this inference I will first discuss theoretical and experimental treatments involving prokaryotes. I will then further illustrate the manner in which HGT affects bacterial evolutionary trajectories by discussing a set of examples indicating adaptive changes via gene acquisition.

5.4.1 Horizontal gene transfer and the fitness of bacteria: Theory

"The power of HGT to transform evolutionary landscapes has become increasingly evident with the number of sequenced genomes . . . with fitness effects that range from fine tuning of enzymatic function to enabling major ecological transitions." This statement from Baltrus (2013) reflects well how most reports of horizontal exchange between prokaryotic lineages approach the topic. In other words, and understandably, the detected gene acquisitions are almost always discussed by referencing their importance for the adaptation and diversification of lineages and clades (e.g., Wisniewski-Dyé et al. 2011). Though as the quote indicates, Baltrus (2013) was keenly aware of the evolutionary significance of horizontal transfers between lineages, his emphasis was reflected in the title of his review: "Exploring the costs of horizontal gene transfer." In particular, he pointed to

evidence indicating "recently acquired regions often function inefficiently within their new genomic backgrounds so that, despite great evolutionary benefits, they can be energetically or physiologically costly" (Baltrus 2013).

A number of theoretical analyses have explored the ways in which the introduction of genetic material by HGT may impact the fitness of the recipient organism; two examples of such studies were those by Raz and Tannenbaum (2010) and Tazzyman and Bonhoeffer (2014). Both of these analyses detected differential fitness effects from the introduction of foreign DNA carried on plasmids and/or the bacterial chromosome, but the results obtained indicated a complexity dependent upon environmental conditions. In the earlier study, the influence of conjugation-mediated HGT (i.e., via plasmid transfer) on the mutation-selection balance was modeled. As expected, HGT had a significant effect on the mean fitness of asexual prokaryotic populations when in the presence of an antibiotic (Raz and Tannenbaum 2010). However, the HGT fitness effect was slightly deleterious in a stable environment. This led to the inference that the acquisition of foreign DNA would be selectively advantageous in dynamic environmental settings by facilitating more rapid adaptation (Raz and Tannenbaum 2010).

Tazzyman and Bonhoeffer (2014) also modeled the development of antibiotic resistance as an indicator of adaptive evolution in bacterial populations. As with the previous analysis, the predicted evolutionary patterns indicated that environmental selection had a profound influence on fitness. Specifically, this study revealed that competition between "nonresistant" and "resistant" plasmids retarded or prevented the spread of antibiotic resistance (i.e., the spread of the "resistant" plasmids; Tazzyman and Bonhoeffer 2014). This was an unexpected finding given the observation that most resistance genes are carried on plasmids, rather than by the bacterial chromosome (Eberhard 1990; Rankin et al. 2011; de Been et al. 2014). To explain the lack of concordance between the outcome of the modeling exercise and what is observed in nature, the authors also invoked environmental selection. Specifically, they argued "that resistant plasmids can benefit by transferring across species boundaries in an environment that has antibiotics, bringing drug resistance to species that otherwise may not have evolved it" (Tazzyman and Bonhoeffer 2014).

5.4.2 Horizontal gene transfer and the fitness of bacteria: Experiments

As with the theoretical treatments, a number of manipulative, experimental and meta-analyses have estimated fitness consequences due to HGT in bacteria. For example, Omer et al. (2010) tested two hypotheses relating to why genes that encode subunits of complexes (e.g., those involved in transcription) are less frequently involved in horizontal transfers. The two alternative hypotheses both related to fitness effects from the incorporation of the foreign genes were: (1) "the failure of a new gene product to correctly interact with pre-existing protein subunits can make its acquisition neutral" and (2) "foreign subunit-encoding genes may reduce the fitness of the new host by disrupting the stoichiometric balance between complex subunits, resulting in purifying selection against gene retention" (Omer et al. 2010). Inserting and expressing the RNA polymerase β subunit from *Bacillus subtilis* into the genome of *Escherichia coli* demonstrated the formation of nonspecific interactions between the introduced and native subunits. However, these novel interactions did not reduce the fitness of the host as inferred by growth. The observation of "neutrality" of effect from the simulated HGT event into *E. coli* led to the conclusion that the foreign genomic material would not be removed by purifying selection, but rather would provide the potential for novel adaptive evolution during environmental perturbations (Omer et al. 2010).

In contrast to the findings of Omer et al. (2010), an analysis of *Methylobacterium extorquens* detected a range of fitness values for various bacterial genotypes resulting from the transfer of novel genes into the *Methylobacterium* genome, once again mimicking HGT (Chou et al. 2011). Natural populations of *Methylobacterium* are capable of growing on methanol through the oxidation of formaldehyde into formate using a tetrahydromethanopterin-dependent pathway, with the genes that are required for this process likely horizontally transferred between this proteobacteria and metabolically similar Archaeal lineages (Chistoserdova et al. 1998). In the experimental analysis by Chou et al. (2011), the genes necessary for growing in a methanol-containing habitat were replaced with a non-ortholog, resulting in the engineered bacteria being dependent upon glutathione. Though still capable of surviving on methanol, these *Methylobacterium* demonstrated greatly reduced fitness as reflected by growing at only one third the rate of non-engineered lineages (Chou et al. 2011). However, after 600 generations of growth in the methanol habitat, an average fitness increase of 66.8% was observed for a number of the lineages. Thus, genotypes analogous to HGT-derived recombinants demonstrated very low fitness initially, but resulted in lineages with high relative fitness following subsequent adaptive evolution (Chou et al. 2011).

Park and Zhang (2012) used a meta-analysis of horizontal gene transfers in both experimental and natural populations in order to test for causal factors in fitness variation in lineages receiving foreign genes thus leading to either success or failure of incorporation of the introduced genes. Among the factors tested in this analysis, expression level was the best predictor of the transferability of genes, or gene classes, in strains of *E. coli* (Figure 5.9). In addition, the highly significant difference in expression levels between transferred and resident genes (Figure 5.9) and thus the effect on patterns of HGT was found to be independent of three other factors thought to contribute to promoting or limiting horizontal exchanges—i.e., protein function, complexity, and GC% (Park and Zhang 2012). Finally, the causal nature of expression level and HGT was also inferred by analyzing genomes of a broad array of prokaryotes. Of the 133 genomes examined, 127 demonstrated a correlation between the genes transferred and their expression level, with the median expression levels significantly lower for the foreign genes relative to those of resident genes (Park and Zhang 2012). Taken together, these results suggested that gene expression has been a major causal factor in promoting and limiting HGT among prokaryotes through the sieve of recombinant fitness. Furthermore, it seems likely that "most successful HGTs are initially slightly deleterious, fixed because of their negligibly low costs rather than high benefits to the recipient" (Park and Zhang 2012).

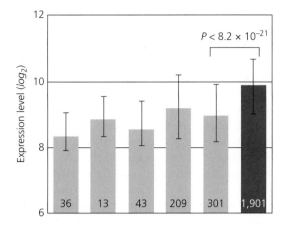

Figure 5.9 Microarray expression levels of horizontally acquired genes were found to be lower than those of resident genes in *E. coli*. The grey bars reflect the expression levels of genes from different strains of *E. coli* and the black bar indicates the expression level of all non-transferred genes (i.e., "resident" genes) studied. The numbers of genes analyzed for each strain (or set of strains) are indicated at the bottom of each bar. The *P* value was derived from a Mann-Whitney *U* test (Park and Zhang 2012).

Notwithstanding the reduced fitness of early-generation recombinant genotypes, prokaryotic adaptive evolution has been greatly facilitated by the horizontal transfer of genomic material among closely and distantly related lineages. To exemplify this observation I will highlight findings for the *E. coli/Shigella* and *Legionella* species complexes.

5.4.3 Horizontal gene transfer and the fitness of bacteria: Shigella

Members of the genus *Shigella* are now recognized as the main cause of bacterial dysentery in humans (Kotloff et al. 1999; von Seidlein et al. 2006). Kotloff et al. (1999) estimated the annual number of cases of shigellosis to be *c.* 165,000,000, with 99% of the persons contracting the disease located in developing countries. Furthermore, the majority of the one million deaths caused annually by *Shigella* infections (i.e., 61%) occurred in children less than five years old (Kotloff et al. 1999). Multiple genomic analyses of *Shigella* have led to the conclusion that this clade is actually a member of the *E. coli* assemblage (Pupo et al. 2000; Sims and Kim 2011; Holt et al. 2012), consisting of strains that have gained adaptations allowing them to cause dysentery in humans by

invading the gut mucosa (Pupo et al. 2000). However, some analyses have also demonstrated that phylogenies constructed from much of the genomes of *Shigella* and *E. coli* isolates resolve members of the two "genera" into monophyletic assemblages (Sims and Kim 2011). Regardless of whether or not *Shigella* lineages should be considered members of the genus *Escherichia*, their impact on human health indicates the importance of defining their origin as a human pathogen.

Several analyses have provided results indicating an extended association of humans and *Shigella*. For example, Pupo et al. (2000) estimated that various *Shigella* clades had evolved from 270,000—35,000 ybp, leading to the inference that shigellosis had been an infectious disease of *H. sapiens* for most, if not all, of its evolutionary history. Furthermore, Holt et al. (2012) documented that the origin and diversification of *Shigella sonnei*, the now dominant cause of bacterial dysentery, occurred in the last 500 years. The origin of *Shigella* in general, and *S. sonnei* in particular, reflect well the degree to which HGT can lead to adaptive evolution in prokaryotes. Hale (1991) reviewed evidence suggesting that genes located both on the bacterial chromosome as well as on plasmids controlled the virulence of *Shigella*. In particular, the ability of this *E. coli* derivative to invade the human gut mucosa was due to the horizontal acquisition of a "virulence plasmid" (Hale 1991).

The recent derivation and global dissemination of *S. sonnei* strains has involved the horizontal transfer and incorporation of genes on both plasmids and the bacterial chromosome that control resistance to multiple antimicrobial drugs (Holt et al. 2012). Additional adaptive evolutionary change in *S. sonnei*, also catalyzed by HGT, has involved the acquisition of plasmids carrying genes that provide a fitness increase in competitive habitats. For example, Holt et al. (2013) detected the addition of plasmids possessing colicin genes that caused "potent bactericidal activity against nonimmune *Shigella* and *Escherichia coli*." Not surprisingly, the strains of *Shigella* carrying these plasmids went to fixation due to a selective sweep (Holt et al. 2013). The colicin genes, along with a number of other genomic components, derived from other gut-inhabiting bacteria such as *E. coli*, *Salmonella*, and *Yersinia* (Holt et al. 2013). Thus, the evolutionary and ecological trajectories of the members of the *Shigella* clade have been, and continue to be, shaped by the horizontal acquisition of genes from a diverse array of donors followed by natural and human-mediated selection.

5.4.4 Horizontal gene transfer and the fitness of bacteria: Legionella

Water-borne pathogens cause numerous health hazards for humans. However, from the perspective of the pathogenic organisms, many potential habitats are extremely hostile due to purification methods employed by *H. sapiens*. Water sources for semiconductor, pharmaceutical, and food-industrial factories are routinely treated to remove prokaryotes (Kulikova et al. 2002). Although purification steps may include filtration, ultraviolet light, heat, and ozonation, the "complete removal of contaminating microorganisms is considered to be nearly impossible." (Kulikova et al. 2002). Because of these extremely restrictive environments, microorganisms that are associated with human-utilized water systems often demonstrate adaptations that allow them to escape from biocontrol measures.

One of the causative agents of Legionnaires' disease, *Legionella pneumophila*, has the ability to survive and reproduce in the incredibly demanding environment of plumbing systems treated with biocides (Chien et al. 2004). Fields et al. (2002) concluded that Legionnaires' disease (i.e., legionellosis), instead of reflecting an "exotic plague," more accurately belonged in the category of severe pneumonia. Similarly, the original report described the disease as "An explosive, common-source outbreak of pneumonia caused by a previously unrecognized bacterium" (Fraser et al. 1977). The 25 years of research on *L. pneumophila*, reviewed by Fields et al. (2002), was initiated after the outbreak of severe pneumonia linked to a July 1976 American Legion convention in Philadelphia. Of the 182 cases, 29 were fatal (Fraser et al. 1977). Although isolates of this genus had been first collected in the 1940s, the taxonomic recognition of the genus *Legionella* also occurred after the 1976 outbreak (Brenner et al. 1979; Fields et al. 2002). Though *L. pneumophila* is now recognized as the major cause of worldwide legionellosis infection, a second species from this genus, *L. longbeachae*, has been shown to be associated with major outbreaks as well—for example, this species results in *c.* 30% and 50% of Legionnaires' cases in South Australia and Thailand, respectively (Yu et al. 2002; Phares et al. 2007).

Notwithstanding Fields et al's. (2002) statement, members of the genus *Legionella* can indeed be considered "exotic" in that they demonstrate unique adaptations to their environments, some of which are due to HGT. For example, the ability of *L. pneumophila* to infect the Legionnaires at the Philadelphia meeting, and its lethality relative to other strains of the same species, was caused partially by gene acquisition (Chien et al. 2004). Included in these acquisitions were (1) F-plasmid *tra/trb* genes (involved in the synthesis of extracellular filaments which establish contact between donor and recipient cells during F-plasmid-mediated conjugation thus leading to the transfer of the F-plasmid; Frost et al. 1994), (2) eukaryotic genes involved in recruitment of substrates (Nagai and Roy 2003) to the vacuole in which *L. pneumophila* resides within the host cell (Sturgill-Koszycki and Swanson 2000), and (3) *dot/icm* and *lvh/lvr* secretion system genes (associated with conjugation and host cell death; Segal et al. 1998, 1999; Vogel et al. 1998).

Chien et al. (2004) have argued that two of *L. pneumophila*'s most remarkable adaptations involve its ability to utilize the organelle trafficking functions of a broad range of host cells (to create and maintain its vacuolar niche), and to exist in extremely harsh environments, such as plumbing systems that are treated with biocides. As mentioned, the first of these adaptations is partially due to horizontally transferred genes belonging to the *icm/dot* clusters. In its natural habitat, *L. pneumophila* is a pathogen of amoeba from multiple genera (Rowbotham 1980). Whether infecting its natural amoebic hosts or humans, this bacterial species requires the type IV secretion system (i.e., encoded by *dot/icm* genes) to transfer a wide array of proteins into the host cells (e.g., Segal et al. 1998; Vogel et al. 1998). However, the horizontally acquired genomic regions that control this secretion system demonstrate host-specific functions in amoebic species, but apparently not in mammals (Figure 5.10; O'Connor et al. 2011). For example, the deletion of even a single genomic region associated with the type IV secretion system in

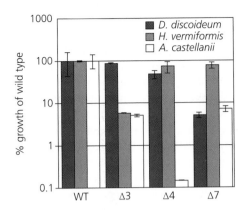

Figure 5.10 The growth characteristics of wild-type and deletion strains of the causative agent of Legionnaires' disease, *L. pneumophila*. Growth of all four strains was observed in three amoeba species, *D. discoideum*, *H. vermiformis*, and *A. castellanii*, and was calculated as a percentage of the wild-type strain (from O'Connor et al. 2011).

L. pneumophila reduced significantly growth within amoeba (Figure 5.10; O'Connor et al. 2011). In contrast, strains lacking several of the genomic islands, and up to 31% of the substrates generated by this system, demonstrated nearly normal growth in mammalian cells. These observations led O'Connor et al. (2011) to conclude, "The host-specific requirements of these genomic islands support a model in which the acquisition of foreign DNA has broadened the *L. pneumophila* host range" (Figure 5.10).

Coscollá et al. (2011) reached a similar conclusion concerning the importance of HGT on the adaptive evolution of *L. pneumophila*. These workers utilized both a phylogenomic and population genomic analysis to infer the proportion of the genome of this species acquired from other prokaryotes as well as eukaryotes. Coscollá et al. (2011) concluded that *c.* 41% of the 1700 genes examined were derived from horizontal transfer events. That such a large percentage of a genome was horizontally acquired suggested that the amoebic niche was a "training ground for pathogenic species" such as *L. pneumophila*, leading to the ability to infect humans as well (Coscollá et al. 2011). A similar degree of evolutionary importance for HGT is likely for other species belonging to this genus as well (e.g., *L. longbeachae*; Cazalet et al. 2010).

L. pneumophila's ability to survive the harsh, human-mediated aquatic environments associated with, for example, air-conditioning systems has also been inferred to result from genetic exchange. The importance of this particular adaptation, specifically for the infection of humans, results from the fact that such environments allow the aerosolization and inhalation of water contaminated with *L. pneumophila* (O'Connor et al. 2011). One of the HGT-derived regions suggested as important for the survival of this species in these types of environments is a 100 kb stretch that includes several genes necessary for processing toxic compounds and heavy metals (Chien et al. 2004). It is possible that the acquisition of this genomic island reflects an adaptation to host organisms that accumulate heavy metals from the environment (Chien et al. 2004). However, regardless of the initial adaptive processes associated with this region, its addition allowed (and allows) *L. pneumophila* to come into contact with human hosts resulting in legionellosis. The production of recombinant *L. pneumophila* via HGT has thus led to adaptive trait transfer resulting in an apparent increase in fitness, especially in the novel environmental settings associated with both the amoebic and human hosts.

5.5 Genetic exchange and fitness: Reassortment and selection in viruses

Viral infections cause millions of deaths annually among humans. For example, in the year 2013, HIV-AIDS led to the death of *c.* 1.5 million people globally. In that same year, there were *c.* 2.1 million new HIV infections recorded with *c.* 35 million people infected worldwide (World Health Organization 2014). Though causing only a fraction of the number of deaths from HIV-AIDS, past and current outbreaks of Ebolavirus have been the subject of much attention (Gire et al. 2014) due to the high mortality rate (*c.* 50–88% for the "Zaire ebolavirus" form; Centers for Disease Control and Prevention 2002), and because of recent media reports concerning the infection and recovery of patients provided with experimental treatments (Centers for Disease Control and Prevention 2014). Indeed, the notoriety of this disease is likely reflected by the 1995 movie "Outbreak" starring Dustin Hoffman, Morgan Freeman, et al. that opens with a July 1967 viral epidemic in the Motaba River Valley, Zaire. The fictional Motaba virus causes mortality in 100% of infected persons. Horrifyingly, the virus causes the liquefaction of internal organs, causing death within three days post-infection (<http://www.imdb.com/title/tt0114069/plotsummary>). The writers of the screenplay must have taken their cue from the origin of the Ebolavirus, first recognized in 1976 from an outbreak in the Sudan, near the border with Zaire. Though not leading to the extreme consequences recorded in the movie, infections from the Ebolavirus result in hemorrhagic symptoms causing internal and external bleeding (Centers for Disease Control and Prevention 2002).

Regardless of their method of transmission, their genomic makeup (i.e., "DNA" or "RNA" viruses), or the percentage mortality they cause, the majority of viral pathogens of humans demonstrate some frequency of reticulate evolution. This inference was reflected in the description of a novel methodology for inferring evolutionary history using HIV, influenza, hepatitis C, dengue, and West Nile isolates as exemplars (Chan et al. 2013). Though the dengue and West Nile viruses showed limited and no evidence of reticulate evolutionary histories, respectively, the other three clades were marked by extensive genetic exchange events (Chan et al. 2013).

A review of the evolution of picorna-like viruses by Koonin et al. (2008) also indicated that non tree-like evolution is well recognized for viral clades. After considering the genomic information gathered for viruses in general, and the positive-strand RNA viruses from the order *Picornavirales* in particular, they concluded that "the notion of monophyly has limited applicability when broad groups of viruses are considered, given the important roles of gene sampling and recombination in the evolution of viruses" (Koonin et al. 2008).

It is likely that, as with prokaryotic and eukaryotic organisms, a number of factors contribute to the fitness of viruses. Furthermore, as observed from studies of bacteria, plants, and animals, genetic exchange between divergent viral lineages would be predicted to result in a variety of recombinant progeny, demonstrating a range of fitnesses relative to the parental types (O'Keefe et al. 2010; Sabehi et al. 2012). One commonly used metric for estimating the fitness of different pathogenic viral genotypes has been their relative ability to infect their hosts, i.e., "virulence." In fact, though

the virulence of pathogens is recognized as a complex "trait," it is a standard metric for estimating the fitness of all classes of disease-causing organisms (e.g., prokaryotic pathogens—Casadevall and Piroski 2001; viral pathogens—Wargo and Kurath 2011). Furthermore, the origin of virulence has been used to demonstrate the potential for genetic exchange to contribute to pathogen evolution (Woolhouse et al. 2005; Lee et al. 2012; Xu et al. 2013). In Chapter 4, I discussed the reticulate evolutionary history associated with the origin and spread of HIV and influenza pathogens. In those cases, viral recombination resulted from naturally occurring variants. In the following sections I will highlight recent reticulate evolutionary events resulting in the origin of novel forms of the causative agents of poliomyelitis and herpes. In particular, I will review evidence that the origin of some of these new lineages has resulted from recombination involving attenuated viruses in vaccines. In the final example, I will discuss findings indicating that, like human pathogens, viruses that infect bacteria (i.e., bacteriophages) likewise show extensive effects from reticulate processes, thus leading to fitness consequences in recombinants and their hosts.

5.5.1 Genetic exchange and viral fitness: Polioviruses

The fight to eradicate poliomyelitis through vaccinations has been very successful, as reflected by the reduction in cases from 350,000 in 1988 to ~2,000 in 2005 (Arita et al. 2006). Notwithstanding the World Health Organization's target of the year 2000 as the date for worldwide eradication, outbreaks of poliomyelitis in various parts of the world are still occurring: e.g., 46 cases in the "Eastern Mediterranean Region" in 2007 (<http://www.emro.who.int/polio>) and 62 cases in Nigeria in 2011 (Burns et al. 2013). Furthermore, though incredible strides have been made through polio vaccination programs—with the apparent extinction of natural populations of Type 2 viruses and the endemism of Types 1 and 3 in only a few African and Asian countries (Savolainen-Kopra and Blomqvist 2010)—there is a growing recognition that web-of-life processes are providing a means of escape from eradication for the poliovirus (Figure 5.11; Savolainen-Kopra

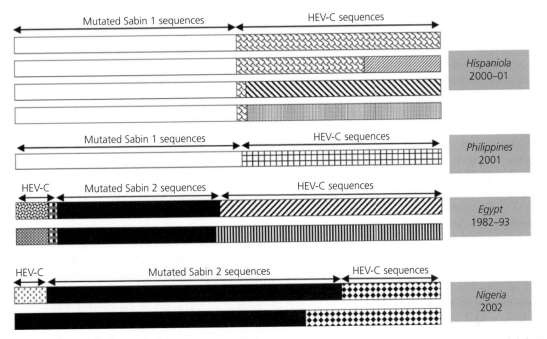

Figure 5.11 Genomic structure of VDPVs associated with poliomyelitis outbreaks in four regions. Vaccine-derived sequences are indicated above the various country isolates (i.e., "mutated Sabin 1 or 2 sequences"). The portions of the VDPV genomes originating from non-polio, species C, human enteroviruses are indicated by the designation, "HEV-C sequences." Different fill patterns within HEV-C genomic regions indicate significant differences between the HEV-C sequences (from Combelas et al. 2011).

and Blomqvist 2010; Combelas et al. 2011; Duintjer Tebbins et al. 2013). Specifically, recombinant, pathogenic, "vaccine-derived poliovirus" (i.e., "VDPV") lineages are now distributed throughout many regions of the world (Figure 5.11; Kew et al. 2005; Agol 2006; Savolainen-Kopra and Blomqvist 2010; Combelas et al. 2011). The production of these novel genotypes appears to be due to the use of "Sabin's live attenuated oral poliovirus vaccine" (i.e., "OPV") as the main tool in the eradication program. OPV includes three divergent serotypes (Sabin 1–3), thus facilitating genetic exchange between these three lineages, as well as between these viral classes and naturally occurring human enteroviruses (Figure 5.11; Savolainen-Kopra and Blomqvist 2010; Combelas et al. 2011; Duintjer Tebbins et al. 2013).

Recombination among strains of poliovirus has been recognized for over 50 years (Savolainen-Kopra and Blomqvist 2010). More recently, viral recombination between poliovirus strains and divergent, human enteroviruses classified as belonging to the same species as the poliovirus (i.e., "HEV-C") has also been detected (Adu et al. 2007;

Jegouic et al. 2009; Savolainen-Kopra and Blomqvist 2010; Combelas et al. 2011). For example, the majority of VDPV genotypes associated with polio outbreaks in Nigeria between 2005 and 2011 derived from poliovirus × "unidentified species C" lineages (Adu et al. 2007; Burns et al. 2013). Likewise, VDPVs associated with outbreaks in the Toliara province of Madagascar during 2001–2002 and 2005 reflect genotypic diversity derived from recombination between divergent species C lineages (Jegouic et al. 2009; Joffret et al. 2012; Razafindratsimandresy et al. 2013). Genome sequences from the 2001 to 2002 and 2005 isolates indicated that novel pathogenic strains had arisen through recombination between serotype 2 and 3 viruses and the divergent coxsackie A human enterovirus (Jegouic et al. 2009; Joffret et al. 2012). Though no cases of polio have been reported from Madagascar since 2005, because OPV application continues to fluctuate in this province, as in other regions of the world (e.g., Romania; Combiescu et al. 2007), the probability of new outbreaks from novel VDPVs will continue to increase. In this regard, Razafindratsimandresy et al. (2013)

detected potentially pathogenic VDPVs in the feces of healthy children living in Toliara province.

The origin of recombinant polioviruses is extremely significant given that vaccination-based extinction was thought possible largely because there was no zoonotic reservoir in which the virus could be maintained and thus elimination in the human host would drive the virus to extinction. In contrast, the tool expected to eradicate the virus has been the cause of some new epidemics and the increase in the number of new polio cases in some areas where the wild virus still exists (Kew et al. 2005; Agol 2006; Arita et al. 2006; Combiescu et al. 2007; Jegouic et al. 2009; Burns et al. 2013; Razafindratsimandresy et al. 2013).

5.5.2 Genetic exchange and viral fitness: Herpesviruses

Herpesviruses that infect humans, and animals that humans depend upon for food, are of major concern globally. For example, viruses belonging to the alphaherpes clade include herpes simplex virus type 1 and type 2 (i.e., HSV-1 and HSV-2), the causal agents of genital herpes (Looker et al. 2008). Likewise, alphaherpesviruses contribute to extensive mortality among domesticated animals such as poultry (Kirkpatrick et al. 2006). Because these classes of viruses have a relatively low rate of mutation via nucleotide substitutions, recombination is viewed as an "essential evolutionary driving force" (Thiry et al. 2005).

Instances of genetic exchange between divergent strains of the same alphaherpesvirus lineage and between different lineages (e.g., HSV-1 and HSV-2), both *in vitro* and *in vivo*, have been documented (Thiry et al. 2005). Furthermore, the use of live attenuated herpesvirus vaccines in both humans and domesticated animals has been widespread, leading to the possibility that new virulent forms would arise through recombination of components of the vaccines (Thiry et al. 2005). Indeed, outbreaks of infectious laryngotracheitis (i.e., ILT), caused by alphaherpesviruses, have been linked to the use of live attenuated vaccines (e.g., García and Riblet 2001), with different vaccines being more or less likely to cause ILT (Lee et al. 2011). However, it was not known whether recombination between the viral components of the live attenuated vaccines was

responsible for the various epidemics in poultry (Thiry et al. 2005).

Recently, a number of ILT epidemics in Australia, in some instances causing up to *c.* 18% mortality, have been linked to the inoculation of chickens with live attenuated vaccines (Blacker et al. 2011; Devlin et al. 2011). These series of outbreaks have provided the first definitive evidence of vaccine-derived recombinants as the causal factors in ILT. Lee et al. (2012) used genome sequencing to determine the relationship of attenuated viruses from three commonly used vaccines (i.e., SA2, A20, and Serva) and viruses isolated from infected chickens (i.e., Class 8 and Class 9). The genome analysis revealed that Class 8 and 9 viruses were nearly identical to the Serva virus genomic sequence. However, there were islands in both the Class 8 and 9 viral genomes that diverged from the Serva sequence, and instead were identical (or nearly so) to either the SA2 or A20 genomes. These findings led to the conclusion that the ILT infections in Australia were due to novel, virulent strains of alphaherpesvirus that had arisen through genetic exchange between divergent viral lineages (Lee et al. 2012). Furthermore, "the rapid emergence of two virulent recombinants suggests that recombination between attenuated herpesvirus vaccines and resultant restoration of virulence may be rare but can bring about a fitness advantage, with severe consequences" (Lee et al. 2012).

5.5.3 Genetic exchange and viral fitness: Bacteriophages

Bacterial viruses, or bacteriophages, are now recognized to be genomically highly variable elements of the environment. The significant effects of these viral elements on their hosts are wide-ranging, but often include contributions to the virulence of bacterial pathogens of humans (e.g., Waldor and Mekalanos 1996; Wagner and Waldor 2002; Chen and Novick 2009). A significant proportion of the genomic variability of bacteriophages has arisen due to viral recombination/HGT (Hendrix et al. 2000; Filée et al. 2005; Silander et al. 2005; Deng et al. 2014), with some of the recombinant genotypes, as in prokaryotic and eukaryotic systems, demonstrating reduced fitness relative to non-recombinants (Rokyta and Wichman 2009; O'Keefe et al. 2010). Yet, the evolutionary history of bacteriophages is now recognized as being reticulate. Indeed, with the advent of genome sequencing, classification of phage became problematic. For example, Lima-Mendez et al. (2008) observed that "an ever increasing number of sequenced phages cannot be classified, in part due to a lack of morphological information and in part to the intrinsic incapability of tree-based methods to efficiently deal with mosaicism. This problem led some virologists to call for a moratorium on the creation of additional taxa . . . in order to let virologists discuss classification schemes that might better suit phage evolution."

Genetic exchange during phage evolution often involves recombination between divergent bacteriophages, likely facilitated by co-infection of the same bacterial host (Waterbury and Valois 1993; Sullivan et al. 2003). However, horizontal transfers also occur between the viruses and their bacterial hosts (Hendrix et al. 2000; Filée et al. 2003). HGT events from the bacteriophages into their host genomes can be prevalent, leading to unique contributions to the pattern and processes associated with bacterial evolution. As mentioned above, effects of phage → bacteria transfers can take the form of introductions of virulence factors. Regardless of the specific phenotypic outcome, the introduction of phage DNA into the bacterial genome often leads to significant increases in genetic diversity, potentially resulting in modifications to the evolutionary trajectory of the host (Wei et al. 2008). Thus, there are many data sets indicating that transfers from bacteriophages to their bacterial hosts have shaped not only the genomic structure of the host, but also their adaptive responses to the environment (Forterre 1999; Wagner and Waldor 2002; Filée et al. 2003; Pedulla et al. 2003; Lindell et al. 2004).

Horizontal genetic exchange from bacteria → bacteriophage has also apparently been widespread during the association of host and virus. Many transfers of host genes have contributed to the multiplication of the resident bacteriophages—including those affecting DNA replication, RNA transcription, and nucleic acid metabolism (Moreira 2000; Chen and Lu 2002; Casjens 2003; Filée et al. 2003; Miller et al. 2003b). In addition, numerous host genes, not specifically involved in multiplication of the viruses, have also been identified within the pathogen

genomes (Figueroa-Bossi and Bossi 1999; Rohwer et al. 2000; Miller et al. 2003a). It has been suggested that some of these may provide increased fitness for the bacteriophage by improving host function prior to lysis (Mann et al. 2003; Lindell et al. 2004).

As the examples illustrate, HGT between phage and bacteria is bidirectional, with potential fitness effects for both host and pathogen. A model system, developed by Lindell and her colleagues, that demonstrates all possible types of transfers as well as their role in the adaptive evolution of both participants, is reflected in the viral pathogens (families, Myoviridae and Podoviridae) of the marine cyanobacteria, *Synechococcus* and *Prochlorococcus* (Mann et al. 2003; Lindell et al. 2004; Avrani et al. 2011; Sabehi et al. 2012). These cyanobacterial genera are of fundamental importance for marine systems, accounting for up to 90% of the primary production of oligotrophic oceanic ecosystems (Liu et al. 1997). HGT between the bacterial pathogens and their cyanobacterial hosts is reflected by the presence of host photosynthesis genes in the bacteriophage genomes (Mann et al. 2003; Lindell et al. 2004; Millard et al. 2004). Lindell et al. (2004) suggested that the occurrence of these host photosynthesis genes was common among the associated bacteriophages. For example, a newly described bacteriophage member of the Myoviridae, found to infect *Synechococcus*, possesses numerous bacterial-like metabolism genes; the genes include photosynthesis, carbon metabolism, and phosphorus acquisition loci. From this observation, Sabehi et al. (2012) concluded, "This suggests a common gene pool and gene swapping of cyanophage-specific genes among different phage lineages." Furthermore, an array of genes transferred into the cyanobacteria genome from bacteriophage results in variation in resistance to phage infections, thus affecting the fitness of both pathogen and host (Avrani et al. 2011).

Although the host species and the bacteriophages that infect them cluster phylogenetically based upon photosynthesis gene sequences (Lindell et al. 2004), the genomic organization of these gene families in the hosts and pathogens is diagnosably different. In the host species the genes are dispersed throughout the genome, while they are clustered in the viral genomes (Hess et al. 2001; Dufresne et al. 2003; Palenik et al. 2003; Rocap et al. 2003; Lindell et al. 2004). The divergence in the host/pathogen gene organization may reflect multiple, independent acquisitions of the genes for photosynthesis (Lindell et al. 2004; Millard et al. 2004; Sabehi et al. 2012). As with the acquisition of genes contributing to resistance (Avrani et al. 2011), the horizontal exchange of the genomic machinery necessary for photosynthesis likely has fitness consequences. Lindell et al. (2004) argued that the presence of highly conserved PSII reaction center and *hli* genes in three different phages infecting the cyanobacterium *Prochlorococcus* resulted from positive selection. In freshwater cyanobacteria the production of bacteriophage progeny is dependent upon continuous photosynthetic activity until just prior to lysis. Given this analogous situation, it was concluded that bacteriophages in the marine environment that contain functional photosynthesis genes might also facilitate their host's metabolism until just prior to lysis. This capability should provide a selective advantage over those bacteriophages that lacked the photosynthesis gene arrays (Lindell et al. 2004). However, as with HGT between phage and host in general (Avrani et al. 2011; Sabehi et al. 2012), the exchange of photosynthesis genes is apparently not unidirectional. For example, phage-to-cyanobacterium exchange of the *hli* multigene family also occurs (Lindell et al. 2004). Furthermore, the evolution of this gene family—via genetic exchange—is postulated to have been adaptive for the recipient cyanobacteria as well (Partensky et al. 1999; Hess et al. 2001; Bhaya et al. 2002; Rocap et al. 2003). Therefore, HGT has been an important factor in alterations to the genomic composition and the adaptive evolution of both components of this host/pathogen system.

5.6 Conclusions

The object of the examples and discussion given in this chapter was to illustrate the inference that hybrid genotypes vary in fitness. Sometimes this variation is independent of the environmental setting, while in other cases it shows indications of being environment-dependent. Thus, hybrids that have low or high fitness across many habitats (i.e., their fitness appears environment-independent) occur. I have emphasized a variety of approaches that allow tests for hybrid fitness. In general, these can be classified into (1) theoretical or simulation studies, (2) experimental, manipulative analyses, and (3) assays for the transfer of known, or inferred, adaptive traits. Theoretical and manipulative analyses provide a detailed, composite fitness not available from the more correlative analyses of natural populations, including those that infer adaptive trait transfers. However, the latter studies provide very strong indications of fitness differentials resulting from the exchange of the genomic components of adaptations. As illustrated, such transfers have been detected by a novel phenotype and/or by the horizontal transfer (by whatever process) of genomic material between divergent lineages. Clearly, my choice of examples for this chapter was designed to point to the fact that hybrid fitness is conceptually and empirically the same measurement as that for any non-hybrid genotype. Furthermore, whether a "hybrid" results from viral recombination, HGT, or introgressive hybridization, the patterns of hybrid fitness found in nature and the laboratory are concordant.

CHAPTER 6

Evolutionary outcomes of genetic exchange

"Successful polyploid complexes apparently arise from recombination through hybridization between genetic types adapted to radically different environments, or possessing different modes of adaptation to the same environment."
(Stebbins 1956)

"Linnaeus and Kerner were the outstanding early exponents of the idea that natural hybrids can be the starting points of new species."
(Grant 1981, p. 245)

"These results support an origin of the bisexual taxon G. seminuda through introgressive hybridization. Interspecific hybridization is potentially an important mode of evolution among western North American fishes, and valid species of hybrid origin may exist in other groups as well."
(DeMarais et al. 1992)

"The key finding from this study is that Lake Malawi cichlids share genetic polymorphism broadly with lineages throughout eastern Africa. The degree of allele sharing across Africa is unexpected under simple neutral models of coalescence and likely requires gene exchange."
(Loh et al. 2013)

"Our results show that in contrast to paternal derived rRNA genes, maternal derived rRNA genes have increased in number in the allopolyploid and appear to have begun to colonize chromosomes derived from the paternal parent."
(Zozomová-Lihová et al. 2014)

6.1 Genetic exchange: Its role in genomic and organismic evolution

In sections of the preceding chapters I have discussed various effects from recombination between divergent genomes. One conclusion that is hopefully clear from the illustrations already provided is that web-of-life processes have contributed significantly to the evolutionary and ecological trajectories of organisms. In the current chapter I will expand on the previous observations by illustrating additional organismic groups and different outcomes that may occur due to natural hybridization. I will thus review instances in which adaptive evolutionary change has led to single evolutionary lineages, and also to entire clades (i.e., adaptive radiations). Likewise, I will discuss the possible effects from whole-genome duplication (WGD)—arising from genetic exchange—in terms of the evolution of (1) whole genomes, (2) genes and gene families, and (3) adaptive radiations. In the current chapter, I will discuss examples from only animal and plant assemblages. Even with limiting the discussion in this chapter to examples from the botanical and zoological literature, the data come from a myriad of studies, including representatives of many plant and animal clades, and once again reflect the role of genetic exchange-mediated evolutionary change.

Divergence with Genetic Exchange. Michael L. Arnold.
© Michael L. Arnold 2016. Published 2016 by Oxford University Press.

6.2 Genetic exchange and organismic evolution: Introgression, hybrid speciation, and adaptive radiations in animals

The question might be asked, given the wealth of examples of reticulate evolution among plant, microbial, and viral lineages (see Arnold, 2006, 2009 and previous chapters), why I often choose to focus much attention on examples from animal clades (e.g., Arnold and Fogarty 2009; Arnold et al. 2015). I have two major motivations for this emphasis. I have illustrated the first motivation in preceding chapters. The predominant viewpoint that came out of the neo-Darwinian synthesis (and which in some quarters has continued to the present day; Coyne and Orr 2004) was that divergence-with-gene-flow was rare and unimportant during the evolutionary history of most animals (Mayr 1963). And, if animals did happen to make the "mistake" of mating with members of another evolutionary lineage, the resulting offspring would demonstrate fitness deficits. For the majority of instances of hybridization, these assumptions have been shown to be inaccurate. Thus, hybrid fitness can vary from less-than to greater-than that of conspecific progeny (see Chapter 5) and rarity of hybrid formation, just as with rarity of point mutations that cause increases in fitness does not indicate unimportance as a catalyst of evolutionary diversification (Arnold and Hodges 1995). An extension of the neo-Darwinian concept of the maladaptive nature of hybridization and introgression in animals (and plants) was the assumption—based upon theoretical considerations (Wright 1931; Slatkin 1985)—that even low-levels of gene flow were sufficient to homogenize gene frequencies between populations. The assumption that such gene flow would disrupt gene combinations, leading to similar gene frequencies among previously divergent populations, bolstered the conceptual framework that introgressive hybridization was a "violation of species integrity."

The second motivation for selecting animals as exemplars of genetic exchange is simply that this domain of life reflects an ever increasing number of illustrations of the various outcomes of reticulate

evolution—e.g., hybrid speciation, adaptive introgression, loss of biodiversity through genetic assimilation, and reinforcement. In the following sections I will once again provide evidence that the major mode of evolutionary divergence in many animal groups is non-allopatric. This conclusion, as demonstrated throughout the previous five chapters (and in the following chapters as well), will be shown to explain patterns of divergence across a wide range of animal complexes.

6.2.1 Genetic exchange and animal evolution: Divergence-with-introgression

Chipmunks

Work by Sullivan and his colleagues on the phylogenetics and population genetics of members of the North American genus of chipmunks, *Tamias*, have identified taxonomically and temporally widespread divergence-with-gene-flow (Good et al. 2008; Hird and Sullivan 2009; Reid et al. 2012). Sullivan et al. (2014) reported that *c.* 16% of all the individuals genotyped possessed mitochondrial DNA from another taxon. This introgression reflected cases of contemporary genetic exchange between subspecies and species as well as ancient introgressive hybridization between groups not presently hybridizing (Sullivan et al. 2014). One of the consistent findings from Sullivan et al.'s (2014) analysis was that the divergence-with-introgression of mitochondrial DNA (mtDNA) often involved lineages that were diverged ecologically and morphologically. For example, Hird and Sullivan (2009) carried out analyses of DNA sequence and morphological variation within a present-day zone of overlap between two subspecies of *T. ruficaudus*. Their observations supported the hypothesis that introgressive hybridization was an important component of evolutionary diversification within this species (Hird and Sullivan 2009), likely producing the type of evolutionary novelty expected when divergent genomes recombine (Arnold 1997, 2006). This latter conclusion resulted from the discovery that hybrid zone populations were genotypically differentiated from those of the parents (Hird and Sullivan 2009).

As discussed by Sullivan et al. (2014), ancient introgressive hybridization has also impacted the

North American chipmunk clade. This inference was drawn from a series of analyses that included broad taxonomic and geographic sampling of the *Tamias* species complex. One such study was that of Good et al. (2008) who tested for genetic exchange between *T. ruficaudus* and *T. amoenus*. By comparing data from nuclear and mtDNA loci, they detected genomic variability indicative of ancient introgressive hybridization; coalescent analyses dated mtDNA introgression between these two species at 1–3 mya (Good et al. 2008). Likewise, Reid et al. (2012) detected widespread introgression among various species pairs, with the majority of the transfer events occurring deep within the phylogenetic assemblages, an inference consistent with ancient genetic exchange (Reid et al. 2012). Each of the cases of ancient introgression (where species relationships were resolved) involved non-sister taxa; four of the six cases involved asymmetric introgression from *T. minimus* into another lineage (Reid et al. 2012). In contrast, both sister and non-sister lineages contributed to recent introgression. Finally, Reid et al. (2012) suggested a primary role for niche partitioning in causing the numerous examples of recent

overlap and introgressive hybridization within the Southern Rocky Mountain clade of *Tamias*.

European, North American, and Asian hares

Introgressive hybridization between native species and between native and introduced species of the hare genus, *Lepus*, has been recognized for at least two decades (Thulin et al. 1997). A series of recent investigations on this complex by Alves, Melo-Ferreira, and their colleagues has provided some of the best evidence for contemporaneous and ancient reticulate evolution in animals (Alves et al. 2003, 2006, 2008; Melo-Ferreira et al. 2005, 2007, 2009, 2011, 2012, 2014a, b; Pietri et al. 2011; Acevedo et al. 2012a, b; Cheng et al. 2014). They have thus described the complex pattern of divergence-with-gene-flow resulting in introgression among a number of Old World and New World species (Alves et al. 2003; Cheng et al. 2014). A number of the instances of interspecific genetic exchange involve transfer from species that no longer exist in the areas of introgression (Figure 6.1; Melo-Ferreira et al. 2014a). For example, Melo-Ferreira et al. (2009, 2012) used population genetic and phylogenetic approaches,

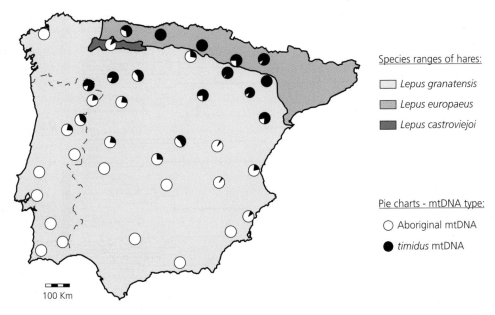

Figure 6.1 Geographic distribution and mitochondrial DNA sequence variation for the hare species, *Lepus granatensis, L. europaeus, L. castroviejoi*, and *L. timidus*. The pie diagrams reflect the frequencies of mtDNA introgressed from *L. timidus* (filled portions of the diagrams) into the Iberian *Lepus* species (from Melo-Ferreira et al. 2009).

respectively, to test for the source and directionality of genetic exchange. Their findings indicated that the arctic/arboreal species *L. timidus*—now extinct from the Iberian Peninsula—had left its genetic signature behind through introgression with several extant, temperate species (i.e., *L. granatensis, L. europaeus, L. castroviejoi, L. corsicanus*; Figure 6.1). Interestingly, as with the examples from the chipmunks, the introgression events reflected in the mtDNA of the Iberian hare species resulted from both ancient and more recent genetic exchange. Melo-Ferreira et al. (2012) reflected this conclusion in the following manner: "Despite the many uncertainties on divergence time estimates and the difficulty of finding paleontological calibration points within *Lepus*, it seems clear that mtDNA introgression occurred at 2 different epochs, first presumably into the ancestor of *L. castroviejoi* and *L. corsicanus* and then more recently into the former, in the Iberian Peninsula."

The Alves/Melo-Ferreira group has recently extended their analyses of *Lepus* species into the North American clade that includes the snowshoe hare (*L. americanus*) and the white- and black-tailed jackrabbits (*L. townsendii* and *L. californicus*, respectively). Cheng et al. (2014) sampled both nuclear and mtDNA loci across the range of the snowshoe hare to determine the number of significant conservation units for this key component of boreal ecosystems. This analysis detected several divergent lineages within *L. americanus*, reflecting populations in (1) the northern, Boreal regions, (2) the Pacific Northwest, and (3) the southwest (Figure 6.2). The detection of relatively divergent subunits within the snowshoe hare was not surprising given its wide geographic and ecological distribution. However, a very unexpected result was that, based on mtDNA sequences, the Pacific Northwest lineage shared a most recent common ancestor with the black-tailed jackrabbit (Figure 6.2). Cheng et al. (2014) concluded that

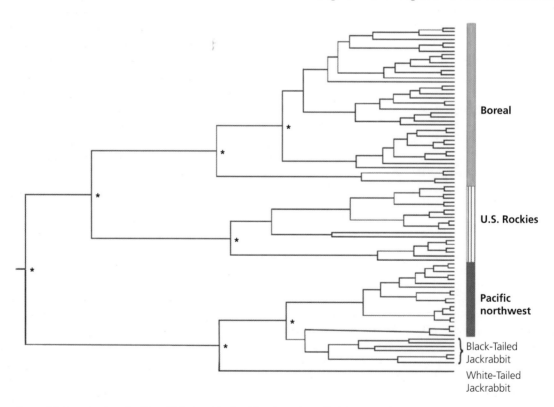

Figure 6.2 Snowshoe hare (i.e., "Boreal," "U.S. Rockies," and "Pacific Northwest") phylogenetic relationships determined using mtDNA sequences from this species as well as the other two North American hare species, white- and black-tailed jackrabbits. Asterisks (*) indicate divisions with ≥95% posterior probability support (from Cheng et al. 2014).

this non-concordant phylogenetic pattern could have been caused by incomplete lineage sorting or introgressive hybridization, but due to their limited sampling of *L. californicus* they could not determine which of these explanations was most likely.

A subsequent study of the North American *Lepus* clade, combining extensive sampling of *L. americanus, L. townsendii*, and *L. californicus* with a coalescent-based phylogenetic approach, allowed a test of whether incomplete lineage sorting or introgressive hybridization was responsible for the phylogenetic discordance observed by Cheng et al. (2014). Melo-Ferreira et al. (2014b) examined sequence data for autosomal, mitochondrial, X chromosome, and Y chromosome loci. Once again, divergent lineages were detected within *L. americanus*. Likewise, some of the mtDNA sequences of the Pacific Northwest samples of snowshoe hares were found to be most similar to *L. californicus* (Melo-Ferreira et al. 2014b). By comparing the observed mtDNA sequence variation with the distributions expected from either incomplete lineage sorting or introgression, it was inferred that the close similarity between snowshoe hares and black-tailed jacket rabbits was a consequence of genetic exchange. Thus, though nuclear introgression among the three species has apparently been rare, there was ancient (*c.* 500,000 ybp) mtDNA introgression from *L. californicus* into *L. americanus* (Melo-Ferreira et al. 2014b).

Similar to the Iberian species, lineages of Asian *Lepus* reflect ancient and contemporaneous genetic exchange. Indeed, as predicted by Alves et al. (2008), the genetic input from *L. timidus* was also detected in Asian hares (Liu et al. 2011). For example, the mtDNA haplotypes present in *L. mandshuricus* and *L. capensis* consisted of those from *L. timidus* and *L. sinensis*. Recent hybridization among a number of Asian *Lepus* species was also detected from the presence of a number of individuals that were heterozygous for species-specific nuclear alleles (Liu et al. 2011). Likewise, an analysis of *L. capensis* and *L. yarkandensis* revealed bidirectional introgression of Y chromosome and mtDNA between these ecologically and morphologically distinct species (Wu et al. 2011). Thus, as the landmark studies of Alves, Melo-Ferreira, et al. predicted, the genus *Lepus* is a paradigm for reticulate evolutionary processes.

European rabbits

Long used as a model system for studying genomic and evolutionary processes (e.g., Biju-Duval et al. 1991; Hardy et al. 1995), members of the rabbit assemblage likewise possess genomic evidence of both ancient and recent introgressive hybridization. In particular, the occurrence of divergence-with-gene-flow among lineages has been well illustrated by the work of Ferrand and his colleagues on domesticated and wild lineages of the European rabbit, *Oryctolagus cuniculus*. Their studies, spanning more than 15 years, have detected numerous introgression events leading to the formation of natural and domesticated rabbit populations (Branco et al. 2000, 2002; Queney et al. 2001; Geraldes et al. 2005, 2006, 2008; Carneiro et al. 2009, 2010, 2011, 2013, 2014a, b).

The recent series of studies by Carneiro et al. (2009, 2010, 2011, 2013, 2014a, b) have yielded detailed insights into the genomic architecture of introgressive hybridization, reproductive isolation, and domestication within the *O. cuniculus* complex. For example, Carneiro et al. (2009, 2010) analyzed sequence variation at autosomal and X linked loci across natural populations of the two parapatrically-distributed subspecies, *O. c. algirus* and *O. c. cuniculus*. Both of these analyses revealed widely varying frequencies of genetic exchange for the loci studied; loci near centromeres and on the X chromosome demonstrated high levels of divergence (i.e., low levels of introgression) between the two subspecies, while those autosomal loci near telomeres generally reflected high levels of introgression (Carneiro et al. 2009, 2010). The widely varying estimates of introgression and haplotype sharing between *O. c. algirus* and *O. c. cuniculus* were consistent with both ancient and contemporaneous genetic exchange and thus a non-allopatric model of divergence. Furthermore, these studies revealed the presence of genomic regions that contributed differentially to reproductive isolation (Carneiro et al. 2009, 2010).

Consistent with the earlier findings, Carneiro et al.'s (2013) analysis of autosomal, X and Y linked loci across the hybrid zone between the two subspecies also detected differential introgression and reproductive isolation. X and Y linked loci showed greatly reduced recombination, likely reflecting selection against the introgression of these loci. In contrast, many genomic regions demonstrated extensive interchange between the two taxa. Carneiro et al.'s (2013) conclusions included the following: "These results imply an old history of hybridization and high effective gene flow and anticipate that isolation factors should often localize to small genomic regions."

In terms of the evolutionary processes resulting in the domestic rabbit, Carneiro et al. (2011) were able to determine the region of origin (France) and the direct ancestor (French populations of *O. c. cuniculus*). Though there was no evidence that *O. c. algirus* populations were used in the domestication process some 1200 years ago, this subspecies did have an impact on the genomic makeup of the domesticated form. Natural introgressive hybridization between the two wild subspecies, predating the domestication event, thus resulted in *O. cuniculus* possessing a mosaic genome made up of elements from both subspecies (Carneiro et al. 2011). Furthermore, Carneiro et al. (2014b) concluded that patterns of genomic variability in domestic and wild populations of *O. cuniculus* were "consistent with polygenic and soft sweep modes of selection . . . that primarily acted on standing genetic variation in regulatory regions of the genome." It is tempting to speculate that some of this standing genetic variation might have arisen through genetic admixture between the two subspecies.

Ducks

Grant and Grant (1992) estimated that *c.* 10% of avian species hybridized in nature. However, they found great variation in frequency between different clades. The highest frequency of hybridizing pairs was found for the order Anseriformes (ducks and geese) with *c.* 42% of all species involved in genetic exchange (Grant and Grant 1992). Analyses utilizing genomic markers have confirmed the extent to which duck and geese taxa have been impacted by introgression. For example, I have discussed elsewhere (Arnold 2006; pp. 161–163) studies that illustrate the conservation issues arising from introgressive hybridization between the mallard, *Anas platyrhynchus* and a relatively rare species, the black duck (*A. rubripes*). Briefly, Mank et al. (2004) demonstrated that the lack of reciprocal monophyly of black duck and mallard samples

reflected the genetic assimilation, through introgressive hybridization, of *A. rubripes* by *A. platyrhynchus*. The mallard was thereby implicated in causing the extinction of the black duck through hybridization.

Given the propensity for interspecific hybridization among Anseriformes lineages, extinction through genetic assimilation appears possible for other taxa as well (e.g., Muñoz-Fuentes et al. 2007; Peters et al. 2014a). However, it is also likely that the widespread genetic exchange occurring within the duck/goose clade has provided evolutionary benefits. In this regard, Kraus et al. (2012) argued that the high frequency of genetic exchange between lineages within this avian group was similar in effect to HGT in prokaryotes. Their conclusion drawn from this analogy was that "observed parallels to horizontal gene transfer in bacteria facilitate the understanding of why ducks have been such an evolutionarily successful group of animals. There is large evolutionary potential in the ability to exchange genes among species and the resulting dramatic increase of effective population size to counter selective constraints" (Kraus et al. 2012). Finally, it seems plausible that, as with natural populations, the biased contribution of, for example, *A. platyrhynchus* to certain domestic duck populations (Hird et al. 2005; Li et al. 2010) might also reflect the evolutionarily beneficial role of introgression—possibly benefiting both the domestic animal and humans.

Though *A. platyrhynchus* has been implicated in the genetic assimilation of other taxa, the genomic makeup of the mallard is also a mosaic of elements derived from other species. For example, Avise et al. (1990) defined two divergent mtDNA lineages and these lineages have been shown to be paraphyletic with respect to a number of other related species (Peters et al. 2014b). Kulikova et al. (2004, 2005) invoked introgressive hybridization as at least a partial explanation for the presence of multiple mtDNA haplotypes among *A. platyrhynchus* populations. Figure 6.3 illustrates a portion of the mallard's geographic distribution and the mtDNA constitution of its populations indicating the role of introgression with other species. The cline in the frequency of the A and B mtDNA haplotypes is consistent with introgression of the Group A mtDNA into mallards

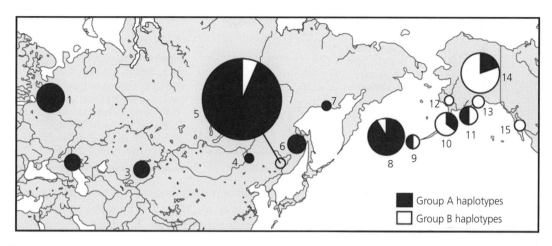

Figure 6.3 Geographic distributions and frequencies of two mtDNA haplotypes ("Group A" and "Group B"; Avise et al. 1990) in *Anas platyrhynchus* (i.e., mallard). Each pie diagram indicates the proportion of individuals with either haplotype (as indicated by the filled and unfilled portions); the sizes of the pie diagrams reflect the relative sample number for each population. The Group A haplotypes have been inferred to have originated through the introgression into mallards of mtDNA from other species such as the Eastern Spotbilled Duck, *A. zonorhyncha* (from Kulikova et al. 2004, 2005).

from related species of ducks from Asia (that only possess Group A haplotypes; Kulikova et al. 2005). Likewise, introgression between mallards and other duck species has also been detected from nuclear loci. For example, Lavretsky et al. (2014) demonstrated that phylogenies for 14 closely related taxa belonging to the species complex containing *A. platyrhynchus* had lower support (i.e., were less likely to reflect shared ancestry) when mallard samples were included. This was consistent with the widespread occurrence of introgressive hybridization between *A. platyrhynchus* and numerous other taxa during their evolutionary history.

Genetic interactions between *Anas* species and subspecies are not limited to cases involving mallards (Peters et al. 2007; Winker et al. 2013). The general interfertility between members of *Anas*, and the likelihood of introgression in nature, has been well established for decades (Johnsgard 1960). Figure 6.4 illustrates the discordant phylogenetic relationships derived from morphological and mtDNA data sets among a series of species known as the wigeons (Peters et al. 2005). As with the mallard studies, similar patterns of non-concordance were also detected between those phylogenies based on either morphological traits or nuclear loci (Peters et al. 2005). Variation at nuclear

loci was consistent with the hypothesis that genetic exchange caused the discordance involving *A. strepera* (gadwall) and *A. falcata* (falcated duck) illustrated in Figure 6.4 (Peters and Omland 2007; Peters et al. 2007).

6.2.2 Genetic exchange and animal evolution: Homoploid hybrid speciation

Homoploid hybrid speciation in animals has become a recognized mechanism in certain clades (Arnold 1997, 2006; Mallet 2007; Mavárez and Linares 2008). From the early work on the cyprinid fish species, *Gila seminuda* (DeMarais et al. 1992), to the recent analyses of the Neotropical butterfly species, *Heliconius heurippa* (Mavárez et al. 2006), there is a growing appreciation of how this reticulate evolutionary process may have contributed to biodiversity (see also Nietlisbach et al. 2013). To illustrate the evolutionary significance and organismic distribution of web-of-life processes, several examples of homoploid hybrid speciation leading to sexually reproducing species were reviewed in Chapter 1. Therefore, I will focus the majority of the discussion in this and the following sections on the processes leading to either asexually reproducing homoploid or allopolyploid animals. However, for

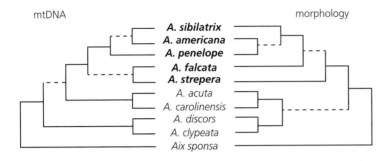

Figure 6.4 Phylogenetic relationships of species from the duck genus *Anas* known commonly as wigeons (indicated by bold type), along with their sister taxa, based on mtDNA sequences and morphological traits. Dashed lines indicate the three regions of the phylogenies that show discordances in species placement (from Peters et al. 2005).

completeness, I will first discuss an additional example of bisexual, homoploid hybrid species formation, in this instance from the order Chiroptera.

Genetic exchange and animal evolution: Homoploid, sexual species

As with a number of other mammalian clades, bats have been a model system for depicting evolution catalyzed by chromosomal rearrangements. In particular, Baker, Bickham, and their colleagues have reported numerous analyses designed to test for evolutionary diversification catalyzed by the structural reorganization of the genome (e.g., Baker and Bickham 1986; Hoffmann et al. 2003; Baird et al. 2009; Sotero-Caio et al. 2013). A recent analysis has also pointed to the effect of genetic exchange on the origin of bat species.

Larsen et al. (2010) collected both molecular (nuclear and mitochondrial loci) and morphological data from mainland South American and Caribbean species of the Neotropical bat genus *Artibeus*, representing all seven recognized Caribbean taxa. The analysis of the island taxa defined a zone of hybridization and introgression among *A. jamaicensis*, *A. planirostris*, and *A. schwartzi*, with the hybrid zone extending across the chain of islands from St. Lucia to Grenada (Larsen et al. 2010). Interestingly, not only did the data support admixture among these three *Artibeus* species (in spite of their divergence from a common ancestor *c.* 2.5 mya), but also led to the inference that *A. schwartzi* was a homoploid hybrid derivative. The findings supporting this latter conclusion were that *A. schwartzi* possessed a nuclear genome made up of alleles from both *A. jamaicensis* and *A. planirostris*, mtDNA haplotypes from another, as yet unidentified, lineage, and a transgressive morphology that did not overlap with that

of either *A. jamaicensis* or *A. planirostris* (Figure 6.5; Larsen et al. 2010). The transgressive phenotypic variation detected for *A. schwartzi* (Figure 6.5) is of particular significance because such a pattern of morphological variability is a common outcome of homoploid hybrid speciation in both plants and animals (Rieseberg 1997; Bell and Travis 2005).

Thus, ongoing and ancient divergence-with-gene-flow characterizes the Caribbean Island lineages of this Neotropical bat genus. Larsen et al. (2010) summarized well the overall processes that likely contributed to the evolution of this group, and other animal complexes in this way: "the evolutionary processes occurring within the genus *Artibeus* perhaps are best conceptualized within the framework of an 'open system,' whereby allopatric diversification, reinforcement during periods of sympatry, and hybrid speciation have contributed to contemporary diversity."

Genetic exchange and animal evolution: Homoploid asexual species

Warramaba

The Australian grasshopper genus, *Warramaba*, consists of a series of taxonomically named and unnamed sexual species (*W. picta*, P125, P169, P196, P152, P188) and the *parthenogenetic* species, *W. virgo* (Key 1976; Kearney and Blacket 2008). Numerous earlier studies by White and his colleagues, and more recent analyses by Kearney et al. have defined the reproductive biology, population genetics, and evolutionary origins of the various species, with particular emphasis given to *W. virgo* (e.g., White et al. 1963, 1977, 1980, 1982; Hewitt 1975; Webb et al. 1978; White 1980; Dennis et al. 1981; White and Contreras 1982; Honeycutt and Wilkinson 1989; Kearney et al. 2006; Kearney and Blacket 2008).

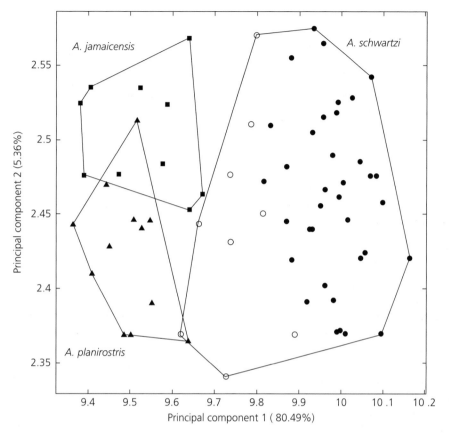

Figure 6.5 Transgressive morphological variation detected in the homoploid hybrid bat species, *Artibeus schwartzi*. Results come from a principal components analysis using 17 cranial and mandibular traits from the putative parental taxa, *A. jamaicensis* and *A. planirostris* as well as *A. schwartzi* collected from Caribbean islands. For *A. schwartzi* samples: ● = specimens collected from St. Vincent and ○ = specimens collected from St. Lucia and the Grenadines (from Larsen et al. 2010).

White (1980) defined the mechanism of reproduction of *W. virgo* as a type of automictic parthenogenesis. This process involves the premeiotic doubling of chromosomes with synapsis limited to sister chromosomes thus ensuring that "all offspring of a given female are genetically identical to one another and to their mother" (White 1980).

Though *W. virgo* reproduces parthenogenetically, White's (1980) conclusion concerning a lack of genetic variation produced by recombination was revised subsequent to a study of allozyme variation (Honeycutt and Wilkinson 1989). The genetic polymorphisms detected in the parthenogenetic lineage reflected multiple occurrences of recombination (Honeycutt and Wilkinson 1989). Furthermore, as with asexual, allopolyploid lineages,

cytogenetic and genetic variation in populations of *W. virgo* were best explained by a model including recurrent formation by hybridization between its bisexual progenitors (White et al. 1977; Webb et al. 1978; Dennis et al. 1981). For example, the pattern of variability at the allozyme loci was indicative of *W. virgo* consisting of many clonal lineages resulting from numerous hybridization events between the sexual forms known as "P196" and "P169." Likewise, though it appeared that only one of the sexual species (i.e., P169) acted as the maternal lineage during the origin of *W. virgo*, there were multiple mtDNA lineages within the parthenogen which was also consistent with separate hybridization events during the formation of asexual populations (Kearney et al. 2006). Analyses of mtDNA in a phylogeographic context led to the further inferences that Plio-Pleistocene environmental changes caused the formation of glacial refugia in regions of northern Australia and the expansion of *W. virgo* from these northern areas into its current, southerly locales (Kearney and Blacket 2008). Finally, in light of its clonal diversity and geographically extensive distribution, *W. virgo* has also been inferred to be fit relative to its bisexual progenitors, with the observation that new genetic variation had arisen since the formation of this parthenogen indicative of a genetically dynamic complex (Honeycutt and Wilkinson 1989).

Pelophylax esculentus species complex

"Hybridogenetic species possess a hybrid genome: half is clonally inherited (hemiclonal reproduction) while the other half is obtained each generation by sexual reproduction with a parental species" (Semlitsch et al. 1996). One of the best investigated *hybridogenetic* species complexes is the *Pelophylax esculentus* clade (e.g., Hotz et al. 1992; Semlitsch and Reyer 1992; Semlitsch 1993a, b; Santucci et al. 1996; Semlitsch et al. 1996, 1997; Plénet et al. 2000, 2005; Schmeller et al. 2005; Som and Reyer 2007; Plötner et al. 2008; Leuenberger et al. 2014; Pruvost et al. 2015). Known as the *Rana esculenta* complex prior to Frost et al. (2006), this Palaearctic water frog continues to be repeatedly formed through natural hybridization between *P. lessonae* and *P. ridibundus* (Hellriegel and Reyer 2000; formerly *R. lessonae* and *R. ridibunda*, respectively; Frost et al. 2006). Initially, three breeding systems were defined for the category *P. esculentus*, as reflected by LE (*lessonae esculentus*), RE (*ridibundus esculentus*), and all-hybrid populations (Figure 6.6; Christiansen et al. 2005). The designations of *lessonae esculentus* and *ridibundus esculentus* indicated population types in which *lessonae* or *ridibundus* acted as the sexual donor, respectively. In contrast, the all-hybrid system reflected those populations in which neither of the progenitors (for the hybridization event that gave rise to *P. esculentus*) was present (Figure 6.6; Christiansen et al. 2005). Finally, there were a number of LE populations into which *P. ridibundus* had been introduced. In these cases, this latter species had replaced (due to asymmetric success of different genotypes) both *P. esculentus* and *P. lessonae*

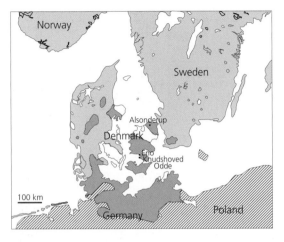

Figure 6.6 Geographic distribution of the Palaearctic water frog, *Pelophylax esculentus*. The portion of the range map covered with dark gray shading indicates regions where this hybridogenetic species does not co-occur with either of its two progenitors (*P. ridibundus* and *P. lessonae*). The hatching indicates regions where *P. esculentus* occurs with its progenitors. The locations of three Danish all-hybrid populations (Alsønderup, Enø, and Knudshoved Odde) from which reproductive data were collected are also indicated (from Christiansen et al. 2005).

(Vorburger and Reyer 2003). In contrast, a recent analysis of populations into which *P. ridibundus* had been introduced concluded that the introduced and native forms were likely to coexist due to the different niche requirements of the three species (Leuenberger et al. 2014).

Though I am discussing the *P. esculentus* clade as an example of a "homoploid asexual species," this designation does not encompass the genetic or evolutionary intricacy of this system. For example, hybridization between the parental species does not only lead to the formation of the hybridogen, but also has resulted in asymmetric mtDNA introgression from *P. lessonae* into *P. ridibundus* (Plötner et al. 2008). In addition, Mikulíček et al. (2014) documented both mtDNA and nuclear introgression from *P. esculentus* into its sexual progenitor, *P. ridibundus*. In the latter instance, the highest degree of introgression occurred in populations containing both parental and hybridogen animals suggesting, "an ongoing and site-specific interspecific genetic transfer mediated by hybridogenetic hybrids" (Mikulíček et al. 2014).

As with using the term homoploid hybrid speciation, applying the term "hybridogenesis" likewise does not indicate the complex breeding system that gives rise to *P. esculentus*. Christiansen and Reyer (2011) reported one example of the evolutionary diversity involving populations of *P. esculentus* located in the Danish archipelago as well as in the adjoining countries of Sweden and Germany. Not only did these populations consist entirely (or nearly so) of hybrid frogs, they also contained both diploid and triploid hybridogenetic individuals. Significantly, the various hybrid types were associated with certain habitats, suggestive of a role for environment-dependent fitness differences in the establishment and maintenance of the array of genotypes (Christiansen and Reyer 2011). Furthermore, the diploid and triploid *P. esculentus* hybrids have the capability to recombine their genomes (Arioli et al. 2010) demonstrating the ability to reproduce sexually. Christiansen and Reyer (2009) thus detected recombination in both experimental and natural triploid hybrids, leading them to conclude: "This direct evidence of sexual reproduction in *P. esculentus* calls for a change of the conventional view of hybridogens as clonally reproducing diploids. Rather, hybridogens can be independent sexually reproducing units with an evolutionary potential."

Poecilia formosa

Hubbs and Hubbs (1932) defined the novelty of their discoveries concerning the fish species, *Mollienisia formosa* (since named *Poecilia formosa*) by stating, "In conclusion, the conditions demonstrated by this study, so far as we know novel in the biology of the vertebrates, are: (1) The abundant occurrence in nature of a form of demonstrated hybrid origin, having nearly all of the characteristics of a natural species; (2) the occurrence of a form as females only, over a wide portion of its range; (3) the consistent and abundant production of wholly female and purely matroclinous young; (4) apparent parthenogenesis in nature." Hence, theirs was not only the first ever description of a unisexual vertebrate species, but also of a hybrid, asexual species. In support of a hybrid origin, Hubbs and Hubbs (1932) described the morphology of *P. formosa* as being "exactly intermediate" between the presumed parents, *M. latipinna* and *M. sphenops*, (now, *P. latipinna* and *P. sphenops*, respectively). Though *P. formosa* is indeed asexual, Hubbs and Hubbs (1932)

suggested that, rather than parthenogenetic, its reproductive system was more accurately defined as *gynogenetic*—i.e., asexual reproduction that requires activation of embryogenesis by sperm from a related, sexual species (Avise et al. 1992). In fact, Hubbs and Hubbs (1932) gave an excellent description of the expected association of a gynogen and its bisexual progenitors in their observation that *P. formosa* always occurred with one of its parental species.

Following its discovery, various analyses were designed to test the specific processes associated with the origin and subsequent evolution of *P. formosa* (Lampert and Schartl 2008). In terms of the origin of this hybrid gynogen, *P. latipinna* was inferred to have acted as the male parent (Turner et al. 1980; Avise et al. 1991). However, in contrast to the original description, it was determined that *P. sphenops* was not involved in the origin of *P. formosa*. Instead, *P. mexicana*, a close relative of *P. sphenops*, was identified as the likely maternal parent of the hybrid species (Turner et al. 1980; Avise et al. 1991). Furthermore, unlike most asexual animal lineages (e.g., *Heteronotia*, *Warramaba*, *Daphnia*, and *Pelophylax*), the origin of hybrid *P. formosa* does not appear to have occurred repeatedly. Evidence for a limited (possibly a single) number of formations of this gynogenetic species has come from multiple experimental crossing arrays and from analyses of genomic data. Both Turner et al. (1980) and Stöck et al. (2010) attempted to produce gynogenetic individuals by experimentally crossing *P. latipinna* and *P. mexicana* individuals from numerous populations. Both of these efforts failed to synthesize asexual progeny. Turner et al. (1980) suggested that the failure to recreate the gynogen might be best explained by the control of gynogenesis by a few genes that were polymorphic in the populations of the progenitors. If this hypothesis were accurate, the lack of any gynogenetic offspring from the crossing analyses by Turner et al. (1980) and Stöck et al. (2010) could have resulted from their using individuals that did not possess the correct genotypes at these few loci.

Consistent with the lack of gynogenetic progeny from crosses of *P. latipinna* and *P. mexicana*, the molecular analyses by Lampert et al. (2005) and Stöck et al. (2010) led to the inference of a single (or at the most a few) episode(s) of hybridization

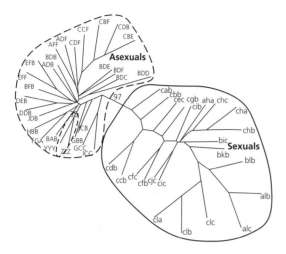

Figure 6.7 The phylogenetic relationships of gynogenetic *Poecilia formosa* clones ("asexuals") and bisexual *P. mexicana* samples ("sexuals") using microsatellite data. The two triploid, gynogenetic clones are YYY and ZZZ (indicated by the small, dashed ellipse within the asexual assemblage). The remaining asexual lineages are diploid gynogens (from Lampert et al. 2005).

resulting in the formation of both diploid and triploid gynogenetic clones of *P. formosa* (Figure 6.7). Thus, sequences from both mtDNA and nuclear loci resolved the asexual lineages into a monophyletic clade relative to *P. latipinna* and *P. mexicana*. The rarity of formation of *P. formosa* led to the inference "that some unisexual vertebrates might be rare not because they suffer the long-term consequences of clonal reproduction but because they are only very rarely formed as a result of complex genetic preconditions necessary to produce viable and fertile clonal genomes and phenotypes" (Stöck et al. 2010).

In addition to the predicted difficulty of formation of *P. formosa*, the maintenance of this gynogenetic form is dependent on the seemingly unlikely success of its sexual parasitism of the diploid progenitors. If there are no bisexual individuals sympatric with the gynogen, or males belonging to the sexually reproducing species do not mate with the *P. formosa* females, the gynogenetic lineage will become extinct. In this regard, results from a series of mate choice experiments have provided evidence that the maintenance of *P. formosa* is likely facilitated by reproductive interactions. Specifically, males belonging to the progenitor species (Gabor

and Aspbury 2008; Alberici da Barbiano et al. 2012) and, surprisingly, even to an unrelated sexual species (Joachim and Schlupp 2012) were attracted to the *P. formosa* females. This suggests that, as long as one of the parental species is present in populations of *P. formosa*, there will be no lack of a sperm donor for gynogenetic reproduction.

The rarity of formation and the hypothesized difficulty in maintaining *P. formosa* belie results indicating an adaptive advantage due to its hybrid origin. For example, an analysis of gene expression at opsin loci provided evidence of "hybrid-sensory expansion" in the hybrid gynogen (Sandkam et al. 2013). The expression analyses demonstrated allelic differences and differential opsin repertoires in the progenitor species, *P. latipinna* and *P. mexicana* (Sandkam et al. 2013). Significantly, *P. formosa* possessed both of the parental variants and an expanded opsin repertoire consistent with hybrid-sensory expansion. Sandkam et al. (2013) argued that the expanded genomic and phenotypic variation in *P. formosa* might indicate the importance of natural hybridization in the evolution of sensory systems in other groups as well.

6.2.3 Genetic exchange and animal evolution: Allopolyploid speciation

Though seemingly much less frequent than in plants, as already discussed, some animal species complexes possess polyploid taxa. A number of hypotheses have been proposed to address the question of why polyploidy is relatively rare in animal taxa. For example, Stebbins (1950) suggested that restrictions for establishment of taxa with whole-genome duplications in animals were caused by a more easily perturbed developmental system. More recently, Orr (1990) argued instead for the disruption of a balance between gene products from the X chromosomes and the autosomes.

Regardless of partial barriers to the formation of allopolyploid animal taxa, WGD events have nonetheless played a significant role in the evolution of the animal clade (Van de Peer et al. 2009). Indeed, polyploidy has been suggested as a causal factor in the radiation of the entire vertebrate clade (Spring 1997; Furlong and Holland 2002; Donoghue and Purnell 2005) and for subclades (such as fish; Le

Comber and Smith 2004) within vertebrates. However, whether or not WGD events caused the diversification of entire clades (Van de Peer et al. 2009), the polyploid events themselves and the genomic consequences for the organisms characterized by these events have been well substantiated (e.g., Grismer et al. 2014; Martin and Holland 2014). In fact, the number of recorded polyploid, animal species, and the taxonomic breadth of the clades in which they occur, continues to increase (Mable et al. 2011; Neaves and Baumann 2011; Bogart and Bi 2013; Montelongo and Gómez-Zurita 2015).

Genetic exchange and animal evolution: Allopolyploid, sexual species

Not only have many allopolyploid animal lineages now been identified, but sexually reproducing polyploid lineages have been recorded as well. Though sometimes reported from endothermic groups (Gallardo et al. 1999, 2013), bisexual, allopolyploid vertebrate species have been most commonly detected in fish and amphibian clades (e.g., Stöck et al. 2009; Cunha et al. 2011; Levin et al. 2012; Luo et al. 2014; see Mable et al. 2011 for additional examples). Polyploidization in general, and the production of bisexual allopolyploid taxa in particular, thus occurred from the earliest divergence of vertebrates through to recent radiations.

Members of the related amphibian genera, *Xenopus* and *Silurana*, illustrate well the evolutionary processes associated with the origin and development of bisexual allopolyploid animals (Figure 6.8). These allopolyploid lineages have been the basis for a number of studies testing for genomic changes

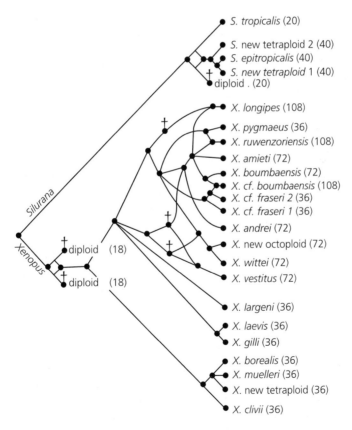

Figure 6.8 Evolutionary relationships among diploid and allopolyploid lineages of the African clawed frog genera, *Silurana* and *Xenopus*. The ploidy level follows each species name including three diploid and three tetraploid taxa (indicated with daggers) predicted, but not observed among extant forms (from Evans 2007). Reproduced with permission from the Genetics Society of America.

that occurred after hybridization and WGD. For both individual genes and gene families, some of the possible molecular evolutionary consequences from allopolyploidy are *non-functionalization* (i.e., *silencing*), *sub-functionalization, neo-functionalization, transcriptomic shock* and *concerted evolution* (e.g., Prince and Pickett 2002; de Souza et al. 2005; Duarte et al. 2006; Zhou et al. 2011; Koh et al. 2012; Gong et al. 2014; Jourda et al. 2014; Xu et al. 2014; Zozomová-Lihová et al. 2014; Douglas et al. 2015). The occurrence of these processes, resulting from hybridization and polyploidization, has been inferred using genomic and expression analyses among lineages of *Xenopus* and *Silurana* (Evans 2007; Chain et al. 2008; Sémon and Wolfe 2008; Anderson and Evans 2009).

Xenopus and *Silurana* are sister complexes within the family Pipidae (De Sá and Hillis 1990; Evans et al. 2004). Various analyses have suggested that extant species belonging to both genera had an ancient origin (c. 64 mya) in the central and/or eastern portions of equatorial Africa (Evans et al. 2004). Chromosome numbers displayed by various taxa suggest the role of polyploidization in the evolution of the African clawed frogs. Thus, *Silurana* includes species with either 20 or 40 chromosomes; the lower number is associated with non-duplicated (i.e., "diploid") taxa and the larger number reflects tetraploidy (Figure 6.8; Evans et al. 2005). Interestingly, no diploid *Xenopus* taxa have been identified. Instead, there are ten tetraploid ($4n = 36$), five octoploid ($8n = 72$), and two dodecaploid ($12n = 108$) species (Figure 6.8; Evans et al. 2005). Furthermore, recent analyses have suggested that *X. laevis sensu lato* actually consists of at least four separate species (Furman et al. 2015).

Additional earlier evidence of the polyploid constituency of these species was reflected by the increasing series of DNA contents in the various chromosomal forms and the detection of polyploid gametes in both natural and experimental hybrids (Kobel and Du Pasquier 1986). Likewise, both phylogenetic discordance obtained using various molecular and morphological data and the identification of contemporary hybrid zones also supported the likelihood of allopolyploid evolution within these genera (Carr et al. 1987; Evans et al. 1997; Fischer et al. 2000; Jackson and Tinsley

2003; Evans et al. 2005). Finally, recent analyses have identified the expected pattern of duplicated, paralogous genes within the genomes of bisexual, allopolyploid *Xenopus* and *Silurana* lineages (Evans 2007; Chain et al. 2008; Sémon and Wolfe 2008; Anderson and Evans 2009). All of the available data for this clade point to a fundamentally important role for genetic exchange, and specifically, WGD in the evolution of the African clawed frogs. The *Xenopus*/*Silurana* clade reflects numerous instances of natural hybridization-mediated, evolutionary diversification (Figure 6.8).

Genetic exchange and animal evolution: Allopolyploid, asexual species

Heteronotia

The Australian gekkonid lizard clade—identified taxonomically as *Heteronotia binoei*—reflects a classic example of the evolutionary sequence, hybridization → allopolyploidy → parthenogenesis. Beginning with the work of Moritz et al. and more recently Kearney and his colleagues, the diploid and polyploid lineages have been defined in terms of their mode of origins, evolutionary and ecological history, molecular evolution, and the relative fitness of the parental diploids and allopolyploid parthenogens (e.g., Moritz 1983, 1991; Moritz et al. 1989; Hillis et al. 1991; Kearney 2003, 2005; Kearney and Shine 2004a, b; Strasburg and Kearney 2005; Strasburg et al. 2007; Roberts et al. 2012).

As with almost all allopolyploid (and diploid; e.g., see Webb et al. 1978; Honeycutt and Wilkinson 1989), parthenogenetic animals, triploid *Heteronotia* parthenogenetic lineages were repeatedly formed (Moritz 1983, 1991; Moritz et al. 1989). Specifically, the allopolyploid forms originated from reciprocal crosses between two of three diploid, sexual races (Figure 6.9; Moritz 1983), with backcrossing between the initial, diploid parthenogenetic forms and the sexual diploids, leading to the formation of the triploid populations (Figure 6.9; Strasburg and Kearney 2005). The putative backcrosses (Figure 6.9) were suggested to explain not only the triploid, parthenogen formation, but also the high level of nuclear-marker polymorphism demonstrated by the parthenogenetic lineages (Moritz 1983; Moritz et al. 1989). Currently, the geographic ranges of the parthenogenetic, hybrid lineages overlap with both

(a)

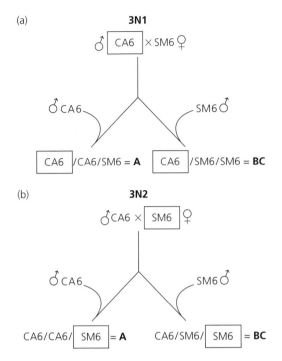

Figure 6.9 Reciprocal formations of the triploid, parthenogenetic races of the gekkonid lizard, *Heteronotia binoei*, designated as (a) "3N1" and (b) "3N2." Backcrossing (indicated at the bottom of the two panels) between the inferred diploid hybrids produced multiple clones ("A" and "BC") within 3N1 and 3N2 (from Strasburg and Kearney 2005).

of their progenitors as well as with a third sexual, diploid race. Mitochondrial DNA sequence variation does not support a contribution from this latter diploid form in the hybrid origin of the 3N1 and 3N2 lineages (Strasburg and Kearney 2005).

It has been postulated that gradual aridification and, more recently, cyclical arid and mesic periods, altered the geographic ranges of both the diploid bisexual and triploid parthenogenetic forms (Strasburg and Kearney 2005; Strasburg et al. 2007). Analyses of mtDNA variation encompassing the present-day range of the two parthenogenetic lineages (Figure 6.9), combined with modeling of past distributions, predicted two separate expansion events. It was thus inferred that the 3N1 and 3N2 lineages (Figure 6.9) expanded from the northwest and far west portions of their present-day distributions, respectively (Strasburg et al. 2007). The timing for the 3N1 geographic extension was estimated at *c.* 0.24 mya, while the

range of 3N2 enlarged more recently at *c.* 0.07 mya (Strasburg et al. 2007).

The occurrence of not only *H. binoei* parthenogens in the Australian arid zone, but many additional asexual animal and plant forms as well, has led Kearney (2003) and Kearney et al. (2006) to argue for a cause-and-effect relationship. For example, the hybrid parthenogenetic animals and plants could be favored due to the uncertainty of finding mates in the harsh environments and/or through the bringing together of variation from two divergent evolutionary lineages. Consistent with the second hypothesis, various analyses of fitness components detected higher, lower, and equivalent estimates for parthenogenetic *H. binoei* relative to the diploid progenitors (Kearney and Shine 2004a, b; Kearney et al. 2005; Roberts et al. 2012). Kearney et al. (2005) found that, in lower temperature environments, the parthenogenetic individuals demonstrated significantly greater endurance, higher maximum oxygen consumption rates, higher maximum aerobic speeds, and greater amounts of voluntary activity relative to diploid animals. These observations led to the conclusion that the *H. binoei* individuals possessed a fitness advantage due to their greater aerobic activity levels (Kearney et al. 2005). Similarly, a recent comparison of physiological traits suggested that the allopolyploid constitution of parthenogenetic *H. binoei* "may have produced a broader overall niche for the species" (Roberts et al. 2012).

Like the sexual, allopolyploid *Xenopus* and *Silurana* species, parthenogenetic *H. binoei* also exemplify the genomic effects from post-WGD molecular evolutionary processes. Hillis et al. (1991) thus detected a role for concerted evolution in the resulting DNA variation among members of the ribosomal RNA gene families found in the parthenogenetic lineages. Their analyses demonstrated the asymmetric replacement of the rDNA variants from the sexual, diploid parent, CA6, by allelic forms of the SM6 progenitor. These data were consistent with biased *gene-conversion*-mediated molecular evolution (Hillis et al. 1991). Thus not only did the initial hybridization events and subsequent backcrossing (Figure 6.9) combine highly divergent genomes into the same nucleus, they also resulted in molecular mechanisms (i.e., concerted evolution) that affected the genetic variation found in the present-day parthenogenetic lineages.

Daphnia

The crustacean genus, *Daphnia*, has been the focus of scientists interested in fields as diverse as taxonomy and climate change (e.g., Benzie 1986; Orsini et al. 2013). However, it is probably best known as a model system for understanding the dynamics of *cyclic parthenogenesis* (Lynch 1984). Both North American and European clades of *Daphnia* have been investigated to test for the evolutionary causes and consequences of the alternating sexual–asexual reproduction, with numerous analyses of both the *D. pulex* and *D. longispina* assemblages reflecting the diversity of outcomes from genetic exchange (e.g., Lynch 1984; Beaton and Hebert 1988; Dufresne and Hebert 1994, 1997; Taylor et al. 1998; Weider et al. 1999; Brede et al. 2009; Thielsch et al. 2009; Vergilino et al. 2011; Tucker et al. 2013). For example, the *D. pulex* species complex encompasses the diversity of evolutionary phenomena detected within *Daphnia*. In particular, genetic exchange is linked to WGD and parthenogenesis. The hybridization

events leading to these outcomes have occurred repeatedly, resulting in the recurrent origin of various hybrid parthenogenetic lineages (Dufresne and Hebert 1994, 1997). Further diversity of outcomes is reflected by the observation of both hybrid diploid and allopolyploid parthenogens (Hebert and Finston 2001) as well as diploid, sexual species (Adamowicz et al. 2002) within the *D. pulex* clade.

Not all *Daphnia* species are the product of reticulate evolutionary processes (Taylor et al. 1998), however, much of the biodiversity contained within this genus does exemplify the role of genetic exchange-generated evolution. In general, the various complexes reflect partial reproductive isolation resulting in both introgression and hybrid parthenogens (Figure 6.10; Weider et al. 1999). Significantly, it has been demonstrated that not only are reticulate events (i.e., hybridization and allopolyploidy) associated with the formation of numerous taxa possessing either cyclic or obligate parthenogenesis (Lynch 1984; Crease et al. 1989; Colbourne et al. 1998; Mergeay et al. 2008), but that the bringing together of the divergent genomes was causal in producing asexuality. For example, both

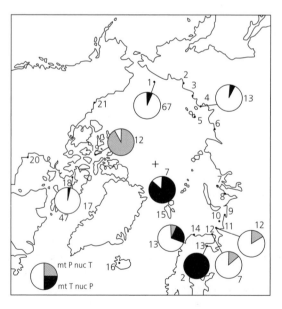

Figure 6.10 The frequency of introgressive hybridization between two members of the *Daphnia pulex* species complex, *D. pulicaria* (P) and *D. tenebrosa* (T). The black portions of each diagram reflect the proportion of parthenogenetic clones in a certain geographical region that possessed the mtDNA ("mt") of the *tenebrosa* lineage, but the nuclear ("nuc") markers of the *pulicaria* lineage. Grey portions of each diagram reflect the reciprocal pattern of mtDNA and nuclear variation. Sample sizes are indicated next to each diagram (from Weider et al. 1999).

Xu et al. (2013) and Tucker et al. (2013) were able to show that the alleles in the *D. pulex* parthenogenetic lineages underlying the loss of meiosis in females originated via introgression from *D. pulicaria*. Furthermore, though predicted by theory, there was no evidence that the relatively young age of the parthenogenetic populations (Tucker et al. 2013) reflected a disadvantage due to their inability to rid themselves of newly arisen, deleterious mutations. Rather, post-formation molecular mechanisms such as gene conversion resulted in a loss of heterozygosity thus uncovering preexisting, recessive deleterious alleles (Tucker et al. 2013).

Various analyses have resulted in inferences concerning the fitness of parental and hybrid *Daphnia* as well as the evolutionary potential of parthenogenetic lineages (Jankowski and Straile 2004; Wolinska et al. 2006; Keller et al. 2007). Observations of differential biogeographic patterning of diploid and allopolyploid lineages (i.e., parthenogenetic), in particular, have been used to infer possible benefits derived from hybridization and polyploidy. Once again the *D. pulex* complex has been used as an exemplar for such inferences. In North America and Europe, members of this species complex show a complex pattern of geographic distributions: polyploids and diploids predominate in the high Arctic and temperate regions, respectively; both diploid and allopolyploid assemblages occur in the low Arctic (Adamowicz et al. 2002).

The combination of selection favoring the allopolyploid parthenogenetic lineages in the high Arctic and the diploid parental forms in temperate regions could have established a selective gradient leading to the clinal variation in the frequency of diploid/polyploid populations seen in North America and Europe (Beaton and Hebert 1988; Adamowicz et al. 2002). However, as discussed, the parthenogenetic lineages in some complexes have been shown to be very young and thus likely under some measure of negative selection (Tucker et al. 2013). Furthermore, clinal distributions for parthenogens (and sexual species) have not been documented in the southern hemisphere. Instead, Adamowicz et al. (2002) detected only allopolyploid parthenogens in samples from temperate zones in Argentina. It was argued that the presence of the parthenogenetic forms in atypical (i.e., temperate)

environments, and the absence of diploid *D. pulex*, could be the result of the allopolyploids being "fortunate founders" (Adamowicz et al. 2002, 2004). Consistent with the hypothesis that contingency may play a significant role in the evolution of sexual and asexual lineages of *Daphnia*, Brede et al. (2009) documented the origin of novel complexes following the transition of two European lakes from human-caused eutrophic conditions back to the pre-perturbation settings. Notwithstanding the role of contingency, the evolution of *Daphnia* in general has been dependent upon natural hybridization. In this regard, Brede et al. (2009) observed: "anthropogenically induced temporal alterations of habitats are associated with long-lasting changes in communities and species via interspecific hybridization and introgression."

6.2.4 Genetic exchange and animal evolution: Adaptive radiations

As discussed in previous chapters, the concept of natural hybridization-generated adaptive evolution and speciation was proposed in the middle of the twentieth century (Anderson 1949; Anderson and Stebbins 1954; Stebbins 1959). However, not until the work of Seehausen (2004) was there an explicit model predicting how genetic exchange could catalyze adaptive radiations. This model reflected the ever-increasing genetic and ecological data from cases of natural hybridization, as well as the conceptual framework of ecological speciation (Schluter 2000). Because natural hybridization often resulted in greatly increased phenotypic variability, the resulting populations have been termed a "hybrid swarm" (e.g., Anderson 1949). Seehausen (2004) thus named his model the "hybrid swarm theory." Tests of the hybrid swarm theory require a combination of data that address several predictions (Seehausen 2004). First, bouts of hybridization must be shown to precede numerous radiations. Second, a majority of the present-day variability must be tied to admixture between divergent lineages. Third, a portion of the diversity originating during natural hybridization must be demonstrably functional (e.g., adaptive phenotypes in hybrid lineages should reflect genomic contributions from multiple progenitors). Fourth, genetic exchange (i.e., in this case, natural hybridization) must increase the likelihood of adaptive radiations (Seehausen 2004). A number of biological examples suggesting the validity of the hybrid swarm theory have now been reported (e.g., Herder et al. 2006; Hudson et al. 2011; Lundsgaard-Hansen et al. 2014). In the following two sections I will discuss data that suggest the predictive efficacy of Seehausen's model for both cichlids and heliconiine butterflies. In later sections, I will discuss evidence that this model accounts for adaptive radiations in some plant clades as well.

Cichlids

In Chapter 1 (Section 1.6.3), cichlids were used as examples of homoploid hybrid speciation in animals. Furthermore, numerous analyses have utilized this fish clade to define both the causal factors for reproductive isolation between different taxa (Salzburger et al. 2006; Stelkens and Seehausen 2009; Stelkens et al. 2009; Dijkstra et al. 2011; Böhne et al. 2014; Santos et al. 2014; Selz et al. 2014b; Theis et al. 2014) as well as both ancient and recent introgressive hybridization (Koblmüller et al. 2007, 2010; Sturmbauer et al. 2010; Nevado et al. 2011; Egger et al. 2012; Schwarzer et al. 2012a). Though excellent examples of introgression, homoploid hybrid speciation and the development of pre- and post-zygotic reproductive isolating barriers, both New World and Old World (i.e., African) cichlid assemblages are known best as models of sympatric speciation and adaptive radiation (Barluenga and Meyer 2004; Kocher 2004; Barluenga et al. 2006; Geiger et al. 2010; Bezault et al. 2011; Kautt et al. 2012; Muschick et al. 2014; Salzburger et al. 2014). For example, the African cichlids are a major contributor to "collectively the earth's most remarkable and species-rich freshwater feature" (Salzburger et al. 2014)—i.e., the East African Rift Lakes. Furthermore, the detection of recent, parallel evolutionary diversification and monophyly within the various lake systems has strongly implicated the role of natural and sexual selection as main drivers of the Old and New World radiations (e.g., Meyer et al. 1990; Kocher 2004; Barluenga et al. 2006; Seehausen et al. 2008; Kautt et al. 2012; Wagner et al. 2012).

Recent analyses of both the New World Crater Lake and African Rift Lake cichlid radiations have defined monophyletic origins for entire New World crater lakes (Geiger et al. 2010) and at least some species within individual African lakes (Wagner et al. 2013). However, particularly for the cichlids existing in the Great Lakes of East Africa, evidence has accumulated supporting the hypothesis (Seehausen 2004) that syngameons formed the basis of radiations of entire assemblages (e.g., Seehausen et al. 2003; Seehausen 2006; Day et al. 2007; Genner and Turner 2012; Brawand et al. 2014). All of the Great Lake cichlid diversifications occurred over a very short period of time following the invasion of the African rift lakes by riverine lineages (Seehausen 2006). Remarkably, given such a short period for divergences to occur, the various endemic species flocks demonstrate extreme morphological and ecological diversity, leading to their separation into hundreds of related species. Earlier studies suggested that the cichlid assemblages from the various lakes were monophyletic (Seehausen 2006). However, this inference likely reflected a lack of appropriate data for testing for the alternative signatures of paraphyly and monophyly (Joyce et al. 2011). In particular, sequences from mtDNA were often the only data sets used to define phylogenetic history. Sequences from a uniparentally inherited, non-recombining genome often provide a less sensitive assay for evolutionary pattern and process. This would likely be the case when an adaptive radiation was founded upon multiple lineages, particularly when hybridization was involved (Arnold 1997, 2006).

Joyce et al. (2011) highlighted the lack of a robust test of the hypothesis of monophyly for the adaptive radiation of cichlids in Lake Malawi. In this instance, the inference of monophyly was drawn using only mtDNA sequence data and without assays of key riverine taxa that could have participated in the founding of the Lake Malawi radiation. Thus, the evolutionary origin of possibly the largest radiation of cichlids in any East African Great Lake (i.e., up to 800 species formed) was only weakly tested. To identify the likely progenitor(s) of the Lake Malawi adaptive radiation, both nuclear (2045 polymorphic AFLP loci) and mitochondrial sequence (control region) data were collected from representative taxa (in terms of ecological associations and morphological characteristics) from the six previously-defined mtDNA clades as well as

17 cichlid populations from river systems of varying distances from Lake Malawi (Joyce et al. 2011). Genotyping of the riverine and Lake Malawi samples detected recombination among lineages belonging to multiple cichlid clades indicating a hybrid origin for this adaptive radiation. For example, the most diverse clade within the lake, that of the "rock-dwelling mbuna" possessed nuclear and mitochondrial genomes from highly divergent progenitor lineages (Joyce et al. 2011).

Consistent with the findings of Joyce et al. (2011), Loh et al. (2013) detected extremely high levels of shared SNP polymorphisms (*c.* 100 out of 200 loci) between Lake Malawi species and cichlid species from across Africa (Figure 6.11). This shared variation supported the inference that the Lake Malawi assemblage was not monophyletic and, further, that riverine cichlid taxa had contributed significantly to the genomic makeup of Malawi species (Figure 6.11; Loh et al. 2013). Indeed, the entire, present-day African Rift Lakes, cichlid assemblage would seem to have originated from admixed palaeolake and riverine cichlid communities (Seehausen et al. 2003; Joyce et al. 2005; Schwarzer et al. 2012b; Nichols

et al. 2015). Therefore, as predicted by the hybrid swarm model (Seehausen 2004), the African Great Lake cichlids were founded in the context of introgressive hybridization followed by the rapid origin of hundreds of species with admixed genomes.

Heliconius

The Neotropical butterfly genus, *Heliconius*, has been a focus of evolutionary and ecological research due to its aposematic wing color patterns that are a paradigm of Müllerian mimicry (Benson 1972; Mallet 1986; Naisbit et al. 2002; Joron et al. 2006; Blum 2008; Chamberlain et al. 2011). The wing color patterns thus provide a greater level of protection for all of the associated *Heliconius* species due to the cumulative effects on predator behavior (Benson 1972). In addition, taxa belonging to the same clade often reflect "mimicry rings" in which the variation in wing patterning also results in some degree of reproductive isolation from closely related species (Kronforst et al. 2007; Rosser et al. 2014). Though additional factors likely contribute to reproductive isolation (Arias et al. 2012; Mérot et al. 2013), associations between wing color variation and both pre- and postzygotic reproductive isolation have been described, with loci affecting wing patterning and levels of natural hybridization being associated within the *Heliconius* genome (Merrill et al. 2011). This genetic linkage should impede the disassociation, through recombination and segregation, of the loci that affect the phenotypic traits (i.e., wing coloration) from those that result in pre- and postzygotic reproductive isolation (Salazar et al. 2010; Merrill et al. 2011).

Over the past several decades, the *Heliconius* species complex has, in addition to being a model of Müllerian mimicry and ecological speciation, become a paradigm for testing the significance of genetic exchange-mediated evolutionary change. For example, *Heliconius* species, like cichlids have become a focus of studies concerning the process of homoploid hybrid speciation (Salazar et al. 2005, 2010; Mavárez et al. 2006; Jiggins et al. 2008). In particular, *H. heurippa* has been identified as a hybrid derivative of *H. melpomene* and *H. cydno*. The original definition of this species recognized its admixed morphological and genomic characteristics (Salazar et al. 2005; Mavárez et al. 2006), both of which were synthesized through experimental,

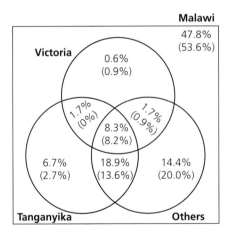

Figure 6.11 Overlapping SNP polymorphisms between Lake Malawi cichlids and other East African assemblages. The numbers in parentheses reflect values for non-CpG SNP loci (i.e., reflecting the elimination of loci that may have inflated observed levels of polymorphism due to increased mutation rates; Loh et al. 2013). 47.8% of Lake Malawi SNP loci are polymorphic only within endemic Malawi species; 8.3% are shared across Malawi, Tanganyika, Victoria, and other river haplochromine cichlid groups. Overall, 52.2% of Lake Malawi SNP loci share polymorphic alleles with cichlid lineages that do not belong to the endemic Malawi flock (from Loh et al. 2013).

introgressive hybridization between *H. melpomene* and *H. cydno* (see Jiggins et al. 2008 for a review). The admixed morphology of *H. heurippa* can be accounted for via adaptive trait introgression involving the alleles causing the red-banded *H. melpomene* phenotype onto the color background of *H. cydno* (Salazar et al. 2010). The genomic mosaic associated with the homoploid hybrid, *H. heurippa*, supports the hypothesis of adaptive transfer, with some of the genetic loci putatively affecting pattern formation of the forewing and thus an important ecological trait (i.e., predation avoidance; Salazar et al. 2010). Significantly, diversification due to genetic exchange has apparently not only affected individual lineages (such as *H. heurippa*), but the entire *Heliconius* complex as well (Kronforst et al. 2006; Dasmahapatra et al. 2012; Pardo-Diaz et al. 2012; Martin et al. 2013; Arias et al. 2014).

The inference of introgression-mediated diversification across the *Heliconius* assemblage resulted from the recognition of the importance of ecological speciation combined with the identification of pervasive ancient and recent genetic exchange (Figure 6.12). In this regard, Kronforst et al. (2006) and Arias et al. (2014) documented sequence variation at both mtDNA and nuclear loci for *H. cydno, H. pachinus*, and *H. melpomene* and *H. cydno, H. timareta, H. heurippa*, and *H. melpomene*, respectively. The gene flow parameters estimated from their data indicated extensive, though often asymmetric, intra- and interspecific introgression (Figure 6.12; Kronforst et al. 2006; Arias et al. 2014). Along with other processes (e.g., geoclimatic events), admixture was seen as a driving force for the biodiversification within and between the different mimetic races and species (Figure 6.12; Kronforst et al. 2006; Arias et al. 2014).

Recent analyses of the genes underlying mimetic phenotypes, in comparison to variation within whole-genomes, have provided additional opportunities to test for the role of introgression in the adaptive radiation of *Heliconius*. For example, Hines et al. (2011) uncovered sequence diversity around the locus controlling the red color-pattern in *H. erato* and *H. melpomene* (i.e., "*optix*") suggesting "that the red-rayed Amazonian pattern evolved recently and expanded, causing disjunctions of more ancestral patterns." Similarly, genomic resequencing demonstrated genetic exchange among *H. melpomene, H. timareta*, and *H. elevatus*; this exchange was particularly well-defined for two genomic segments containing loci affecting mimicry patterns (Dasmahapatra et al. 2012). Taken together, the analyses of *Heliconius* spanning many decades have provided evidence in support of the hybrid swarm hypothesis. In particular, the loci underlying the adaptive Müllerian mimicry traits demonstrate a pattern of "promiscuous exchange" facilitating the radiation of this New World butterfly complex (Kronforst et al. 2006; Dasmahapatra et al. 2012; Pardo-Diaz et al. 2012; Martin et al. 2013; Arias et al. 2014).

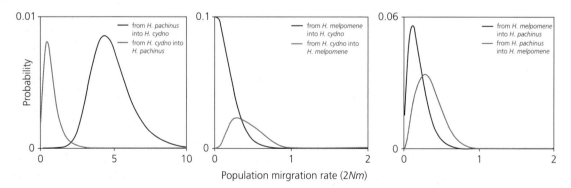

Figure 6.12 Estimates of introgression between three species of New World *Heliconius* butterflies; *H. pachinus, H. cydno*, and *H. melpomene*. The methods of Nielsen and Wakeley (2001), Hey and Nielsen (2004), and Won and Hey (2005) were used to calculate the probability of past genetic exchange between each pair of species. The results are consistent with extensive introgression from *H. pachinus* into *H. cydno* (left panel). Lower frequencies of genetic exchange were detected from *H. cydno* into *H. pachinus* (left panel), and from *H. cydno* and *H. pachinus* into *H. melpomene* (middle and right panel, respectively) (from Kronforst et al. 2006).

6.3 Genetic exchange and organismic evolution: Introgression, hybrid speciation, and adaptive radiations in plants

Unlike animal clades, very few arguments have been constructed suggesting that genetic exchange among plant lineages is an unimportant evolutionary phenomenon (but see Wagner 1970; Mayr 1992). Instead, plants have been used as exemplars for all of the sub-processes and diverse outcomes predicted by the web-of-life metaphor. These include prokaryotic-like, horizontal gene transfer events (e.g., Bergthorsson et al. 2003; Yoshida et al. 2010; Li et al. 2014; see also Section 1.7.3), introgressive hybridization, hybrid speciation, and adaptive radiations. The last three of these categories will be discussed later. One cautionary note that should be sounded is that these "categories" are not necessarily discrete in terms of their genomic and/or phenotypic effects. Thus, introgressive hybridization often accompanies hybrid speciation, even in the instances where allopolyploid lineages—presumed to be instantaneously-formed and reproductively isolated from their progenitors and other polyploid lineages—originate (e.g., Wendel et al. 1995; Williams and Arnold 2001; Bendiksby et al. 2011; De Hert et al. 2011; Lipman et al. 2013). Likewise, if natural hybridization underlies adaptive radiations then, by definition, the burst of diversification reflects a form of hybrid speciation (Seehausen 2004). In the following sections I will highlight the same types of processes detected in the animal taxa discussed in the preceding sections. Though some of the classes discussed will not be covered (e.g., the origin of hybrid, asexual, plant lineages; but see van Dijk 2003 and Beck et al. 2011 for discussions of such processes in plants), the range of outcomes included will hopefully reflect the extraordinary complexity of evolutionary products originating from genetic exchange among divergent plant taxa.

6.3.1 Genetic exchange and plant evolution: Divergence-with-introgression

The evolutionary histories of a majority of plants fit a model of divergence-with-gene-flow. This may seem like too strong a statement, but if it is accurate

that most plants and all angiosperms contain genomes marked by hybridization-derived polyploidy of some type (Soltis and Soltis 2009), then the majority of plants have indeed been affected by genetic exchange during their evolutionary history. With the increasing evidence from genome sequencing exercises, it is unlikely that the estimates of WGD due to hybridization are overestimated. However, even if they are somewhat inflated, the plant clade *sensu lato* is demonstrably characterized by hybrid speciation and introgressive hybridization (Anderson and Stebbins 1954; Arnold 1992, 1997, 2006; Rieseberg 1997; Doyle et al. 2008). I have previously summarized work on various plant clades that demonstrate divergence accompanied by introgression with related lineages (e.g., annual sunflowers and Louisiana irises). In this section, I will briefly discuss results from studies leading to the inference of genetic exchange during the divergence of tree species belonging to the genera *Populus* (poplars and aspens), *Picea* (spruces), and *Fraxinus* (ashes).

Populus

A survey of *c.* 38,000 SNP loci, across the genomes of the naturally hybridizing species *Populus alba* and *P. tremula*, allowed Stölting et al. (2013) a means of testing "the genomic landscape of divergence in taxa with 'porous' species boundaries." With regard to the present topic, these data were the basis of analyses describing patterns of genomic divergence associated with introgression. Figure 6.13 illustrates estimates of the "porosity" of different genomic regions based on Wright's F_{st} values (Wright 1951; Holsinger and Weir 2009; i.e., reflecting levels of differentiation between *P. alba* and *P. tremula*). The distribution of F_{st} estimates (Figure 6.13) was consistent with the hypothesis that introgression between these species had differentially affected loci scattered across the two genomes. For example, regions found on linkage groups VI, VIII, and X were demarcated by extremely high F_{st} values in certain blocks (Figure 6.13). In fact, the average multilocus F_{st} value (i.e., 0.634) between the two species reflected highly divergent genomes (Stölting et al. 2013).

In contrast to both the average [high] levels of divergence between *P. alba* and *P. tremula* and those regions of their genomes demonstrating even greater

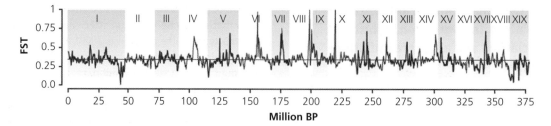

Figure 6.13 Interspecific genomic divergence, as estimated using Wright's F_{st} values (Wright 1951) between *Populus alba* and *P. tremula*. Grey and white vertical blocks indicate chromosomes I–XIX (from Stölting et al. 2013).

differentiation, were segments characterized by low F_{st} values (i.e., Chromosomes I, III, XVIII, XIX; Figure 6.13). Surprisingly, one of the most marked regions of low differentiation—indicative of high frequencies of introgression—involved the incipient sex chromosome XIX (Figure 6.13; Stölting et al. 2013). This result was unexpected given the general finding from both plants and animals of greater interspecific divergence at sex chromosome loci, reflected in a greater contribution by these loci to reproductive isolation, relative to genes located on autosomes (e.g., Brothers and Delph 2010; Ruegg et al. 2014; Yoshida et al. 2014). Though Stölting et al. (2013) discussed additional processes (such as selection favoring similar alleles on the sex chromosomes of both species) to account for the greater than expected interspecific similarity of loci on the *Populus* sex chromosome, the genomic data supported the occurrence of "rampant introgression" along this incipient sex chromosome. Notwithstanding the various contributing factors, it is clear that divergence-with-introgression has characterized the evolutionary histories of *P. alba* and *P. tremula*.

Fraxinus

Species of ashes (genus *Fraxinus*) span the temperate and subtropical regions of North America, Asia, and Europe, with most taxa occurring in the North American and eastern Asian portions of the distribution (Wallander 2008). Introgression and polyploidization within the genus have been repeatedly identified leading to numerous difficulties in delimiting taxa (Wallander 2008; Hinsinger et al. 2013). For example, a recent analysis of both AFLP and ribosomal RNA loci for the species belonging to the

section *Fraxinus* (*F. angustifolia, F. excelsior, F. nigra,* and *F. mandshurica*) detected phylogenetic discordance (Hinsinger et al. 2014).

The lack of phylogenetic resolution reported for the section *Fraxinus* species was not unexpected given previous genomic analyses indicating ongoing introgressive hybridization (e.g., Jeandroz et al. 1995; Fernandez-Manjarres et al. 2006). What *was* novel was the inference of multiple, ancient genetic exchange events between various members of this clade, including currently non-overlapping European and Asian species (*F. angustifolia / F. excelsior* and *F. mandshurica*) and European and North American species (*F. angustifolia / F. excelsior* and *F. nigra*; Hinsinger et al. 2014). The various reticulation episodes occurred during the Miocene and Pliocene, and continue in present-day hybrid zones (Hinsinger et al. 2014).

Picea

As with the analysis of *Populus*, a survey of SNP variation at 290 nuclear loci in the spruce species, *Picea glauca* and *P. engelmannii*, provided evidence of long-term introgressive hybridization. By genotyping individual trees from both allopatric populations of the two species and plants from a hybrid zone, De La Torre et al. (2014a) detected "extensive admixture and introgression . . . in the contact zone, with most alleles being shared by white spruce, Engelmann spruce and their hybrids." This evidence of high levels of introgression stood in contrast to the well-differentiated nature of *P. glauca* and *P. engelmannii*, suggesting that though divergence-with-gene-flow had contributed greatly to the genomic constitution of individuals in contact zones across time, the two taxa were indeed well-defined species (De La Torre et al. 2014a).

One of the most significant findings derived from the study of De La Torre et al. (2014a)—with regard to testing for divergence accompanied by introgressive hybridization—was that genetic exchange among the diverging *P. glauca* and *P. engelmannii* occurred over an extended time period. Thus, though the two species are thought to have diverged from a common ancestor *c.* 5 mya (i.e., during the Pliocene; Lockwood et al. 2013), they likely exchanged genetic material during repeated periods of overlap, with the most recent occurring across the last 21,000 years (Lockwood et al. 2013; De La Torre et al. 2014a). Such an extended period of introgressive hybridization, not only between these spruce species, but others as well (e.g., *P. likiangensis* and *P. purpurea*; Du et al. 2011), is a likely contributor to the lack of monophyletic relationships for species such as *P. engelmannii* (Lockwood et al. 2013). Therefore, as with *Populus* and *Fraxinus*, the ash genus demonstrates yet another example of divergence in the face of repeated episodes of introgression. In the following section, I will again use *Picea* to illustrate the web-of-life metaphor, in this case, the outcome of homoploid hybrid speciation.

6.3.2 Genetic exchange and plant evolution: Homoploid hybrid speciation

As reflected by Yakimowski and Rieseberg (2014), "while homoploid hybridization was viewed as maladaptive by zoologists, the possibility that it might play a creative role in evolution was explored and debated by botanists during the evolutionary synthesis." More recently, the occurrence of hybrid speciation resulting in reticulate taxa at the same ploidal level as the parental lineages has been confirmed for numerous plant and animal taxa (Arnold 1997, 2006; Rieseberg 1997; Mallet 2007; Yakimowski and Rieseberg 2014).

Various mechanisms have been suggested to help in the maintenance of a new homoploid, hybrid species in the face of potential introgression with parents and other hybrid lineages. Both intrinsic (e.g., chromosome structural differences, gene-gene interactions, mating system polymorphisms) and extrinsic (e.g., niche differences) processes have thus been implicated in the origin and stabilization of hybrid taxa (Stebbins 1959; Grant 1981;

Mallet 2007; Yakimowski and Rieseberg 2014). Various findings for two plants systems that have been used in defining the processes associated with the evolution of homoploid hybrid species have been presented in previous chapters; the two complexes were the Tibetan pines (Section 1.6.4) and the annual sunflowers (Section 4.6.2). An additional example of apparent homoploid hybrid speciation associated with the Tibetan ecosystems involves the spruce species, *Picea purpurea*.

The genus *Picea* shares a number of characteristics with the ashes clade: (1) spruce species are widely distributed, occurring in North America, Europe, and Asia; (2) the largest number of species is found in Asia; and (3) this genus has been problematic for those interested in producing taxonomic treatments (see Ran et al. 2006 for a discussion and additional references). Once again, at least a partial role must be assigned to reticulate events for the difficulty experienced when attempting to assign various samples of *Picea* into named categories. In this regard, *P. purpurea* has been the focus of much debate due to conflicting phylogenetic signals resolved from different data sets. For example, Li et al. (2010) examined the phylogenetic placement and nuclear, genomic constitution of *P. purpurea* in order to test the hypothesis that it originated from hybridization between *P. likiangensis* and *P. wilsonii*. Though results from this analysis did indeed suggest a role of reticulate processes in the evolutionary history of *P. purpurea*, Li et al. (2010) favored a model including the derivation of this taxon from a common ancestor shared with a third species (i.e., *P. schrenkiana*), followed by episodes of introgressive hybridization with *P. likiangensis* and *P. wilsonii*. Likewise, analyses of both mtDNA and cpDNA haplotypes present in allopatric and sympatric populations of *P. purpurea* and *P. likiangensis* found patterns consistent with the introgression of the cytoplasmic genomes in present-day hybrid zones, but not into areas of allopatry (Du et al. 2011). Though agreeing with a model of introgression-mediated sharing of cytoplasmic haplotypes between *P. purpurea* and *P. likiangensis*, Zou et al. (2012) also detected genomic variation suggestive of long-distance introgression of cpDNA from the former species into allopatric populations of the latter taxon.

A recent study by Sun et al. (2014) once again attempted to test the hypothesis of a homoploid hybrid origin for *P. purpurea*. Unlike the previous studies, Sun et al. (2014) utilized sequence information from nuclear, mitochondrial, and chloroplast loci for *P. purpurea*, *P. schrenkiana*, *P. likiangensis*, and *P. wilsonii*. In addition, they utilized environmental data and ecological modeling in order to define the niches occupied by each of the four species. The combination of the genomic and ecological data sets for each species allowed a multifaceted examination of the predictions provided by the homoploid hybrid speciation model. Consistent with this model, Sun et al. (2014) defined 69% and 31% of the nuclear genome of *P. purpurea* as originating from *P. likiangensis* and *P. wilsonii*, respectively. Following its origin from natural hybridization between these two species, asymmetric introgression of the organellar genomes from *P. wilsonii* resulted in *P. purpurea* possessing both mtDNA and cpDNA from *P. wilsonii* (Sun et al. 2014). Finally, the ecological niche modeling involving the two apparent progenitors and the putative homoploid hybrid found that they each occupied unique habitats. It would appear that the formation of *P. purpurea* did likely involve a more complex evolutionary history than would be suggested by a single reticulate evolutionary event resulting in a homoploid hybrid species. Instead, there were multiple genetic exchange events contributing to the unique genomic and ecological characteristics displayed by this hybrid spruce species.

6.3.3 Genetic exchange and plant evolution: Allopolyploid speciation

Whole-genome duplication events have been detected in the evolutionary history of all eukaryotic lineages (e.g., see Arnold 2006 for a review and references). Thus, as reflected by Doyle et al. (2008), "it is now known that flowering plant genomes are fundamentally polyploid." Furthermore, it is now accepted that the majority of plant clades likely underwent multiple rounds of allopolyploidy (Soltis and Soltis 2009; Symonds et al. 2010). In fact, the evolutionary history of the model "diploid" plant, *Arabidopsis thaliana*, included a minimum of two WGD events, the first apparently involving hexapolyploidy (Bowers et al. 2003; Jaillon et al. 2007). Findings such as those for *A. thaliana* indicate the degree to which plant speciation reflects a reticulate, rather than a purely divergent, evolutionary pattern (Soltis and Soltis 2009). This reticulate pattern of evolution is illustrated well by instances of both ancient and recent allopolyploid speciation—termed "palaeopolyploids" and "neoallopolyploids," respectively—with neoallopolyploids commonly being formed multiple times from hybridization events between the same diploid parents (see Doyle et al. 2008 and Soltis and Soltis 2009 for further discussion and references). I have used a number of plant examples in previous chapters (e.g., cotton, Chapter 1; *Tragopogon*, Chapter 2) to illustrate the process of allopolyploid speciation. This evolutionarily ubiquitous process is also well illustrated by findings from the plant genus *Glycine*, particularly those from the research groups of Doyle and Brown.

Since the 1980s Doyle, Brown, and their colleagues have reported on a detailed series of investigations into the evolutionary history of the perennial sister clade to the cultivated soybean, genus *Glycine* (Sherman-Broyles et al. 2014). This complex belongs to the subgenus *Glycine* and was originally recognized as consisting of 12 diploid ($2n = 40$) and three allopolyploid ($2n = 80$) species (Doyle et al. 1990a, b; Figure 6.14). The "diploid" lineages were also recognized as ancient palaeopolyploids (forming *c.* 15 mya), based on the chromosome number for the legume clade to which *Glycine* belongs being $2n = 20, 22$ (Doyle et al. 2004). Analyses of both nuclear and cytoplasmic (i.e., cpDNA) variation among the diploid progenitors and the allopolyploid derivatives detected the typical pattern of multiple origins of the neoallopolyploid taxa, with their appearances occurring within the past several hundred thousand years (Doyle et al. 1990b, 1999; Bombarely et al. 2014; Figure 6.14). Given that the diploid taxa all occur within the confines of Australia and Papua New Guinea (González-Orozco et al. 2012), it is most parsimonious to assume that the multiple origins of the allopolyploid species occurred in this region (Doyle et al. 1999). Recent treatments have refined the taxonomy of the subgenus *Glycine* complex to include > 25 diploid ($2n = 38, 40$) and eight allopolyploid ($2n = 78, 80$) species (Figure 6.14; see

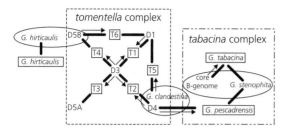

Figure 6.14 The *Glycine* diploid/allopolyploid species complexes of *G. tomentella*, *G. tabacina*, and *G. hirticaulis*. The boxes denoted by dashes and dashes/dots enclose the *G. tomentella* and *G. tabacina* clades, respectively. The *G. hirticaulis* complex is illustrated at the left of the *G. tomentella* box. The letter/number designations within the *G. tomentella* complex indicate *Glycine* races (i.e., species not recognized taxonomically). Those not enclosed within the boxes are diploid taxa; allopolyploid races fall within a box. Likewise, for the *G. tabacina* complex, the boxes indicate allopolyploid taxa. Similarly, the diploid and allopolyploid *G. hirticaulis* complex races are indicated by the absence or presence of boxes. Ovals surround diploid taxa that possess the same diploid genome. Solid lines connect allopolyploids with their diploid progenitors and arrows indicate the cpDNA donor species (from Doyle et al. 2004).

Harbert et al. 2014 for references). Furthermore, the recent studies have provided even greater resolution of the patterns and effects of hybridization and WGD on the genomic, biogeographic, and phenotypic constitutions of the neoallopolyploids.

Coate et al. (2012) documented transgressive phenotypes for photosynthetic-related traits in the recently derived allopolyploid, *G. dolichocarpa*. In particular, 17 of 21 traits measured in this allopolyploid were found to exceed those of either of the diploid progenitors (i.e., *G. tomentella* and *G. syndetika*; Coate et al. 2012). Consistent with an adaptive photosynthetic phenotype deriving from its allopolyploid origin, a subsequent study demonstrated that *G. dolichocarpa* possessed a higher capacity for photoprotection under high light intensity (Coate et al. 2013). Also consistent with an adaptive effect from allopolyploidization were the findings by Harbert et al. (2014) from climate niche modeling for five diploid progenitors and their four, neopolyploid derivatives. In this case, though the allopolyploid species did not demonstrate uniformly greater geographic distributions, all four did occur in niche space "not climatically available to their progenitors" (Harbert et al. 2014). Taken together, the results from studies of the perennial *Glycine* species complex reflect the complex nature of

allopolyploidization, both in terms of the diversity of origins (Figure 6.14) and the evolutionary/ecological trajectories of the hybrid lineages.

6.3.4 Genetic exchange and plant evolution: Adaptive radiations

In Section 6.2.4 I discussed two examples of animal adaptive radiations underlain by genetic exchange, in particular, introgressive hybridization. As Abbott et al. (2013) inferred, "promotion of adaptive divergence as a result of introgression may be much more common and have the potential to lead to increased reproductive isolation between populations." However, as they also observed, "systematic tests which conclusively distinguish introgressed alleles from shared polymorphisms are needed extending beyond cases where there are initial phenotypic clues (such as in butterfly wing patterns) and specifically addressing the role of introgression in adaptive radiation." There are numerous methodologies that allow relative rigor in the assignment of shared polymorphisms to the processes of coalescence or genetic exchange (see Chapter 3, Section 3.2) thus providing a solution for that portion of Abbott et al.'s (2013) systematic tests. More difficult by far is the resolution of whether a particular adaptive radiation has been catalyzed by genetic exchange between divergent lineages. In this topic area, evolutionary zoologists have led the way in designing both models and procedures for testing those models (see Seehausen 2004; Abbott et al. 2013; Seehausen et al. 2014, as well as Section 6.2.4). There are, however, exemplars from the botanical literature as well that provide support for the role of genetic exchange *sensu lato* in contributing to adaptive radiations. I will review data from two such groups, both components of island floras, the first from the Hawaiian archipelago and the second from Lord Howe Island located in the Tasman Sea.

Hawaiian silversword assemblage

The Hawaiian silversword assemblage, as with the African Rift Lake cichlids, is known as a paradigm of the process of adaptive radiation (Witter and Carr 1988; Robichaux et al. 1990; Baldwin and Sanderson 1998). This clade includes 30 species belonging to three genera that are endemic to six of the eight main Hawaiian islands. These species occur across widely varying ecological settings from lava flows to bogs, and demonstrate widely varying growth forms including cushion plants, shrubs, trees, and lianas. The sister clade to the Hawaiian silversword complex is the North American tarweeds. Unlike the diploid tarweed progenitors, silversword taxa possess polyploid genomes (Barrier et al. 1999, 2001). At a minimum, for the adaptive radiation of the silverswords to be consistent with Seehausen's hybrid swarm model, the derivation of the ancestral lineage(s) that invaded the Hawaiian archipelago would necessarily need to have been allopolyploid (i.e., hybrid). Furthermore, support for a causal association of natural hybridization and adaptive radiation in this group would also require evidence that genomic changes could have affected biological characteristics leading to the diversification of the many lineages into divergent niches.

Like other adaptive radiations (e.g., the Cichlidae and *Heliconius*), silversword species have the ability to form natural hybrids. In fact, 35 different combinations of interspecific natural hybridization have been reported (Carr and Kyhos 1981; Carr et al. 1989). Furthermore, hybrid speciation and/or widespread introgression among various members of the silversword genera are necessary to explain patterns of phylogenetic discordance (Baldwin et al. 1990). For example, Lawton-Rauh et al. (2007) detected differential introgression between two species of *Dubautia* (i.e., *D. ciliolata* and *D. arborea*) in comparisons of structural genes and floral regulatory loci. Alleles at the regulatory loci introgressed asymmetrically, from *D. arborea* into *D. cilioata*, with bidirectional introgression detected at a structural locus (Lawton-Rauh et al. 2007). In a second study, Friar et al. (2008) likewise inferred genetic exchange-affected evolution of a *Dubautia* species. By comparing nuclear and cpDNA sequence data to those from cytogenetic studies, Friar et al. (2008) examined the evolutionary history of *D. scabra* to test its means of origin and its relationship to other members of this silversword genus. Results from their analysis of the genomic data led to the conclusion that this species was of ancient hybrid origin. Interestingly, the derivation of this hybrid taxon reflects homoploid hybrid speciation between lineages that, being members of the silversword

alliance, were polyploid. The specific steps leading to the origin of *D. scabra* posited by Friar et al. (2008) included: (1) the origin of one progenitor on the island of Kauai; (2) the dispersal of this progenitor onto younger islands of the Hawaiian archipelago; (3) hybridization between this progenitor and a second related species resulting in the homoploid hybrid, *D. scabra*; and (4) the widespread dispersal of the homoploid hybrid across Maui Nui and Hawaii.

Given the propensity for species of this complex to hybridize, and the detection of hybrid species such as *D. scabra*, it is not surprising that the Hawaiian silversword genera are the products of allopolyploid speciation. In particular, the silversword lineages arose from a hybridization event between North American species of tarweeds from the genus *Raillardiopsis*, and possibly also *Madia* (Baldwin et al. 1991; Barrier et al. 1999). Barrier et al. (1999, 2001) and Lawton-Rauh et al. (2003) designed analyses that allowed tests of whether the adaptive radiation of this Hawaiian endemic clade might have been catalyzed by the initial allopolyploidy event. Barrier et al. (1999, 2001) and Lawton-Rauh et al. (2003) thus examined sequence variation at a set of floral homeotic gene loci to construct phylogenetic relationships among silversword and tarweed lineages and to define patterns of molecular evolution. The findings from the first analysis identified 2–3 tarweed lineages that apparently contributed to the extant silversword complex thereby confirming an allopolyploid hybrid speciation event at the base of this adaptive radiation (Barrier et al. 1999). The latter two analyses addressed the molecular evolutionary patterns of the same floral homeotic genes studied by Barrier et al. (1999). Barrier et al. (2001) detected a signature of accelerated evolution at these regulatory loci following the formation of the hybrid lineages. Specifically, they found an increase in nonsynonymous versus synonymous base pair substitutions in these homeotic genes following allopolyploidization (Barrier et al. 2001). Lawton-Rauh et al. (2003) likewise detected allelic variability at these loci indicative of non-neutral protein evolution. Given the low levels of genetic differentiation detected between members of the silversword species (e.g., Witter and Carr 1988), changes at such regulatory loci may be of prime importance in their diversification (Lawton-Rauh et al. 2003). These findings, though not conclusive concerning a role for the hybridization event in the adaptive radiation of the silversword alliance (Lawton-Rauh et al. 2003), are suggestive of such a causal link (Barrier et al. 2001). Likewise, the rapidity of the ecological diversification observed for this group is also consistent with that predicted for genetic exchange-mediated adaptive diversification (Seehausen 2004).

Lord Howe Island flora

Though its total area does not exceed 1500 hectares (c. 11 km long and 2.8 km wide), Lord Howe Island supports 100s of endemic plant and animal species, with this rich biodiversity having accrued in less than seven million years—i.e., since its origin through a volcanic eruption (Department of Environment and Climate Change, New South Wales, Australia 2007). Of relevance for defining processes predicted by the web-of-life metaphor, the flora of this Tasman Sea island have become models of divergence-with-gene-flow. For example, Savolainen et al. (2006) and Babik et al. (2009) provided substantial evidence for the sympatric divergence of two species of palm *Howea belmoreana* and *H. forsteriana*. Papadopulos et al. (2011) extended this initial test for sympatric diversification in an analysis of phylogenetic, cytogenetic, and ecologic data from the entire flora of Lord Howe Island. This latter analysis provided evidence that 10% of the endemic plant species had originated while exchanging genes with other lineages (Papadopulos et al. 2011). Furthermore, c. ten indigenous plant lineages were categorized as hybrid species based on the presence of admixed nuclear genomes (Papadopulos et al. 2011).

Recently, additional analyses of the Lord Howe Island flora have focused on the role of environment-dependent selection on the diversification of the various components, thereby providing a test for genetic exchange-catalyzed adaptive radiations within this assemblage. To this end, Papadopulos et al. (2014) defined "the contributions of isolation by environment (IBE) and isolation by community (IBC)" to the genomic and evolutionary trajectories of 19 plant species representing a diverse array of families. Significantly, all of the species demonstrated signatures of divergence caused by ecological selection. These indicators of ecological speciation included such aspects as

divergence in spite of a lack of geographical separation, the detection of divergent selection at specific loci, and the competitive exclusion of congeners (Papadopulos et al. 2014). The central conclusion taken from this study was that ecologically driven, sympatric speciation was prevalent in this floral complex. This reflects one of the predictions of the web-of-life metaphor, that divergent adaptive evolution can occur in the presence of ongoing genetic exchange. Papadopulos et al. (2013) extended such a prediction to the case of adaptive radiations of some of the Lord Howe Island plant groups (*à la* Seehausen 2004). They thus stated: "Colonization of new niches, partly fuelled by the rapid generation of new adaptive genotypes via hybridization, appears to have resulted in the adaptive radiation in *Coprosma* – supporting the 'Syngameon hypothesis.'" Once again, it is apparent that not only individual lineages, but also whole clades and groups of clades can originate via genetic exchange.

6.4 Conclusions

As illustrated in this chapter, genetic exchange is responsible for the origin of novel evolutionary lineages. The preceding examples reflect the taxonomic diversity of animals and plants that have originated from web-like processes. However, it is equally important to recognize that the diversity of processes underlying the multiplication of lineages is equally great. From whole-genome duplications to adaptive radiations, the number of categories of genetic exchange-mediated lineage multiplication continues to increase. It seems a reasonable prediction that as the number of genomic and ecological analyses continues to multiply so will the recognized mechanisms leading to web-like processes. However, regardless of whether or not additional processes contributing to reticulate evolution are detected, the pattern already identified in nature is indicative of a fundamentally important role for the well-defined processes that lead to genetic exchange. As shown, this is true whether we are considering plant or animal groups. Furthermore, as demonstrated in previous chapters, this is likewise an accurate conclusion for viral and prokaryotic clades as well (e.g., Nelson-Sathi et al. 2015).

CHAPTER 7

Genetic exchange and conservation

"Owing to its much smaller gene pool, the Black Duck is vulnerable to eventual swamping through hybridization and introgression" **(Johnsgard 1967)**

"Hybridization between species or between disparate source populations may serve as a stimulus for the evolution of invasiveness." **(Ellstrand and Schierenbeck 2000)**

"Controversy has surrounded the setting of appropriate conservation policies to deal with hybridization and introgression. Any policy that deals with hybrids must be flexible and must recognize that nearly every situation involving hybridization is different enough that general rules are not likely to be effective." **(Allendorf et al. 2001)**

"Hybridization . . . is intrinsically neither good or bad . . . It is best treated as one of many natural evolutionary processes that have played an important role in shaping the biodiversity that conservation aims to protect" **(Detwiler et al. 2005)**

"Although Salix has a relatively high level of interspecific hybridization, this may not sufficiently explain the near complete failure of barcoding that we observed: only one species had a unique barcode . . . The most likely explanation for the patterns we observed involves recent repeated plastid capture events, aided by widespread hybridization and long-range seed dispersal, but primarily propelled by one or more trans-species selective sweeps." **(Percy et al. 2014)**

7.1 Genetic exchange and the conservation and restoration of endangered organisms

It is often the case that studies considering the effect of genetic exchange on the conservation of taxa emphasize a range of outcomes perceived as negatively impacting the survival of endangered forms (Smith et al. 2014). Some outcomes do indeed reflect risks to biodiversity, such as the reticulate evolution of highly virulent pathogens capable of decimating native populations and the presence of numerically or reproductively superior species within the ranges of rare congeners (Rhymer and Simberloff 1996; Wolf et al. 2001; Roberton et al. 2006; Burgess et al. 2008; Price and Muir 2008; Randi 2008; Farrer et al. 2011; Coleman et al. 2014; Balao et al. 2015; Kovach et al. 2015). However, I would argue that some of the potential outcomes from genetic exchange are less clearly negative. For example, Derr and his colleagues have argued for a number of measures to reduce the amount of domestic cattle genes in the various populations of the North American bison (Ward et al. 1999; Halbert et al. 2005; Halbert and Derr 2007; Hedrick 2009; Derr et al. 2012). The conceptual framework expressed by Derr et al., as in other cases to be discussed shortly (e.g., Felids), thus proposes maximizing genomic "purity/ integrity" (Ward et al. 1999; Halbert et al. 2005; Derr et al. 2012). I must admit to feeling conflicted concerning such a framework. On the one hand, I appreciate the care that needs to be taken to minimize the potential effects on the genotypic and phenotypic constitution of native species due to human-mediated genetic exchange with introduced forms. Indeed, *Bison bison* populations exemplify just such an outcome, with genomic and phenotypic alterations having occurred due to introgression from domestic cattle (Figure 7.1; Derr et al. 2012). Thus, if the conservation goal is to maintain a "bison" genomic and phenotypic bauplan or gestalt, the presence of the genes from a divergent taxon has had unacceptable consequences.

An alternative approach would be to involve recognizing the widespread occurrence of genetic exchange in all organismic clades and emphasizing (as I do) the potential adaptive benefits from genomic transfers. From this perspective, the recent introduction of cattle genes into bison could be argued to reflect the latest in a series of ancient and recent reticulate events, many not anthropogenically derived, impacting bison- and cattle-like lineages (e.g., Verkaar et al. 2004; Achilli et al. 2008; Zeyland et al. 2012; Decker et al. 2014). In this chapter, I will attempt to point to both real and perceived perils for endangered taxa from genetic exchange with related taxa. However, I will also illustrate why genetic exchange can be seen as both natural and often beneficial for rare forms (e.g., providing *genetic rescue* for rare forms; Whiteley et al. 2015). In the following discussion I will thus highlight cases in which introgression led to the genetic replenishment of populations with limited genetic variability or alternatively to the genomic and phenotypic assimilation of rare forms by more numerous congeners. Furthermore, I will argue that because evolutionary diversification is often reticulate, it is not relevant whether members of one evolutionary lineage exchange genes with another when attempting to determine a value for conservation. Rather, it should be considered whether genetic exchange can help or hinder the conservation and restoration

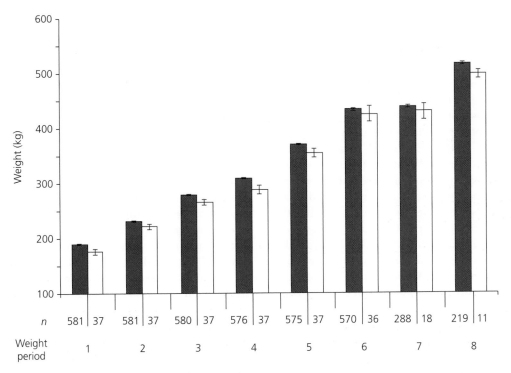

Figure 7.1 The mean weight (with SE bars) of young North American bison males in a single feedlot, measured at intervals of 90–120 days (*n* = sample size). The males possessed mitochondrial DNA from either bison or domestic cattle (shaded and unshaded bars, respectively). All differences in weight between animals with bison mtDNA and cattle mtDNA were statistically significant, with those bison possessing bison mtDNA significantly heavier than those bison males with the mitochondrial DNA from cattle (from Derr et al. 2012).

of manageable units (i.e., taxa; Sampson and Byrne 2012). Finally, without wanting to minimize the importance of understanding the effects of genetic exchange on native taxa, including those that are endangered (e.g., see Allendorf et al. 2004; Beebee 2005; McDonald et al. 2008; Ryan et al. 2009; Kraus et al. 2011), I would argue that the main factors threatening biodiversity revolve around human-caused habitat loss. Therefore, if we wish to conserve lineages and indeed entire assemblages, we must reduce the occurrence of anthropogenic modifications to ecosystems (Seehausen et al. 2006; Lerdau and Wickham 2011).

7.2 Introgressive hybridization and the conservation of endangered animals

Allendorf et al. (2001, 2004, 2005) have suggested a number of factors that should be taken into account for making conservation management decisions when hybrid organisms are present. In particular, they emphasized the need to decipher whether or not the production of admixed individuals was due to "natural" or "anthropogenic" causes. Furthermore, they argued that even in cases in which hybridization was due to human activities (e.g., translocation, habitat modification), it was necessary to carefully consider whether or not the hybrid individuals contained the last vestige of the genomic and phenotypic identity of the rare taxon (Allendorf et al. 2001, 2004, 2005). Yet, for most situations in which there remained "pure" populations of the endangered taxon, they recommended that if the genetic exchange was due to human-mediated factors, the hybrids should not be protected and should be removed from the populations to prevent their contributing to further admixture (Allendorf et al. 2004).

As with the North American bison example, the conclusions of Allendorf and his colleagues reflect important considerations enumerated by scientists who are passionate about protecting rare and endangered organisms. Thus, I cannot easily dismiss their conclusions regardless of the fact that they are based on a conceptual framework to which I do not hold. I do, however, believe that data from genomic analyses, including those involving taxa considered to be endangered (Ouborg et al. 2010; Milián-García et al. 2015),

suggest that (1) genetic exchange is likely to have been a part of the evolutionary history of most organisms and (2) human-catalyzed admixtures in organisms as diverse as viruses to mammals should not be considered "unnatural." The last of these conclusions is based upon not only the data indicating the widespread nature of reticulate evolutionary processes, but also on philosophical grounds, with biological and conservation implications. If we dismiss human-mediated exchange events as being of less importance, we run the real risk of being seen as suggesting that humans are not part of the natural world.

This argument (i.e., "2") can have enormous ramifications if governments, businesses etc. desire rationales for not conserving organisms. To illustrate this general danger, it is only necessary to consider the following quote from O'Brien and Mayr (1991) concerning the so-called hybrid policy instituted as a part of the United States government's Endangered Species Act: "Their opinions, referred to here as the Hybrid Policy, concluded that protection of hybrids would not serve to recover listed species and would likely jeopardize that species' continued existence." O'Brien and Mayr (1991) pointed to specific instances in which there was anthropogenically caused hybridization to illustrate that the hybrid policy would lead to no protection for the endangered, "hybrid" forms (e.g., the Florida panther). However, the ancient admixture of divergent lineages, with no human interference, often has outcomes that are identical to those from anthropogenic events (e.g., adaptive trait introgression, genetic swamping, variable hybrid fitness). This does not support the contention of some biologists that ancient and recent genetic exchanges reflect qualitatively different processes. For example, Placyk et al. (2012) and Orozco-Terwengel et al. (2013) detected cases of ancient introgression and genetic swamping between North American gartersnakes and Madagascar tomato frogs, respectively. As with studies that implicated human-mediated factors resulting in introgression and genetic assimilation, the findings (particularly for the Madagascar species) suggested a role for environmental perturbations as catalysts of the reticulate evolutionary processes (Orozco-Terwengel et al. 2013). Likewise, the following examples reflect the similarity of outcomes from genetic exchange obtained whether or not there has been anthropogenic input, suggesting the value of the continuing debate concerning conservation efforts involving admixed individuals, populations, and taxa.

7.2.1 Introgressive hybridization and the conservation of endangered animals: North American and Asian bears

A recent study of variation at Y-chromosome loci resolved patterns supporting "the emerging understanding of brown and polar bears as distinct evolutionary lineages that started to diverge no later than the Middle Pleistocene, at least several hundreds of thousands years ago" (Bidon et al. 2014). Likewise, an analysis of nuclear loci from *Ursus maritimus* (i.e., polar bears) and *U. arctos* (i.e., brown/grizzly bears) also found these two species (and the North American black bear, *U. americanus*) to be reciprocally monophyletic, with a divergence time of *c.* 600,000 ybp (Hailer et al. 2012). These findings provide further support for the genomic and phenotypic distinctiveness of these ursine species and for the continued necessity of conservation efforts. Yet, these taxa also reflect the significant role of both ancient and recent reticulate evolutionary processes (Figure 7.2; Edwards et al. 2011; Miller et al. 2012; Cahill et al. 2013).

Divergence-with-introgression among polar bears and brown bears was first demonstrated based upon mtDNA loci. Edwards et al. (2011) analyzed mtDNA sequence variation for a collection of individuals including samples of ancient and extant populations from across the ranges of *U. maritimus* and *U. arctos*. The phylogeographic modeling approach applied by Edwards et al. (2011) detected the effect of climate change on the past and present-day distributions of these two species. Furthermore, the pattern of sequence variation indicated that the mtDNA present in contemporaneous polar bear populations originated through ancient introgressive hybridization with Irish brown bears (Figure 7.2; Edwards et al. 2011). In addition to the replacement of the mitochondrial genome of *U. maritimus* by that of *U. arctos*, both autosomal and X-linked loci in these two species lead to the inference of ancient introgressive hybridization. In this regard, Cahill

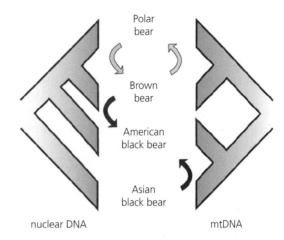

Figure 7.2 Hypotheses for ancient introgressive hybridization between species of North American and Asian bears to account for phylogenetic discordance between trees constructed using either mitochondrial or nuclear sequences. Arrows indicate putative introgression events between the different taxa involving either nuclear (left diagram) or mitochondrial loci (right diagram) (from Kutschera et al. 2014).

et al. (2013) concluded that brown bears located on the Alaskan ABC islands had mosaic genomes consisting mostly of brown bear alleles, but with <1% autosomal and *c.* 6.5% X-chromosomal alleles from polar bears. In order to account for the mosaic genomes of the ABC island individuals, Cahill et al. (2013) proposed that these populations were derived from polar bears isolated by receding ice sheets at the end of the last maximum glacial period followed by unidirectional, male-biased introgression from brown bears.

As discussed, the polar and brown bear assemblage is a clear example of ancient reticulate evolution resulting in present-day individuals possessing mosaic genomes. Furthermore, the recurrent introgression between *U. maritimus* and *U. arctos* has apparently been catalyzed by environmental fluctuations. However, these two taxa are not isolated examples of ursine lineages impacted by reticulate evolution. For example, analyses of autosomal loci by Miller et al. (2012) and Kutschera et al. (2014) detected patterns of phylogenetic discordance and estimates of ancient admixture indicating extensive introgression during the divergences of various North American and Asian species (Figure 7.2). Both studies confirmed previous analyses in their definition of ancient introgression between the polar bear and brown bear (Miller et al. 2012; Kutschera et al. 2014). In addition, these studies identified haplotype sharing and phylogenetic distributions suggesting that, subsequent to the divergence of the lineage leading to *U. americanus* and that resulting in *U. maritimus* and *U. arctos*, there was an extended period of introgression (Figure 7.2). Likewise, various introgressive hybridization events involving the North American black bear, the Asian black bear (i.e., *U. thibethanus*), and the sloth bear (*Melursus ursinus*) were hypothesized to account for the complex array of gene trees derived from nuclear (autosomal, Y-chromosomal) and mtDNA loci (Figure 7.2; Miller et al. 2012; Kutschera et al. 2014).

Some authors have suggested dire consequences for arctic species such as the polar bear due to climate change-catalyzed overlap and hybridization with their brown bear congener, thus resulting in the loss of species-specific genomic and phenotypic characteristics (Kelly et al. 2010). The findings from the above studies would agree that climate change has likely played an important role in shaping the evolution of the ursine clade. In fact, the evolutionary processes created by past environmental fluctuations, like present-day perturbations, have resulted in reticulate rather than bifurcating evolutionary patterning. Furthermore, the conclusion of Edwards et al. (2011) reflects well the dilemma faced by those who wish to conserve the bauplan perceived as species specific because "interspecific hybridization not only may be more common than previously considered but may be a mechanism by which species deal with marginal habitats during periods of environmental deterioration."

7.2.2 Introgressive hybridization and the conservation of endangered animals: Sharks

Marine organisms, from viruses to mammals, are well known for their propensity for genetic exchange (Arnold and Fogarty 2009; Lancaster et al. 2010; Attard et al. 2012; Ladner and Palumbi 2012; Luttikhuizen et al. 2012; Amaral et al. 2014; Gaither et al. 2014; Sakowski et al. 2014). Though often the genetic exchange is detected throughout the evolutionary history of the lineages involved (Riginos and Cunningham 2007; and see Arnold

and Fogarty 2009 for additional examples), conservation concerns may arise due to the risk of losing genotypic and phenotypic variation through alterations in habitats and/or human exploitation (Lancaster et al. 2007; Roberts et al. 2010; Attard et al. 2012).

A series of studies, involving shark species belonging to the genus *Carcharhinus*, illustrate well the need for understanding both past and contemporaneous introgressive hybridization for not only deciphering evolutionary processes, but also in order to make sound management decisions. For example, the analysis of mtDNA sequence variation in the pig-eye shark, *C. amboinensis*, from northern Australian coastal waters revealed two mitochondrial lineages (i.e., eastern and western forms) that likely reflected isolation and divergence during the Pleistocene epoch caused by sea level changes in the Torres Strait (Tillett et al. 2012). In contrast, sequence variation at nuclear loci led to the inference of "unrestricted genetic mixing" between the divergent mtDNA lineages (Tillett et al. 2012). Similarly, extensive genetic exchange (i.e., 45% of samples were identified as hybrid genotypes) between the blacktip whaler species, *C. tilstoni* and *C. limbatus*, has been detected along a 2000 km stretch of the eastern Australian coastline (Morgan et al. 2012). The extensive introgression between these two species (and likely other species, as well; Ovenden et al. 2010) occurs in spite of divergent reproductive ecologies reflected in length at maturity, asynchronous parturition, birth size, and the relative frequencies of newborns in coastal nurseries (Harry et al. 2012).

A variety of management issues, arising from the observation of introgression among the *Carcharhinus* species, were considered in the papers cited. One that was common to each study was the question of whether or not the taxonomic designation of the various species was accurate given that there were successful matings in nature. Furthermore, as pointed out by Morgan et al. (2012), if commonly occurring hybrid genotypes between the commercially exploited *Carcharhinus* species demonstrate reduced fitness, then estimates of population productivity would be inflated thus leading to overfishing. Alternatively, they argued that relatively high fitness of hybrids would lead to genetic assimilation and the concomitant reduction of biodiversity in this shark genus. It is likely that, as found in almost all other organismic groups studied, hybrid fitness will vary across habitats and generations (Arnold and Hodges 1995; Arnold and Martin 2010; Chapter 5 of this book). Thus, given the high frequency and extensive geographic distribution of admixed *Carcharhinus* sharks, hybrid fitness becomes a critical issue for conservation management. Morgan et al. (2012) reflected this conclusion when stating: "Now that hybrids can be identified, obtaining life history measurements to assess their capacity to reproduce, their susceptibility to parasites and general health compared to the parental species will assist in understanding whether the two species will be maintained or combined into one 'hybrid' species in the future."

7.2.3 Introgressive hybridization and the conservation of endangered animals: Felids

In discussing the conservation outlook for the South China tiger (*Panther tigris amoyensis*), Xu et al. (2007) reflected that even if populations remained in the wild, "fragmentation, habitat destruction, and prey shortages call the sustainability for any such remnant wild populations into question." Significantly, like the management steps instituted decades earlier for the Florida panther (*Puma concolor coryi*; O'Brien and Mayr 1991; Johnson et al. 2010), Xu et al. (2007) argued for genetic remediation through introgressive hybridization with a related taxon. As with *P. concolor*, the South China tiger is critically endangered and suffers from inbreeding depression. This observation, along with the detection of previously introgressed alleles from the northern Indochinese tiger, *P. t. corbetti*, provided impetus for the conservation plan that included further genetic enrichment of *P. t. amoyensis* from its congener (Xu et al. 2007). Though many (if not all) of the native felid species have critically reduced population numbers, conservation efforts do not often include proposals for human-mediated introgression with related lineages as instituted for the Florida panther. Instead, native × native and native × domestic introgressive hybridization is usually viewed as detrimental to the preservation of the genetic and phenotypic characteristics of wild felids (Oliveira et al. 2007).

Neotropical felids

Johnson et al. (1999) argued for the use of phylogenetic relationships to determine when to restrict genetic exchange between neotropical cat species for conservation purposes. Yet, natural introgression among these species—belonging to the genus *Leopardus*—has led to extensive and complex patterns of genomic mosaicism. Trigo and her colleagues have generated this latter inference through a series of studies that surveyed morphological traits and sequence variation at Y-chromosome, X-chromosome, autosome, and mitochondrial loci (Trigo et al. 2008, 2013, 2014). Included in their results were the observations of (1) introgression resulting in *L. tigrinus* possessing its own mtDNA haplotypes along with haplotypes from *L. geoffroyi* and *L. colocolo* (Trigo et al. 2008), (2) cytonuclear discordance in northern populations of *L. tigrinus* caused by ancient introgressive hybridization with *L. colocolo* (Trigo et al. 2013), (3) high levels of recent admixture in hybrid zones between *L. geoffroyi* and *L. tigrinus* (Trigo et al. 2013), and (4) complex, mosaic genomes in an area of sympatry between *L. geoffroyi* and *L. guttulus*, but with many parental-like phenotypes indicating advanced generation hybrids (Trigo et al. 2014). Furthermore, the discontinuity of the genotype and phenotype of many hybrid individuals, reflected in many hybrids that were indistinguishable from the parental species, suggested the action of selection that favored a parental, adaptive phenotype (Trigo et al. 2014).

Based on their genetic analyses, Trigo et al. (2008) suggested a number of conservation recommendations for the *Leopardus* species, including not utilizing captive bred individuals collected from the hybrid zone between *L. geoffroyi* and *L. guttulus* in restoration programs. However, given that some of the reticulate evolutionary history of this genus was inferred to be ancient and thus not associated with human activities (Trigo et al. 2013), it seems significant to also consider the possible innovative aspect of divergence-with-introgression in this complex. Therefore, conserving the evolutionary pattern (i.e., taxonomic units) of this species complex should also include conserving the processes that produced said pattern. One of these processes was introgressive hybridization.

Domestic cats and their wild progenitors

"Introgression is an important evolutionary force, which can lead to adaptation and speciation on one hand, but on the other hand also to genetic extinction. It is in the latter sense that introgression is a major conservation concern, especially when domestic species reproduce with their rare wild relatives." Thus, Nussberger et al. (2014) reflected specifically on the risk of losing the distinctive genotypic and phenotypic makeup of European wildcats (*Felis silvestris silvestris*) through introgressive hybridization with their domestic congener (*F. s. catus*). As with other wild and domesticated assemblages, the conservation concerns associated with that of *F. s. silvestris* and *F. s. catus* reflect a combination of anthropogenic effects (Randi 2008). First, humans have fragmented, and in some cases eradicated, wild *F. silvestris* populations through habitat modification and trapping efforts (Randi 2008). Second, domestic cats have been introduced across the geographic regions occupied by the endangered wildcats thus allowing for hybridization resulting in introgression between the two subspecies (Randi 2008; Nussberger et al. 2014).

In spite of the real risks from genetic assimilation of wild forms by their domesticated relatives, Driscoll et al. (2007), in an analysis of domestic cat evolution, pointed to the domestication of wild lineages in general as one of the most successful "biological experiments" accomplished by human populations. Sequence variation from a sample of 979 felids, including domestic cats, European wildcats, Near Eastern wildcats (*F. s. libyca*), central Asian wildcats (*F. s. ornata*), southern African wildcats (*F. s. cafra*), and Chinese desert cats (*F. s. bieti*) suggested a Near Eastern origin for *F. s. catus*. Furthermore, the haplotype diversity at mitochondrial loci among the various lineages indicated that the derivation of *F. s. catus* involved multiple maternal lineages. Also, like other domestication events (see Arnold 2009 for examples), domestic cat evolution was apparently accompanied by bouts of introgression with wild taxa. Figure 7.3 illustrates the phylogenetic associations of domestic cat lineages and their wild congeners from different geographic regions (based on nuclear and mitochondrial sequences; Johnson and O'Brien 1997;

Beaumont et al. 2001; Randi et al. 2001; Pierpaoli et al. 2003; Lecis et al. 2006; Driscoll et al. 2007; Oliveira et al. 2007). Significantly, *F. s. catus* samples from different regions were most closely related to wild *F. silvestris* from the same geographical area.

The findings reflected by Figure 7.3 suggest the contribution of ancient and recent introgression between the wild and domesticated forms of *F. silvestris*. In terms of contemporaneous genetic exchange, and thus conservation issues, Oliveira et al. (2007) examined the genetic variation present in both domestic cats and European wildcats from Portugal. As with earlier studies (Beaumont et al. 2001; Randi et al. 2001; Pierpaoli et al. 2003; Lecis et al. 2006) as well as the more recent study by Nussberger et al. (2014), Oliveira et al. (2007) detected significant introgression between *F. s. silvestris* and *F. s. catus*. In particular, a statistical analysis of microsatellite variation assigned 6 of 34 "wildcats" to the "domestic cat" genotypic category (Oliveira et al. 2007). This led to the conclusion common to other analyses that "hybridization is of major concern for the appropriate implementation of wildcat conservation strategies" (Oliveira et al. 2007).

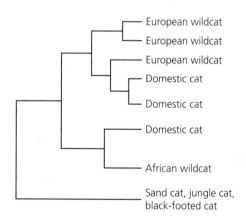

Figure 7.3 Phylogenetic relationships among various lineages of wildcats and their domesticated derivatives sampled from various geographical regions. The phylogenetic tree reflects a compilation of results from sequence analyses of nuclear and mitochondrial loci (Johnson and O'Brien 1997; Beaumont et al. 2001; Randi et al. 2001; Pierpaoli et al. 2003; Lecis et al. 2006; Driscoll et al. 2007; Oliveira et al. 2007; Nussberger et al. 2014). The closer association of domestic lineages with geographically proximate wildcat samples reflects a role for both ancient and contemporaneous introgressive hybridization.

The conservation concerns associated with genetic assimilation of native wildcats through introgression with the domestic cat appear to be well founded. For example, Driscoll et al. (2007) observed that introgression between *F. s. silvestris* and *F. s. catus* had resulted in the common occurrence of mtDNA from the latter in populations of the former. In addition, the presence of (1) feral, domestic cats, (2) wildcats with feral cat morphological traits, and (3) domestic cats with alleles typical of wildcats reflects past and recent introgression between wildcats and domestic cats. Though this introgression appears to be asymmetric (i.e., contemporaneous genetic exchange occurs mostly from domestic into wild populations), the *F. s. catus* lineages are nonetheless being impacted by wildcat genes and thus reflect not only an origin, but also a continued evolutionary trajectory affected by reticulation. Thus, if the conservation of genotypic and phenotypic traits important to cat breeders (Montague et al. 2014) is taken into account, the protection of wildcat gene pools is of increased significance. This may seem a counterintuitive argument, but such concerns drive much of the efforts to identify and conserve, for example, wild landraces from which crop plants have been derived (e.g., Pallotta et al. 2014).

7.2.4 Introgressive hybridization and the conservation of endangered animals: Trout

Fish clades, both freshwater and marine, have provided excellent examples of the role of reticulate evolutionary processes (e.g., Seehausen 2004; McDonald et al. 2008; Arnold and Fogarty 2009; Meraner et al. 2013; Seehausen and Wagner 2014; Sousa-Santos et al. 2014). Thus, Hubbs (1955) stated "During the first ten years of my intensive studies of the freshwater fishes of North America, from 1919 to 1929, I gathered strong circumstantial indications that the species lines are rather often crossed in nature, and during the following fifteen years, from 1929 to 1944, I was able, with the constant aid of Mrs. Hubbs, to confirm these indications." Hubbs (1955) reached this conclusion on the basis of their analysis of *c.* one million specimens each of freshwater and marine fishes, as well as from results of studies by other fish biologists. In addition to illustrating the web-of-life metaphor in

general, fish taxa likewise reflect the need for concern over possible genetic assimilation of endangered forms, particularly due to the translocation of congeners into the ranges of rare, native lineages.

Trout species (including New World species belonging to the genus *Oncorhynchus* and Old World members of the genus *Salmo*) represent a significant, worldwide industry that produces large amounts of revenue for both countries and local municipalities. For example, 20 states within the USA gained US$74.9 million from sales of trout eggs, fingerlings, etc. (National Agricultural Statistics Service 2007). Likewise, trout are utilized extensively in Europe as well. A study by Champigneulle and Cachera (2003) reported one example of their large-scale usage for the sport fishing industry. These authors estimated the effects from human mediated releases of ~500,000 trout annually on replenishment for angling along a 24 km stretch of the River Doubs. Though Champigneulle and Cachera (2003) concluded that natural recruitment, rather than artificial stocking, provided the majority of fish caught through angling, the possible effects on native species, in the face of such introductions, is concerning.

As with the sport fishing industry in Europe, that of North America has also resulted in repeated translocations of non-native species into the ranges of native congeners. One example of the possible outcomes of this type of management program, specifically with regard to genetic assimilation, is reflected by repeated introductions of the rainbow trout, *Oncorhynchus mykiss*, from Canada to Mexico (Allendorf et al. 2001, 2004; Escalante et al. 2014; Kovach et al. 2015). In particular, this species has been introduced into numerous regions within western North America, resulting in widespread introgression into previously isolated populations of the native cutthroat trout, *O. clarkii*. Often, extensive hybrid populations have been reported as a result of the human mediated overlap between these two species. For example, Metcalf et al. (2008) detected hybrid swarms caused by the introduction of *O. mykiss* into watersheds previously occupied only by either greenback or Colorado River cutthroat trout (i.e., *O. clarkii stomias* and *O. clarkii pleuriticus*, respectively). Yet, findings from other analyses have suggested that though admixture can indeed be extensive, the proportion of introgression in populations depends on various factors, including environmental setting (Figure 7.4; Yau and Taylor 2013). This latter observation suggests that the fitness of hybrid and parental genotypes varies across different niches (Figure 7.4), an inference supported by both variation in metabolic traits and the detection of "super invasive alleles" introgressing from *O. mykiss* into *O. clarkii* (Rasmussen et al. 2012; Hohenlohe et al. 2013). On the one hand, environment-dependent selection that favors certain hybrid genotypes suggests a benefit from hybridization due to the occupation of novel niches relative to the parental forms. However, as discussed, if the goal of conservation efforts is to prevent any degree of genotypic or phenotypic impact on endangered forms, the increased survivorship of any hybrids will be viewed as a threat.

7.3 Introgressive hybridization and the conservation of endangered plants

From a consideration of the conservation significance of plant populations located at the margins of a species range, JD Thompson et al. (2010) recommended that not only the peripheral populations, but also populations of rare forms that are introgressed with alleles from more numerous congeners should be seen as sources "of novel diversity that may have adaptive potential." Specifically, these workers found that the rare plant *Cyclamen balearicum* overlapped and hybridized on Corsica, with peripheral populations of the widespread congener *C. repandum*. The admixture between these two species resulted in genomically- and phenotypically-enriched

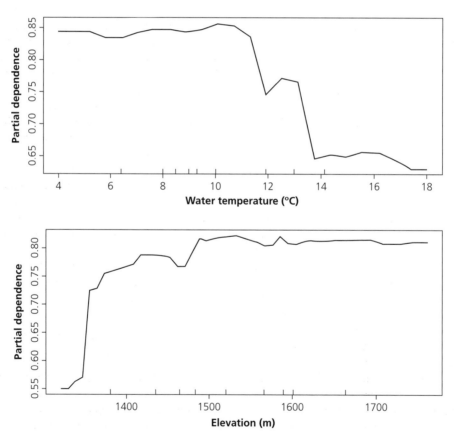

Figure 7.4 "Partial dependence plots" illustrating the correlation between the level of introduced rainbow trout × native cutthroat trout admixture and water temperature (top panel) and elevation (lower panel) across 58 sampling localities in southwestern Alberta, Canada. The plots indicate that both water temperature and elevation are associated with admixture levels, with introgression between the species being lowest in cooler streams and at higher elevations (from Yau and Taylor 2013).

populations relative to the parental forms. In concert with the *Cyclamen* observations, JD Thompson et al. (2010) pointed to the evidence for past climate change resulting in the overlap of species belonging to other Mediterranean plant clades, resulting in hybrid speciation. Given the effects of reticulate evolutionary processes across these clades, it was argued that the Corsican introgressed forms should be included in a conservation strategy to allow their adaptive potential to be assessed (JD Thompson et al. 2010).

Though not arguing for an adaptive benefit from genetic exchange, Conesa et al. (2010) did not identify a threat of genetic assimilation of the endangered, island endemic *Lotus fulgurans*, due to ongoing introgressive hybridization with a widespread congener (i.e., *L. dorycnium*) on the island of Minorca. An analysis of nuclear and cpDNA loci, along with morphological traits of collections from allopatric and sympatric populations of the two *Lotus* species, detected asymmetric introgression between the two species. However, the directionality of the gene flow was from the endangered *L. fulgurans* into the widespread *L. dorycnium* (Conesa et al. 2010). Thus, if there was a threat to the "genetic purity" of one of the species it was to the widespread, rather than the endemic, form. This observation suggested that the conservation of *L. fulgurans* would not be negatively affected by its overlap and introgression with *L. dorycnium* (Conesa et al. 2010).

In contrast to the potential value assigned to introgression by Thompson et al. (2010), and the lack of an apparent threat in the *Lotus* example (Conesa et al. 2010), most plant conservation biologists perceive the process as having largely (or solely) negative effects on endangered lineages. For example, Prentis et al. (2007) found evidence that asymmetric hybridization produced a significant advantage for the introduced *Senecio madagascariensis* relative to the Australian native *S. pinnatifolius*, with the potential of leading to the decline and eventual extinction of the native form. Similarly, Maschinski et al. (2010) determined that widespread introgression was occurring in the state of Florida between two endangered varieties of *Lantana depressa* and the invasive *L. strigocamara*. The extent of the admixture was the catalyst for a proposed conservation program that included the removal of *L. strigocamara*

individuals from mixed populations, the prevention of sales of this exotic species to the public, and the promotion of sales of plants confirmed genetically as belonging to the native varieties (Maschinski et al. 2010). Finally, Zhang et al. (2010) reflected on the potential for introgression in *ex situ* collections of plant species, extinct in nature, to prevent restoration efforts. Specifically, seeds sired by individuals of the naturally extinct, Chinese endemic, *Sinojackia xylocarpa*, in a living garden collection, included an average of 32.7% hybrid progeny from crosses with *S. rehderiana* (Zhang et al. 2010). This led to the warning that "Such extensive hybridization in ex situ collections could jeopardize the genetic integrity of endangered species and irrevocably contaminate the gene pool if such hybrids are used for reintroduction and restoration" (Zhang et al. 2010).

In recognition of the general viewpoint by plant conservation biologists, that genetic exchange has negative impacts on rare taxa, I will review two additional examples of research programs investigating the implications of introgression on native plant species. These two examples come from the poplar and cordgrass complexes (*Populus* and *Spartina*, respectively), and reflect both similarities and differences in the evolutionary processes and the conservation implications that have arisen. For poplar, I will concentrate on the effects detected from the establishment of exotic plantations on populations of native species. In the case of *Spartina*, I will emphasize the role that genetic admixture may have on the evolution of invasiveness. In addition, I will use results from analyses of invasive cordgrass to provide examples of the types of effects these exotic taxa can have on both congeners and entire communities.

7.3.1 Introgressive hybridization and the conservation of endangered plants: Poplar

Plant evolutionary biologists have voiced apprehensions concerning the creation of industrial-scale (exotic) plant populations in the vicinity of native congeners. They have repeatedly warned of the effect that genetic exchange may have on both the genomic and phenotypic constitution of the native taxa (e.g., Unger et al. 2014), particularly with regard to the potential for the introduction of so-called transgenes, thereby spreading poten-tially destructive traits into natural ecosystems (e.g., Ellstrand and Hoffman 1990; Snow et al. 2003; Ellstrand et al. 2013). Though many instances of transgene introgression into native plant popula-tions have been recorded, the predicted negative effects have apparently occurred in only a minority of cases (Ellstrand et al. 2013). Notwithstanding this observation, the question remains as to whether the large-scale propagation of exotic species has cata-lyzed genetic exchange with native congeners thus impacting the genomic and adaptive characteristics of the native lineages.

Both Meirmans et al. (2010) and SL Thompson et al. (2010) examined the potential effects of intro-gression from poplar plantations containing exotic cultivars into native populations of North American *Populus deltoides* and *P. balsamifera*. The plantations have been present since the nineteenth century and consist of both European *P. nigra* and complex hy-brids containing genomic components mostly from *P. nigra*, *P. trichocarpa*, and *P. maximowiczii* (Meir-mans et al. 2010; SL Thompson et al. 2010). Given the relatively weak reproductive barriers between poplar species (e.g., Eckenwalder 1984), some level of introgression among the hybrid cultivars and the native species would be expected. Furthermore, be-cause some of the plantations occurred in a natural hybrid zone between *P. deltoides* and *P. balsamifera* (SL Thompson et al. 2010), there was an added dimension of possible genomic and phenotypic complexity.

A number of patterns emerged from the genomic assays of hybridization and introgression between the two North American species, their natural hy-brids, and *P. nigra* and the multi-species hybrid cultivars. First, natural populations of *P. deltoides* and *P. balsamifera* were found to consistently harbor genomic material from *P. nigra*, *P. trichocarpa*, and *P. maximowiczii*, indicating advanced hybrid gener-ations produced by introgression with not only *P. nigra*, but also with the admixed, hybrid cultivars (Meirmans et al. 2010). In terms of the directionality of gene movement, introgression predominantly oc-curred into the *P. balsamifera* genomic background; this pattern was detected both for hybridization between the native species and between the native and exotic taxa (SL Thompson et al. 2010). Though

introgression between *P. deltoides* or *P. nigra* was not detected, the occurrence of hybrid individuals con-taining genomic material from both of the North American species and *P. nigra* indicated a bridge be-tween the two native taxa and the European species (SL Thompson et al. 2010).

In general, the results from both of the poplar studies indicated the widespread occurrence of genetic admixture between the exotic and native species. This observation suggested the high prob-ability of transgene introgression from the cultivar lineages into populations of native species and their natural hybrids. Furthermore, the determination that smaller populations of native taxa were dispro-portionately impacted by introgression from exotic lineages suggested that guidelines for the introduc-tion of exotic cultivars with novel trait combinations should take into consideration the possibility of re-productive access to numerically inferior popula-tions of native congeners (Meirmans et al. 2010; SL Thompson et al. 2010).

7.3.2 Introgressive hybridization and the conservation of endangered plants: Cordgrass

Along with the concerns discussed, with regard to the potential for the introgression of "exotic" genes from non-natives into native congeners, evolution-ary biologists have also come to recognize the role hybridization may have in catalyzing invasiveness in animals and plants (Ellstrand and Schierenbeck 2000; Wolfe et al. 2007; Verhoeven et al. 2011; He-redia and Ellstrand 2014). For example, Turgeon et al. (2011) and Czypionka et al. (2012) used gen-omic life history and gene expression data to dem-onstrate the effects of admixture on the production of invasive forms of insects and fish, respectively. Likewise, Keller and Taylor (2010) and Travis et al. (2010) detected increased fitness in hybrid inva-sive lineages, relative to their progenitors, in the plant genera *Silene* and *Typha*. Figure 7.5 illustrates the results from an analysis of fitness across North American and European populations of the inva-sive forms of *Silene*. By using genomic markers and observations of fruit production, it was possible to infer the degree of fitness between genotypes be-longing to two categories of admixture (i.e., "lower" and "higher"; Figure 7.5; Keller and Taylor 2010).

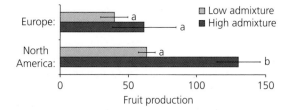

Figure 7.5 Relationship of admixture between divergent *Silene vulgaris* lineages (as assayed using nuclear loci) and fitness (as estimated from fruit production). Mean (± SEM) fruit production by hybrid genotypes characterized as possessing either "high" or "low" admixture frequencies in European and North American populations is illustrated. Bars sharing letters do not differ significantly (from Keller and Taylor 2010).

Though results from this analysis detected no significant difference in fitness between the categories among older European populations, the more recently established North American forms did indeed reflect higher fitness in those lineages possessing greater admixture (Figure 7.5). Keller and Taylor (2010) used the observed differences between the European and North American populations to infer that the heterosis possessed by early generation hybrid genotypes, though transient, would likely "substantially boost the fitness of introduced genotypes, potentially leading to the rapid emergence of invasiveness."

As with other invasive taxa, members of the plant group *Spartina* play key roles in their native habitats, but when translocated elsewhere demonstrate the capacity to affect both the genomic and phenotypic characteristics of native congeners and entire ecological communities. In their 2013 review, Strong and Ayres reflected this conclusion when they stated that members of this genus are "powerful ecological engineers that are highly valued where they are native" and yet when translocated "they overgrow native salt marsh and open intertidal mudflats, diminish biota, increase costs of managing wildlife, and interfere with human uses of estuaries." In terms of the current topic under discussion, the paradoxical biological characteristics of the cordgrass assemblage are likely at least partially catalyzed by the reticulate origin of the species within this genus (Figure 7.6; Ainouche et al. 2003; Hall et al. 2006; Strong and Ayres 2013).

All of the members of *Spartina* are allopolyploid, with species possessing tetraploid to dodecaploid

genomes (Figure 7.6; Baumel et al. 2002; Strong and Ayres 2013). Significantly, each of the major invasive events by cordgrasses has involved introductions of the heptaploid *S. densiflora*, the hexaploid, *S. alterniflora*, or hybrids between different hexaploid species (Ainouche et al. 2003; Strong and Ayres 2013). For example, *S. densiflora*, an invasive taxon in both European and North American coastline habitats, formed through ancient allopolyploid hybrid speciation from crosses between tetraploid *S. arundinacea* and hexaploid *S. alterniflora* (or their ancestral lineages; Fortune et al. 2008). This taxon has been introduced into non-native habitats since the sixteenth century, resulting in competitive exclusion of native congeners by the exotic species, hybridization between *S. densiflora* and native lineages and, finally, hybridization between *S. densiflora* and other introduced *Spartina* species (Ayres et al. 2008; Strong and Ayres 2013). Each of the outcomes from the introduction of *S. densiflora* (and other *Spartina* as well; Figure 7.6) represents a potential conservation issue. For example, hybridization between *S. densiflora* and the native *S. foliosa* has resulted in a new invasive hybrid that has expanded throughout the intertidal niches in San Francisco Bay (Sloop et al. 2011). Therefore, not only is this region impacted by the original introduction of the invasive

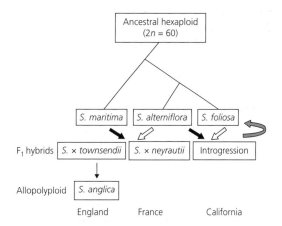

Figure 7.6 Introgressive hybridization and allopolyploidy reflected by a portion of the species in the plant genus *Spartina*. Each of the events was caused by the introduction (into Europe and the west coast of North America) of the eastern, North American species *S. alterniflora*. Arrows indicate hybridization between different species, with unfilled arrows indicating the maternal genome donors for the various crossing episodes (from Ainouche et al. 2003).

hybrid *S. densiflora*, but it faces an additional ecological challenge from the newly arisen hybrid lineage. This series of events—allopolyploid species formation, human-mediated translocations, hybridization with native taxa, thus resulting in major habitat modifications by both the original exotic and the subsequently formed hybrid lineages—has repeatedly occurred with species of *Spartina* (Strong and Ayres 2013). In general, this plant clade reflects well the evolutionary processes associated with the web-of-life, including those with potentially profound, and negative, effects on native plants and whole ecosystems (Ainouche et al. 2003; Hall et al. 2006; Ayres et al. 2008; Fortune et al. 2008; Sloop et al. 2011; Strong and Ayres 2013).

7.4 Conclusions

The main conclusion that I hope readers will draw from the discussion is that any efforts toward the conservation of endangered plants and animals will necessarily be a complex, multipartite process.

This is not, however, a new inference. I began the section on conservation of animals in the context of hybridization by discussing the management recommendations of Allendorf and his colleagues (Allendorf et al. 2001, 2004, 2005). Figure 7.7 reflects one of the major considerations that Allendorf et al. (2001) identified as essential for deciding whether or not a group of organisms should be placed under conservation protection. From this figure one can easily see that there are many outcomes from what these authors termed "natural" or "anthropogenic" hybridization. Furthermore, deciphering whether the admixture present was due to human mediated causes was not necessarily straightforward. Thus, they concluded, "Hybridization . . . provides an exceptionally tough set of problems for conservation biologists. The issues are complex and controversial, beginning with the seemingly simple task of defining hybridization" (Allendorf et al. 2001; Figure 7.7).

The strict application of the web-of-life rather than the tree-of-life metaphor to conservation, preservation, and restoration helps remove some

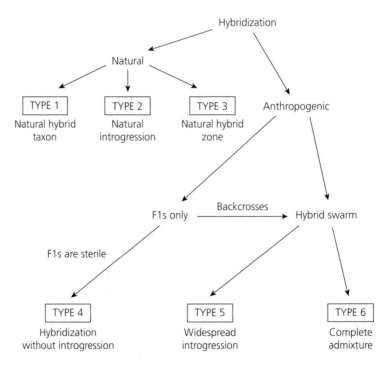

Figure 7.7 Illustration of the potential outcomes of "natural" and "anthropogenic" hybridization. This flowchart was suggested as a mechanism for defining whether admixed individuals and populations should be included in conservation efforts (from Allendorf et al. 2001).

of the issues seen as hurdles for appropriate management decisions. For example, if divergence-with-introgression, rather than purely allopatric diversification, has been the mode of evolutionary change for most eukaryotic (and of course all prokaryotic and viral) lineages and clades, then determining whether a specific example of hybridization was caused by "natural" or "anthropogenic" effects (Figure 7.7) is of less importance. This line of reasoning would follow from the conclusion that human-mediated genetic exchange reflects a continuation of what organisms have experienced episodically (e.g., during previous environmental perturbations) throughout their evolutionary history. Similarly, as discussed, the designation of anthropogenic catalysts of natural hybridization (and indeed genetic exchange *sensu lato*) as effectively "unnatural" obfuscates the fact that the human species is not only itself an admixed taxon (Arnold 2009; Chapter 8 of this book), but also has an ecological setting that, like all other organisms, it modifies simply by existing (Levins and Lewontin 1985, p. 99). To put it another way, if humans are part of the natural world and genetic exchange is common to evolutionary history, then human-mediated perturbations and natural hybridization should be interpreted as prosaic rather than extraordinary.

Though the recommendation to utilize a web-of-life prism to interpret when, where, how, and to what organisms one should apply conservation management efforts does indeed remove some obstacles to conservation programs, it unfortunately simultaneously contributes to others. One of the most important problems derived from such an approach relates to the fact that humans in general are often driven by a desire to place organisms into pigeonholes. One need look no further than the species concept debate discussed in Chapter 2 to see a major outworking of this predilection for ordering and naming. It is of even more practical importance that the persons in control of governmental and private sources of funds for conservation work, like systematists and taxonomists (e.g., those applying barcoding to catalogue biodiversity), likewise demand clean differentiation of units before they will consider providing financial support. Furthermore, and reflecting the desires of many conservation biologists as well, the fund-providers apparently expect that the units being preserved, conserved, and/or restored be kept genomically and phenotypically untouched by genetic exchange with related lineages. Thus, the embracing and communicating of a paradigm that illustrates evolutionary processes leading to admixed lineages could cause problems if the focus of conservation programs continues to emphasize what is tantamount to racial purity.

There does seem to me to be one approach that might circumvent the logjam created by the contrasting arguments that (1) rare forms must be completely protected from making the "mistake" of mating with members of divergent lineages versus (2) that such matings are not mistakes, but rather evolutionary business as usual. This approach would involve the replacement of the proposal that the management of endangered lineages must prevent genetic exchange—which leads for example to constant surveillance and culling of hybrids (e.g., the North American red wolf restoration program; Bohling et al. 2013)—with one that communicates the message that organisms of interest will exchange genes at times with related taxa, but that this can help the endangered forms to adapt to changing environments. I realize that such an approach will necessitate a drastic change of mindset for many fund managers and conservation biologists. In particular, adopting this mindset will require a reexamination of what the important conservation unit is. Basically, humans who are dealing with the protection of endangered lineages will have to decide if keeping the "pure" gestalt and bauplan of organisms, requiring vast expenditures of funds while simultaneously running the risk of extinction from a lack genetic and phenotypic variability, is more appropriate than allowing the occurrence of the commonplace processes associated with the web of life.

I suspect that this chapter and, in particular this Conclusions section, will leave many colleagues unsatisfied and some frustrated. However, I fear that evolutionary biologists risk causing serious damage to conservation efforts if they continue to neglect the incorporation of the paradigm shift that has already occurred, which reflects our understanding that organisms most often evolve in the presence of some degree of genetic exchange with related taxa. The damage will likely occur when educated fund providers read accounts in the popular press (e.g., Velasquez-Manoff 2014) that reveal the message from conservation biologists—"endangered taxa are genomically and morphologically pure relative to related taxa"—to be inaccurate and naïve. This will then lead to the obvious, and understandable, question of "Why should we provide funds to keep something pristine when even the *New York Times* knows that organisms exchange genes naturally?" Clearly formulating and enunciating proposals that accurately reflect a modern understanding of the wonderful complexity of the evolutionary process could encourage, rather than discourage, support for conservation efforts.

CHAPTER 8

Genetic exchange and humans

"This [horse] got a name?" he asked.
The old cowboy replied with a well-worn sentiment.
"Don't like naming things I might have to eat."
(Johnson 2009)

"Subsequently, the compatible alleles emerged and provided an opposing force to hold the differentiated populations together. These alleles might have been favored by selection gradually, likely because of higher reproduction rates in the hybrids than the other two allelic groups."
(Du et al. 2011)

"These results suggest different genetic dynamics within natural and anthropogenic hybridization contexts that carry important implications for primate evolution and conservation."
(Malukiewicz et al. 2014)

8.1 Genetic exchange and the evolution of *Homo sapiens*

At the beginning of 2009, Oxford University Press published another of my books, this volume titled *Reticulate Evolution and Humans—Origins and Ecology*. One of the inferences I drew from the various data available at that time (e.g., Garrigan et al. 2005; Templeton 2005; Hayakawa et al. 2006; Plagnol and Wall 2006; Cox et al. 2008) was that *H. sapiens* and the organisms making up its "ecological setting" possessed reticulate evolutionary histories. By the time this book appeared, the results from numerous genome-sequencing analyses had been published for organisms such as bacteria, plants, and even humans, gorillas, and chimpanzees. Though evolutionary biologists, in general, were willing to accept the arguments concerning the reticulate nature of the vast array of organisms making up the ecology of the human species (e.g., influenza, cotton, and dogs), there was a measure of resistance to the idea that humans had participated in

natural hybridization with other species of *Homo* (but see Holliday 2010). For example, Jolly (2009) argued that "By insisting, against the weight of evidence, that *Homo sapiens* 'must have' interbred with other human species, Arnold misses the opportunity to discuss the interesting paradox in these findings and the many questions arising from it." Given the subsequent explosion of genomic data that supported the reticulate evolutionary history of humans, Jolly's conclusion can now be seen as inversely prophetic. However, to be fair, I too was completely naïve regarding the imminent revolution to be wrought by genomic studies of both extinct and extant lineages of *Homo*. These studies began appearing shortly after the publication of my 2009 book—beginning with Green et al. (2010) and Reich et al. (2010)—and would lead to the falsification of the null hypothesis of divergence-without-introgression during the diversification of *Homo*.

In the first sections of this chapter, I will present information from analyses of genomic and phenotypic traits that provide a test of the various models

Divergence with Genetic Exchange. Michael L. Arnold.
© Michael L. Arnold 2016. Published 2016 by Oxford University Press.

of human evolution. I will begin this discussion by considering analogous cases of reticulate evolution in a limited set of non-hominine taxa. I will then proceed to a consideration of data for the clade containing *Gorilla, Pan,* and *Homo.* The data sets for the hominin primates derive from numerous genomic studies and a more limited set of morphological analyses. However, the diverse types of data lead to the same inference—all hominine lineages analyzed to date likely possess genomes consisting of segments of DNA derived from multiple, divergent lineages.

8.2 Genetic exchange and the evolution of New World primates: Howler monkeys and marmosets

It is now accepted that, like other animal clades, primates reflect numerous examples of introgression and hybrid speciation (Jolly 2001; Arnold and Meyer 2006; Arnold 2009; Zinner et al. 2011; Alves et al. 2012; Arnold et al. 2015). Furthermore, reticulate evolutionary processes have been inferred across all of the various sub-clades within the order Primates, including taxa belonging to either Old or New World assemblages (Zinner et al. 2011; Arnold et al. 2015). Though it had been suggested that natural hybridization might be less frequent in New World primates (Cortés-Ortiz et al. 2007), this inference likely reflected the lack of suitable data to test alternative hypotheses (Zinner et al. 2011). For example, recent analyses of both genomic and morphological traits have provided the extensive data sets necessary to test for divergence-with-gene-flow in the Neotropical, or Platyrrhine, monkeys (Kelaita and Cortés-Ortiz 2013; Fuzessy et al. 2014; Malukie-wicz et al. 2014). Two clades that have been the focus of numerous analyses include the howler monkey and marmoset genera *Alouatta* and *Callithrix.*

8.2.1 Howler monkeys

The genus *Alouatta* has a geographic distribution extending throughout both Meso- and South America, with *c.* 10 and 19 recognized species and subspecies, respectively. Though an analysis of mtDNA sequence variation defined reciprocal

monophyly for species from the two geographic regions (Cortés-Ortiz et al. 2003), as in other groups of Platyrrhine species—and primates in general (Arnold and Meyer 2006; Zinner et al. 2011)—ancient and present-day introgression has been detected among various howler monkey species. This inference has been derived from (1) discordant phylogenies typified by individuals grouping not with members of their own species, but instead in clades occupied by other species (Figure 8.1; Cortés-Ortiz et al. 2003), (2) the production of viable hybrid progeny in captivity (de Souza Jesus et al. 2010), and (3) the detection of natural hybrid zones between various combinations of species (Aguiar et al. 2007, 2008; Cortés-Ortiz et al. 2007; Agostini et al. 2008; Bicca-Marques et al. 2008). As discussed in previous chapters, the discordant phylogenetic signals detected in studies such as those involving the *Alouatta* species (Figure 8.1) could result from incomplete lineage sorting. Furthermore, there is also the possibility that a lack of sufficient sequence variation may have contributed to the patterns detected (Cortés-Ortiz et al. 2003). However, the identification of overlap zones between howler monkey species containing hybrid individuals argues for a contribution of genetic exchange in the production of the phylogenetic discordance observed.

Well-defined hybrid zones containing numerous advanced-generation hybrid individuals, between *A. palliata / A. pigra* and *A. caraya / A. guariba,* have been surveyed for genomic, morphological, and behavioral variability (Aguiar et al. 2007, 2008; Cortés-Ortiz et al. 2007; Agostini et al. 2008; Bicca-Marques et al. 2008; Kelaita and Cortés-Ortiz 2013; Ho et al. 2014). However, because many of the studies of areas of sympatry between howler monkey species have been based solely on morphological traits (e.g., those involving *A. caraya / A. guariba*), it is likely that introgressive hybridization has been underestimated. Support for this conclusion comes from analyses of howler monkeys from an area of sympatry between *A. palliata* and *A. pigra.* These populations have been defined genomically using a combination of mitochondrial, autosomal, and Y-chromosome loci (Cortés-Ortiz et al. 2007; Kelaita and Cortés-Ortiz 2013). Of 128 hybrid individuals, only 12% were found to have relatively equivalent numbers of alleles from each species (i.e., were

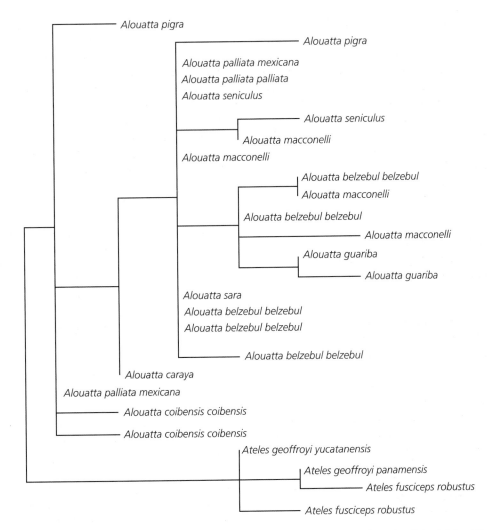

Figure 8.1 Phylogeny of howler monkey species (genus, *Alouatta*) based on the sequence variation at the nuclear *Calmodulin* locus (from Cortés-Ortiz et al. 2003).

"intermediate"), with the vast majority possessing genotypes indicative of advanced-generation backcrosses (Kelaita and Cortés-Ortiz 2013). By comparing the morphological and genomic variability for the same individuals, Kelaita and Cortés-Ortiz (2013) demonstrated that phenotypic traits were much less reliable in identifying multigenerational backcross hybrids. This observation led them to suggest that, given the numerous sympatric distributions of *Alouatta* species and the fact that morphological traits have been used mainly to define variation in such zones, genetic analyses might

"reveal that hybridization is more common in howler monkeys than initially considered and may serve to identify genus wide hybridization patterns" (Kelaita and Cortés-Ortiz 2013).

8.2.2 Marmosets

As with other members of the Platyrrhini (e.g., tamarins: Cropp et al. 1999; da Cunha et al. 2011; wooly monkeys: Ruiz-García et al. 2014; and squirrel monkeys: Lynch Alfaro et al. 2015; Mercês et al. 2015), the clade containing the marmoset assemblage displays genomic and morphological variation indicative of reticulate evolution. For example, a phylogenetic analysis detected cases of paraphyly and a lack of resolution for the placement of several recognized forms (Tagliaro et al. 1997). A portion of the non-concordance was suggested to be a result of poor taxonomy. However, the placement of individuals from well-characterized species into clades occupied mainly by other taxa supported an inference of past introgressive hybridization (Tagliaro et al. 1997). In addition to the hypothesis of introgressive hybridization, Tagliaro et al. (1997) also used their phylogenetic analysis to test whether one of the *Callithrix* species (i.e., *C. kuhli*) might have originated from hybridization between other marmoset species. The placement of different individuals of *C. kuhli* into separate clades was indeed consistent with the hypothesis of hybrid speciation, with the potential parental lineages identified as *C. penicillata* and *C. jacchus* (Tagliaro et al. 1997).

Recent analyses of phenotypic and genomic variability in marmoset hybrid zones have further defined the outcomes of admixture between *Callithrix* species. Fuzessy et al. (2014) assayed both color variation and morphometric traits in hybrid and parental individuals from within (hybrid animals) or outside (parental animals) a hybrid zone between *C. penicillata* and *C. geoffroyi*. The findings from this analysis included the detection of differential pelage patterns (i.e., facial coloration) in hybrid versus parental individuals and some degree of decoupling between the pelage and morphometric traits in hybrid animals (Fuzessy et al. 2014). Both of these observations indicate the potential importance of genetic admixture in generating evolutionarily significant, phenotypic variation in primates.

Novel combinations of genotypic and phenotypic variability have also been detected nearby and within two hybrid zones between the marmoset species *C. penicillata* and *C. jacchus* (Malukiewicz et al. 2014). One of the hybrid zones was located in an area outside of the distributions of both parental species and thus reflected the introduction of these taxa by humans and subsequent introgressive hybridization. The second zone occurred in an area of natural overlap at the edges of the ranges of *C. penicillata* and *C. jacchus* (Malukiewicz et al. 2014). Significant differences in the levels and patterns of

phenotypic and genotypic variation were detected in these two hybrid zones (Malukiewicz et al. 2014). Given the divergent patterns, it was inferred that hybrid fitness within the anthropogenic zone was likely lower than parental populations, but that the fitness of hybrid genotypes/phenotypes in the natural zone of overlap would be equivalent to that found in parental populations (Malukiewicz et al. 2014). In addition, the human mediated hybrid zone contained much less genetic variation than the natural hybrid zone. This observation led to the concern that anthropogenic habitat fragmentation causing hybridization might result in the loss of genetic variation and a concomitant reduction in mean fitness (Malukiewicz et al. 2014). However, Malukiewicz et al. (2014) emphasized that within this genus, "hybridization is a geographically widespread phenomenon," thus necessitating additional studies to decipher the evolutionary consequences of admixture between marmoset lineages.

8.3 Genetic exchange and the evolution of Old World primates: Langurs and leaf monkeys

Old World primates, unlike their New World counterparts, have been recognized for decades as paradigms of reticulate evolution (see reviews by Jolly 2001; Arnold and Meyer 2006; Arnold 2009; Haus et al. 2014; Arnold et al. 2015). Introgression (both ancient and contemporaneous) and hybrid speciation have thus been detected in groups as divergent as lemurs, baboons, macaques, gibbons/orangutans, and colobines (Muir et al. 2000; Roos et al. 2011; Liedigk et al. 2012, 2014; Matsudaira et al. 2013; Fan et al. 2014; Zhou et al. 2014). The last of these groups—the subfamily Colobinae—includes species belonging to radiations in both Africa and Asia (Roos et al. 2011).

One of the clades within the Asian colobine radiation includes species of langurs and leaf monkeys. Taxonomic assignments of the various taxa identified by these common names have been confusing, likely due to the use of "plastic morphological characters such as coat color" (Nag et al. 2011b). In addition, species belonging to the genera of langurs and leaf monkeys (i.e., *Presbytis, Semnopithecus,*

Trachypithecus) demonstrate genomic and morphological variation indicative of reticulate evolutionary processes. In fact, a lack of phylogenetic resolution, reflecting extreme discordance between inferences based upon different traits, would seem to even bring into question the assignment of the various genera to the categories of "langurs" or "leaf monkeys" (Ting et al. 2008; Roos et al. 2011). Furthermore, as illustrated in Figure 8.2, various genomic data sets have suggested that at least some of the members of the genus *Trachypithecus* (e.g., the "capped" and "golden" langurs) originated through natural hybridization between leaf monkeys and langurs possibly belonging to the genera *Semnopithecus* and *Presbytis* (Karanth 2008, 2010; Karanth et al. 2008; Roos et al. 2011).

In addition to the inferred ancient hybrid derivation of some *Trachypithecus* species, population and phylogenetic analyses using morphological characteristics and mitochondrial and nuclear sequences have detected past and contemporary introgression (Ting et al. 2008; Nag et al. 2011a, b). For example, Roos et al. (2011) and Ashalakshmi et al. (2015) provided support for instances of divergence-with-gene-flow among Asian colobines (and for African colobines as well; Roos et al. 2011). In the first analysis, a comparison of sequence variation at autosome, X-chromosome, Y-chromosome, and mitochondrial loci resulted in the inference of introgression between *Semnopithecus* and *Trachypithecus* potentially catalyzed by male behavioral

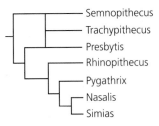

Figure 8.2 Phylogenetic relationships among Asian langur and leaf monkey genera (*Presbytis, Semnopithecus, Trachypithecus*) inferred from sequence variation at nuclear and mitochondrial loci. The outgroup taxa (*Rhinopithecus, Pygathrix, Nasalis, Simias*) also belong to the Asian colobine clade. The horizontal line drawn between *Semnopithecus* and *Presbytis*, that forms the base of the *Trachypithecus* lineage, indicates a hybrid origin for the latter genus (from Roos et al. 2011).

differences (Roos et al. 2011). In the more recent study, sequence variation at both nuclear and mitochondrial loci was also analyzed, in this instance involving species within the genus *Semnopithecus*. Discordance between the accepted taxonomy of these species and the gene trees, particularly those derived from the mtDNA sequences, suggested introgression events between *S. hypoleucos/S. entellus* and *S. priam/S. johnii* (Ashalakshmi et al. 2015).

Taken together, the various data indicate that the evolutionary histories of Asian langur and leaf monkey lineages have included introgressive hybridization and hybrid speciation. Though some instances of recent introgressive hybridization are likely due to anthropogenic habitat modifications (Nag et al. 2011a), many cases of reticulation among these primates reflect ancient events (Karanth 2010; Roos et al. 2011). As with all other major primate assemblages, the colobines thus support a web-like pattern of evolutionary diversification.

8.4 Genetic exchange and the evolution of Old World primates: Baboons

The baboon genus, *Papio*, has been used extensively to test various evolutionary hypotheses, with many of the analyses defining patterns indicative of reticulate evolutionary processes (e.g., Shotake et al. 1977; Shotake 1981; Jolly et al. 1997; Alberts and Altmann 2001; Ackermann et al. 2006; Bergman et al. 2008; Keller et al. 2010; Zinner et al. 2013; Boissinot et al. 2014). Among the various studies have been those primarily undertaken to elucidate phylogenetic relationships, both between baboons and members of sister clades (Zinner et al. 2009a; Guevara and Steiper 2014; Liedigk et al. 2014) and within *Papio* itself (Wildman et al. 2004; Sithaldeen et al. 2009; Zinner et al. 2013; Boissinot et al. 2014). Other analyses have focused on population genetics, life history traits, mating behaviors displayed by parental and hybrid individuals, and the relative fitness of hybrid animals (Hapke et al. 2001; Dirks et al. 2002; Colmenares et al. 2006; Tung et al. 2008; Jolly et al. 2011; Ackermann et al. 2014).

A recent study by Zinner et al. (2013) exemplifies well the results from numerous phylogenetic analyses involving members of *Papio*. In this analysis, entire mitochondrial genomes were sequenced from individuals that encompassed the six baboon species and the seven mtDNA haplogroups found within these taxa (defined previously by Zinner et al. 2009b). Figure 8.3 illustrates the observation that phylogenetic clustering was often not correlated with the taxonomic designation of the various individuals sequenced. Thus, the three animals morphologically identified as *P. anubis*, and possessing mtDNA haplogroups D, F, and G, did not group together as sister lineages. Instead, the *P. anubis* individuals possessing haplogroups F and G were more closely related to *P. papio* and *P. hamadryas*/*P. cynocephalus*, respectively (Figure 8.3; Zinner et al. 2013). Likewise, the two *P. cynocephalus* animals included in this analysis belonged to the B and G haplogroups and were thus placed within clades containing either *P. ursinus* and *P. kindae* (haplogroup B) or *P. anubis* and *P. hamadryas* (haplogroup G; Figure 8.3). Zinner et al. (2013) inferred that introgressive hybridization, rather than incomplete lineage sorting, was likely the primary causal factor in producing the phylogenetic discordance because geographically closely allied lineages often clustered together.

As mentioned, a wide variety of studies have analyzed baboon reticulate evolution from a population-level perspective. For example, given the complex behavioral attributes of these primates, the effects of species–specific behaviors on the process of natural hybridization have been of particular interest (Alberts and Altmann 2001; Hapke et al. 2001; Bergman and Beehner 2004; Beehner et al. 2005; Bergman et al. 2008; Jolly et al. 2011). In this regard, Charpentier et al. (2012) used nuclear microsatellite loci to test for the effects of differential behavioral characteristics on the genetic architecture of a hybrid zone between yellow (i.e., *P. cynocephalus*) and anubis (i.e., *P. anubis*) baboons located in southern Kenya. The pattern of genetic variability at the nuclear loci assayed in this analysis indicated a significant contribution from the behavioral traits on the structure of the baboon populations within and adjacent to the *P. cynocephalus* × *P. anubis* hybrid zone. In particular, Charpentier et al. (2012) discovered asymmetric introgression involving genetic exchange from *P. anubis* into *P. cynocephalus* within the area of overlap. This differential genetic exchange was consistent with previous findings of biased migration of anubis individuals into the range of yellow baboons (Tung et al. 2008), the earlier reproductive maturation of anubis baboons relative to yellow baboons (Charpentier et al. 2008), and the advantage detected in courtship success

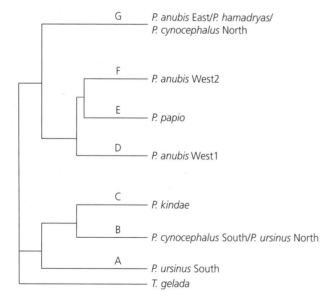

Figure 8.3 Phylogenetic relationships of six baboon species (genus *Papio*) and the seven mtDNA haplogroups (A-G; Zinner et al. 2009b) based upon entire mitochondrial genome sequences. The outgroup taxon used was *Theropithecus gelada* (from Zinner et al. 2013).

for anubis-like males over yellow-like competitors (Tung et al. 2008). Taken together, these behavioral differences between *P. anubis* and *P. cynocephalus* were sufficient to explain the strongly biased introgression from the former into the latter species (Charpentier et al. 2012). Furthermore, it is likely that the reticulate evolution described in this section, observed both within *Papio* and between species of *Papio* and related lineages, has often been shaped by the divergent behaviors of the hybridizing taxa.

8.5 Genetic exchange and the evolution of Old World primates: Gorillas, chimpanzees, and humans

The field of evolutionary biology has been repeatedly challenged, and sometimes catalyzed, by analyses of genomic data. Early studies of biological and genomic evolution utilizing reassociation kinetics, protein electrophoresis, and amino acid sequencing (Hubby and Lewontin 1966; Britten and Kohne 1968; Barnabas et al. 1972; Hamrick and Allard 1972) provided the conceptual framework for the current analyses of RNA and DNA sequences. Regarding the evolution of the chimpanzee/gorilla/human clade, a comparison of DNA reassociation kinetics and protein sequences resulted in the inference of very low divergence (i.e., >95% similarity) among the three lineages (Hoyer et al. 1972; King and Wilson 1975). The extreme similarity in protein (i.e., "coding") sequences relative to morphological, behavioral, etc. differences between chimpanzees and humans resulted in the conclusion that "evolutionary changes in anatomy and way of life are more often based on changes in the mechanisms controlling the expression of genes than on sequence changes in proteins" (King and Wilson 1975). Similarly, Davidson and Britten (1979) argued that given the presence of the same genes across tissues and stages of development, but with large differences in the messenger RNA populations, non-coding sequences were the drivers of gene expression. The observations of both King and Wilson (1975) and Davidson and Britten (1979) thus result in the conclusion that mutations in regulatory loci are likely the basis of the evolutionary

divergence between *Homo* and its nearest living relatives (King and Wilson 1975).

The landmark paper by Sarich and Wilson (1967), in which they reported levels of divergence among blood serum proteins from primates, resulted in a revolutionary inference. This inference was that humans and their closest, extant sister taxa had evolved from a common ancestor only five mya (Figure 8.4). Sarich and Wilson (1967) also concluded that the relative branching order of *Pan, Gorilla*, and *Homo* could not be determined from their data. Figure 8.4 reflects how they illustrated this latter conclusion using an unresolved trichotomy (Sarich and Wilson 1967). Though a furious debate ensued concerning both of these conclusions, after nearly a half a decade the recent sharing of a common ancestor by chimpanzees, gorillas, and humans has been repeatedly demonstrated and is now accepted as fact. Furthermore, there is also now a longstanding consensus that the genera *Homo* and *Pan* shared a common ancestor after separating from the lineage leading to *Gorilla* (e.g., see Avise 1994, pp. 329–331).

In contrast to the well-accepted viewpoint that humans and chimpanzees shared a more recent common ancestor, relative to gorillas, nearly every study testing for the relationships among *Pan, Homo*, and *Gorilla* has resulted in phylogenetic discordance (Figure 8.5). As I have indicated in the previous discussions, one widely supported explanation for such non-concordance is genetic exchange resulting in the differential transfer of loci. In terms of the possibility that introgressive hybridization occurred between the three lineages giving rise to *Pan, Gorilla*, and *Homo*, it should be kept in mind that only *c.* five million years has passed since divergence from a common ancestor. Both Prager and Wilson (1975) and Fitzpatrick (2004) estimated 2–4 million years as the average amount of time since divergence that is necessary for the development of reproductive isolation between mammalian lineages. Applying this metric to chimpanzees, gorillas, and humans results in the conclusion that introgression was possible for 40–80% of the time period since their divergence from a common ancestor. If these taxa did not diverge in complete allopatry, genetic exchange likely occurred (Arnold and Meyer 2006; Arnold 2009). I will thus begin the description of studies that tested for reticulate

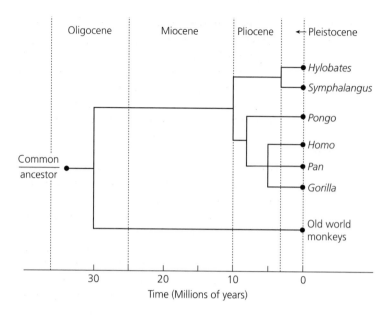

Figure 8.4 Phylogenetic relationships, and times of divergence, among various Old World primates. The depiction of the *Homo, Pan,* and *Gorilla* lineages as forming an unresolved trichotomy reflects the conclusion that their divergence from a common ancestor was essentially simultaneous (from Sarich and Wilson 1967).

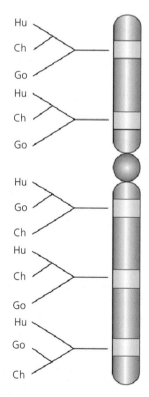

Figure 8.5 Schematic of the phylogenetic discordance observed from analyses of genomic data, reflective of alternate allelic sharing between chimpanzees ("Ch"), gorillas ("Go"), and humans ("Hu") at different loci (from Pääbo 2003).

evolutionary processes within the human/chimpanzee/gorilla clade by discussing evidence of divergence-with-gene-flow among the ancestors of the three genera. I will then proceed to a discussion of evidence of introgressive hybridization between taxa within each of the genera.

8.5.1 Genetic exchange and the evolution of Old World primates: Gorilla × Homo × Pan

Before reviewing evidence that the evolutionary history within the clade containing the chimpanzee, gorilla, and human lineages involved genetic exchange, it is important to note that a number of authors have favored a non-reticulate model for this species complex (e.g., Deinard and Kidd 1999; Pääbo 2003; Innan and Watanabe 2006). A number of these studies were generated as responses to two studies of genomic divergence among *Pan, Gorilla,* and *Homo.* The first of these studies (Navarro and Barton 2003b) was not specifically designed to test for genetic exchange, but instead addressed the genomic footprints predicted if speciation was affected by both chromosomal rearrangement differences between the diverging lineages and accompanying introgression. Navarro and Barton (2003a) had derived a model that predicted greater genomic divergence in the two lineages within chromosomal regions possessing rearrangements relative to segments not marked by structural differences (i.e., that were "collinear"). Navarro and Barton (2003b) did indeed observe elevated sequence divergence between regions of the chimpanzee and human genomes characterized by rearrangements. From this they concluded that chromosomal structural changes likely played a role in the speciation of *Pan* and *Homo* (Navarro and Barton 2003b). In the context of reticulate evolutionary processes, their findings also supported the inference of introgressive hybridization between the collinear portions of the chromosomes possessed by proto-humans and proto-chimpanzees (Navarro and Barton 2003b; Rieseberg and Livingstone 2003). Though a number of subsequent studies took issue with the conclusion of Navarro and Barton (2003b) concerning the role of chromosomal rearrangement differences in the divergence of *Pan* and *Homo* (e.g., Hey 2003; Lu et al. 2003; Marquès-Bonet et al. 2004), few did so with respect to the inference of divergence-with-introgression (but see Zhang et al. 2004). Thus, a representative conclusion contained in the various studies designed to test Navarro and Barton's (2003b) hypotheses was presented by Lu et al. (2003): "We wish to emphasize that this comment does not contradict the elegant model of parapatric speciation [i.e., divergence accompanied by genetic exchange between chimpanzees and humans] driven by differential adaptation."

As with the study by Navarro and Barton, conclusions drawn from an analysis of sequence divergence among humans, chimpanzees, gorillas, orangutans, and macaques led to a flurry of rebuttal papers. Specifically, Patterson et al.'s (2006) analysis of genomic regions spanning the autosomes and X-chromosome detected a range of *c.* four million years between the most ancient and the most recent divergence times between humans and chimpanzees, as estimated by sequence divergence at the loci sampled. A model incorporating differential introgression between the genomes of the proto-chimpanzees and proto-humans predicted such a large discordance among loci (Patterson et al. 2006). Though consistent with data subsequently gathered through whole-genome sequencing, the inference of speciation accompanied by

introgression (Patterson et al. 2006) was argued against by a number of workers. In particular, the observation of much lower than expected sequence divergence between *Homo* and *Pan* at X-chromosomal loci—inferred by Patterson et al. (2006) as evidence of introgression—was subsequently demonstrated to have several different possible causes (Wakeley 2008; Presgraves and Yi 2009; Yamamichi et al. 2012). These objections to a reticulate model of evolutionary change within *Pan / Gorilla / Homo* clade are well reasoned and thus reflect the necessity to closely review all of the data that allow tests of the null hypothesis of allopatric divergence. In fact, though it is likely that some of the range of coalescence times estimated by Patterson et al. (2006) were the result of incomplete lineage sorting among the human, chimpanzee, and gorilla lineages (Deinard and Kidd 1999; Pääbo 2003; Salem et al. 2003; Yamamichi et al. 2012), the conclusion that introgression caused at least a portion of the patterns seen have been supported by a number of analyses. Indeed, two earlier studies by O'hUigin et al. (2002) and Osada and Wu (2005)—involving assays of *c.* 60 and 490 genomic segments, respectively—resolved the effects from both retained ancestral polymorphisms and genetic exchange.

As predicted by previous studies, O'hUigin et al. (2002) found that the highest number of informative base pair substitutions (*c.* 50%) supported a phylogenetic arrangement in which humans and chimpanzees shared the most recent common ancestor. In contrast, 16% of the substitutions indicated that *Gorilla* and *Homo* diverged from a common ancestor. Furthermore, 30% of the informative substitutions reflected shared variation between *Pan* and *Gorilla*, relative to *Homo*, suggesting the sharing of a common ancestor by chimpanzees and gorillas (O'hUigin et al. 2002). A proportion of the nonconcordance was attributable to incomplete lineage sorting. However, as much as 50% was not estimated to reflect coalescent processes (O'hUigin et al. 2002). Likewise, Osada and Wu (2005) reported the results from an analysis of genic and intergenic sequences from *Pan, Gorilla*, and *Homo* designed to test a purely allopatric model of speciation. With regard to the comparison of sequence variation between chimpanzees and gorillas, Osada and Wu (2005) detected a high frequency of allelic sharing (also found in comparisons between *Pan* and *Homo*). They argued that the shared biogeographic history of *Pan* and *Gorilla* would have facilitated introgression between individuals belonging to the lineages leading to both of these extant forms. Osada and Wu (2005) also utilized several methodologies to test the null hypothesis that chimpanzees and humans diverged while allopatric. These analyses included estimates of the time since divergence of alleles and tests of congruence between the phylogenetic relationships inferred from various loci. In both types of analyses, the allopatric paradigm was falsified, while a model of divergence with some level of gene flow between the lineages leading to *Pan, Gorilla,* and *Homo* was not rejected. Both these authors and O'hUigin et al. (2002) thus provided evidence consistent with a reticulate evolutionary history for the clade containing humans, chimpanzees, and gorillas.

Though whole-genome data are still lacking for most New World and Old World primates (Rogers and Gibbs 2014), the most recent comparisons among the *Pan, Gorilla,* and *Homo* lineages have involved the collection of many gigabases of sequence, reflecting a majority of the genomes of interest. A range of inferences has resulted from these sequence analyses concerning the likelihood of divergence-with-gene-flow between and within these three genera (see Sections 8.5.2 to 8.5.4 for estimates of introgression within each of these genera). In particular, Mailund et al. (2012) found a strong signal of introgressive hybridization between many of the primate lineages, including those resulting in *Pan* and *Homo*. Figure 8.6 illustrates the observation that comparisons of sequence divergence between humans and both *P. paniscus* (i.e., bonobo) and *P. troglodytes* (i.e., chimpanzee) resolved a pattern indicating genetic exchange. In fact, in a statement reminiscent of Osada and Wu (2005) and Patterson et al. (2006), Mailund et al. (2012) concluded, "human-chimpanzee speciation has been the focus of considerable attention in previous studies, most of which have assumed a simple (allopatric) speciation model . . . we find evidence favouring a non-allopatric model, in which the initial divergence was followed by gene flow for an extended period."

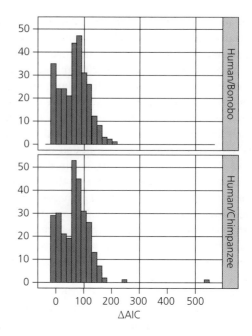

Figure 8.6 Histograms illustrating the distribution of Akaike Information Criteria (i.e., AIC) differences for the allopatric and divergence-with-introgression models for the divergence of the human lineage from the bonobo (upper graph) and chimpanzee (lower graph) lineages. Values <0 are consistent with an allopatric divergence model, while those >0 indicate support for the divergence-with-introgression model. Most values support the latter model with regard to the evolutionary history of *Homo* and *Pan* (from Mailund et al. 2012).

8.5.2 Genetic exchange and the evolution of Old World primates: Gorilla

The genus *Gorilla* was initially considered to consist of one species divided into two subspecies, *Gorilla gorilla gorilla* in the western portion of the range and *Gorilla gorilla beringei* occupying an eastern distribution (Walker et al. 1975, p. 477). Recent taxonomic work has recognized the distinctiveness of the eastern and western forms, providing them with the specific names, *Gorilla beringei* and *Gorilla gorilla*, respectively (see Clifford et al. 2004 for a discussion). A number of previous analyses of both mitochondrial and nuclear loci have also detected subdivisions within the *G. gorilla*, indicative of subspecific status. Clifford et al. (2004) detected mtDNA variation reflecting a subdivision between populations located from eastern Nigeria to southeastern

Cameroon relative to all other populations of *G. gorilla*, with additional genetic diversification within this latter group of populations as well.

Notwithstanding the genetic distinctiveness within *G. gorilla* and between *G. gorilla* and *G. beringei*, a number of earlier studies discovered patterns consistent with introgressive hybridization (e.g., Kaessmann et al. 2001; Clifford et al. 2004; Anthony et al. 2007). The overall findings of these studies can be illustrated using an analysis by Thalmann et al. (2007). This analysis included the collection of *c.* 14,000 base pairs of sequence data from 16, nongenic, autosomal loci from 15 *G. gorilla* individuals (including representatives of both subspecies) and three *G. beringei* animals. These sequence data supported neither an allopatric model, nor a model in which the *Gorilla* ancestral population was in panmixia until very recently. Instead, the sequence variation was consistent with repeated admixture between the *G. gorilla* subspecies and introgression between the *Gorilla* species after their divergence from a common ancestor (i.e., 0.9–1.6 mya) until *c.* 80,000 ybp (Thalmann et al. 2007). Finally, the distribution of sequence variation led to the inference that the genetic exchange between *G. beringei* and *G. gorilla* was asymmetric with introgression occurring mostly from the former into the latter (Thalmann et al. 2007).

A set of more recent analyses—using both morphological and genomic data sets—have further refined the understanding of the reticulate evolutionary history of members of *Gorilla* (Mailund et al. 2012; Scally et al. 2012; Das et al. 2014; Soto-Calderón et al. 2014; McManus et al. 2015). For example, Ackermann and Bishop (2010) compared cranial variation from representatives of each species and subspecies with previously reported mtDNA sequence data. The inclusion of both phenotypic and genetic information provided evidence of recent genetic exchange between the subspecies belonging to either *G. beringei* or *G. gorilla* as well as between the two species (Ackermann and Bishop 2010). Furthermore, this analysis resolved hybrid phenotypes in geographic regions predicted by a model including refugia and riverine barriers. Each of these observations reflected the likelihood of vicariance-mediated, episodic admixture among the various subspecies and species (Ackermann

and Bishop 2010). Das et al. (2014), based on an analysis of the entire mitochondrial genomes of *G. beringei* and *G. gorilla*, likewise emphasized the roles of environmental fluctuations and tectonic episodes in the catalysis of extended admixture between the gorilla species. Finally, Scally et al. (2012) and Soto-Calderón et al. (2014) utilized nuclear sequence data and the insertion of nuclear translocated copies of mitochondrial DNA (i.e., *numts*), respectively, to test for genetic exchange between the *Gorilla* lineages. These two analyses provided evidence of an extended period of introgression (e.g., ~0.5 million years; Scally et al. 2012) following the divergence of *G. beringei* and *G. gorilla* from a common ancestor.

The studies of *Gorilla* indicate that lineages within this genus form a network rather than a simple bifurcating tree. Once again primates are seen to be an excellent example of evolutionary diversification with genetic exchange. For the current discussion, reticulate evolution within *Gorilla* also provides a robust analogy for the genus *Homo*. Section 8.5.3 will illustrate that reticulate evolutionary processes characterize the genus *Pan* as well.

8.5.3 Genetic exchange and the evolution of Old World primates: Pan

As with the studies of gorilla species and subspecies, an extensive series of genomic analyses within the genus *Pan* has identified signatures of past reticulate evolutionary processes. This genus is now considered to consist of two species, *P. troglodytes* ("common chimpanzee") and *P. paniscus* ("pygmy chimpanzee" or "bonobo"). The common chimpanzee species is further subdivided into three subspecies: *P. t. troglodytes*, *P. t. verus*, and *P. t. schweinfurthii* (see discussion of systematics in Morin et al. 1994 and Deinard and Kidd 2000). Divergence times for the two species and the subspecies within the common chimpanzee have been inferred to be *c.* 1 mya and 0.5 mya, respectively (Pesole et al. 1992; Kaessmann et al. 1999; Stone et al. 2002; Yu et al. 2003; Caswell et al. 2008).

Various analyses of mitochondrial, X-chromosome, Y-chromosome, and autosomal loci led to the inference of introgression between the bonobo and common chimpanzee lineages (Deinard and Kidd 1999, 2000). For example, an examination of variation at an X-chromosome locus (i.e., Xq13.3) led Kaessmann et al. (1999) to argue that due to hybridization during the divergence of *P. troglodytes* and *P. paniscus*, "certain loci . . . may have crossed the 'species barrier' much later than other loci." The majority of recent analyses comparing large portions of the genomes of *P. troglodytes* and *P. paniscus* have inferred no significant introgression between these lineages subsequent to their divergence from a common ancestor (Won and Hey 2005; Caswell et al. 2008; Hey 2010; Mailund et al. 2012; Prüfer et al. 2012). However, not only did Wegmann and Excoffier (2010) infer an earlier divergence time for the bonobo and common chimpanzee (*c.* 1.6 mya), they also detected evidence of introgression between the lineages leading to these two species. Indeed, the history of genetic exchange between *P. troglodytes* and *P. paniscus* inferred by Wegmann and Excoffier (2010) included genetic exchange before and after the divergence of the subspecies of common chimpanzee.

In contrast to the conflicting inferences regarding introgression between the ancestors of bonobos and chimpanzees, genomic comparisons have consistently supported a model of divergence-with-gene-flow for the subspecies of *P. troglodytes* (Stone et al. 2002; Won and Hey 2005; Becquet et al. 2007; Caswell et al. 2008; Hey 2010; Wegmann and Excoffier 2010; Gonder et al. 2011; Prado-Martinez et al. 2013). For example, Becquet et al. (2007) determined the genotype of chimpanzee and bonobo individuals at 310 nuclear loci. Genetic divergence within the *P. troglodytes* samples suggested the presence of three lineages corresponding to *P. t. troglodytes*, *P. t. verus*, and *P. t. schweinfurthii* (Becquet et al. 2007). These authors found that 2 of the 51 chimpanzees (i.e., 4%) possessed hybrid genotypes. Similarly, a fourth subspecies, *P. t. ellioti*, whose geographic distribution extends from southern Nigeria to western Cameroon, is thought to form a contemporary hybrid zone with *P. t. troglodytes* (Gonder et al. 2011).

In addition to the contemporaneous introgression detected by Becquet et al. (2007; see also Stone et al. 2002) and Gonder et al. (2011), numerous authors have reported evidence for ancient gene flow as well. Thus, ancient migration (i.e., introgression) events between the eastern (*P. t. schweinfurthii*) and central (*P. t. troglodytes*), and the central and western

(*P. t. verus*) subspecies have been inferred (Won and Hey 2005; Caswell et al. 2008; Wegmann and Excoffier 2010). Likewise, introgression between the central and eastern subspecies has been inferred (Wegmann and Excoffier 2010). Interestingly, the ancient introgression detected between *P. t. troglodytes* and *P. t. verus* was asymmetric, with genes moving from the latter into the former subspecies (Figure 8.7; Won and Hey 2005; Wegmann and Excoffier 2010), a result agreeing with inferences by Hey (2010). However, Hey's (2010) inference was that the asymmetric introgression had actually involved the western subspecies and the ancestor of the eastern and central subspecies. If correct, this episode of divergence-with-gene-flow might explain both the western–central and western–eastern genetic exchange detected by Wegmann and Excoffier (2010). In any case, as with each of the other chimpanzee analyses, these results reflect the role of introgressive hybridization during the diversification of the *P. troglodytes* assemblage.

8.5.4 Genetic exchange and the evolution of Old World primates: Homo

As discussed at the opening of this chapter, until very recently, evolutionary biologists in general accepted the paradigm that our species expanded out of Africa with little or no admixture with conspecific taxa (Cann et al. 1987; Kaessmann and Pääbo 2002; Fagundes et al. 2007; Belle et al. 2009). The majority of support for the no-introgression model came from studies of mitochondrial DNA variation in both *H. sapiens* and extinct species of *Homo* (Cann et al. 1987; Currat and Excoffier 2004; Serre et al. 2004; Lalueza-Fox et al. 2006; Orlando et al. 2006; Green et al. 2008). Furthermore, notwithstanding the current strong support for reticulation, there are some evolutionary biologists who have recently argued that the evidence for admixture between species of *Homo* was weak and instead inferred the role of deep coalescence (Lowery et al. 2013) and/or ancestral population structure (Blum and Jakobsson 2011; Eriksson and Manica 2012; but see Sankararaman et al. 2012, Yang et al. 2012, Lohse and Frantz 2014) in producing patterns suggestive of introgressive hybridization. I will now address evidence from a variety of sources

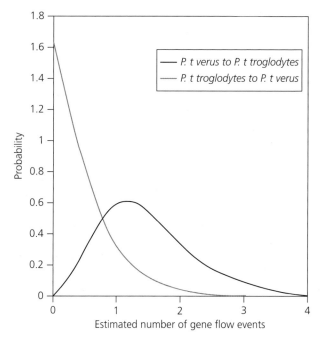

Figure 8.7 Probability distribution of gene flow events between the central and western subspecies of the common chimpanzee (i.e., *P. t. troglodytes* and *P. t. verus*, respectively). The peak shapes and positions indicate an average of one gene flow event from *P. t. verus* into *P. t. troglodytes*, but no introgression from the latter into the former subspecies. This finding suggests that introgression has occurred between the central and western subspecies, but that it is asymmetric (i.e., occurs only from *P. t. verus* into *P. t. troglodytes*; from Won and Hey 2005).

that indicate (1) *H. sapiens* overlapped spatially and temporally with congeneric species, (2) admixture occurred not only outside of Africa, but also within the confines of Africa, and (3) a portion of the introgression from archaic species into *H. sapiens* produced adaptive effects that have carried through into extant populations.

Fossil remains: Evidence of the co-existence of Homo species

Some of the greatest interest regarding the potential role of reticulate evolution in the development of the chimp/gorilla/human web has centered around the question of whether or not different *Homo* species coexisted in space and time, thereby facilitating hybridization (Swisher et al. 1996; Garrigan et al. 2005; Templeton 2005, 2007; Jolly 2009; Stringer 2014). Though it has been well established for some time that our species and archaic forms did indeed occupy broadly defined geographic regions (e.g., *H. sapiens* and *H. neanderthalensis* in present-day Europe), at roughly similar time periods (Figure 8.8), this was not sufficient evidence for contact leading to interspecific matings. One methodology applied to test for coexistence has involved

stratigraphic analyses of fossils in sites characterized by remains of multiple species. The question addressed by these studies has been whether there was evidence of "interstratification" of remains typical of either extinct forms or *H. sapiens*. Figure 8.9 reflects such an analysis of the Grotte des Fées de Châtelperron site in France containing remains from populations of both *H. sapiens* and *H. neanderthalensis* (Gravina et al. 2005). At this site, *H. neanderthalensis* artifacts occurred in sediments located both earlier and later than those of *H. sapiens* (Figure 8.9; Gravina et al. 2005). This finding was consistent with the "chronological coexistence—and therefore potential demographic and cultural interactions—between the last Neanderthal and the earliest anatomically and behaviourally modern human populations in western Europe" (Gravina et al. 2005). Though criticized by a subsequent set of workers (see Zilhão et al. 2006), Mellars et al. (2007) found support for the conclusions of Gravina et al. (2005) with further analyses of the excavations. At this site, *H. sapiens* and *H. neanderthalensis* thus overlapped both spatially and temporally (Figure 8.9), providing the opportunity for introgressive hybridization.

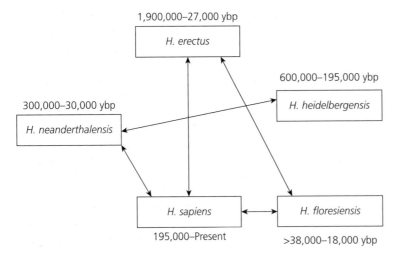

1,900,000–27,000 ybp

H. erectus

600,000–195,000 ybp

H. heidelbergensis

300,000–30,000 ybp

H. neanderthalensis

H. sapiens

195,000–Present

H. floresiensis

>38,000–18,000 ybp

Figure 8.8 Estimates of times of existence for five *Homo* species. Boxes connected by lines reflect species that co-occurred in geographical regions. Data used to construct this figure were found in Antón and Swisher (2004), Morwood et al. (2004), Pääbo et al. (2004), Futuyma (2005), and Finlayson (2005).

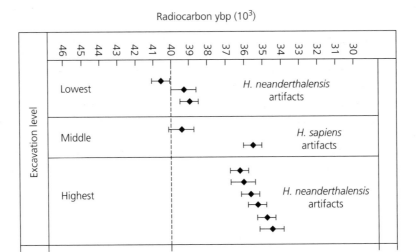

Figure 8.9 Distribution of radiocarbon dates for artifacts collected from three excavation levels at the Grotte des Fées de Châtelperron site in France. Artifacts dated were typical for populations of either *H. neanderthalensis* or *H. sapiens* (from Gravina et al. 2005).

In addition to studies of single sites such as that of Grotte des Fées de Châtelperron, analyses of the dates of habitation of entire geographic regions by *H. sapiens* and *H. neanderthalensis* have allowed defined estimates of temporal and spatial overlap between these species. For example, Pinhasi et al. (2011) inferred a limited period of overlap in the Caucasus through the dating of a number of *H. neanderthalensis* samples. In particular, they estimated that the extinction of this species *c.* 39,000 ybp left only a very narrow time period for interactions

with the expanding *H. sapiens* populations (Pinhasi et al. 2011). However, in a recent analysis incorporating a much more extensive sampling of sites across Europe, this same research group found evidence for an extended period (i.e., 2600–5400 years) of co-existence for *H. sapiens* and *H. neanderthalensis* (Higham et al. 2014). They thus concluded, "Rather than a rapid model of replacement of autochthonous European Neanderthals by incoming AMHs [anatomically modern humans], our results support a more complex picture, one characterized by

Fossil remains: Evidence of introgression among Homo species

The debate over whether or not the available fossil material reflects introgressive hybridization between different species of *Homo* continues, with some authors arguing against a model of admixture (e.g., Tattersall and Schwartz 1999; Schwartz and Tattersall 2010). However, inferences of admixture among hominin taxa, based upon fossil characteristics, have been made by a number of evolutionary anthropologists. For example, remains indicating temporal overlap between *H. erectus* and *H. sapiens* in Central Java led Swisher et al. (1996) to the conclusion that "features shared by the two species are either homoplastic or the result of gene flow." The inference that fossil evidence supports even earlier cases of reticulate evolution, in this case among species of *Australopithecus*, has also been drawn (Holliday 2003; Ackermann 2010). In the case of *Australopithecus*, Ackermann (2010) pointed to the occurrence of anomalous patterns of dental development that have been found frequently in other mammalian and primate hybrids (Ackermann et al. 2006, 2010, 2014; Ackermann 2007). Likewise, though referred to as a "transitional phase" from

H. erectus to *H. sapiens* (SG Larson et al. 2007), the sharing of certain skeletal traits between *H. erectus* and *H. floresiensis* could also be reflective of divergence-with-gene-flow between these species. Furthermore, Ackermann (2010) recently discussed the observation that *H. floresiensis* possessed "full-sized bilateral rotated maxillary premolars," a trait once again found in mammalian hybrids.

As concluded by Holliday (2003), "The most famous case of possible hybridization in *Homo* is that of interbreeding between *H. sapiens* (modern humans) and *H. neanderthalensis* ('Neandertals')." Therefore it is significant that, as with *Australopithecus* species, *H. erectus*, and *H. floresiensis*, traits found among fossils of *H. sapiens* and *H. neanderthalensis* are also consistent with a model of assimilation (i.e., introgressive hybridization and hybrid speciation—Holliday 2003; Trinkaus 2007; Walker et al. 2008; Wolpoff 2009). For example, Trinkaus (2007) compared the morphological variation of European *H. sapiens* fossils from 33,000 ybp to Neanderthal samples as well as more recent human samples (Figure 8.10). This analysis included a wide range of skeletal characters (i.e., those associated with the skull, dentition, upper body, and digits) and resolved a mosaic of Neanderthal and modern human traits in individual, 33,000 year-old *H. sapiens* fossils (Figure 8.10; Trinkaus 2007). Likewise, though not

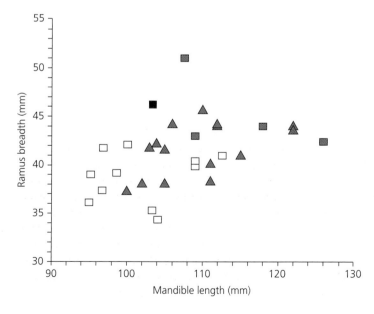

Figure 8.10 Plot of mandibular traits found in the earliest European *H. sapiens* fossils (black square), earliest African *H. sapiens* fossils (gray squares), more recent *H. sapiens* fossils (open squares), and *H. neanderthalensis* (gray triangles) (from Trinkaus 2007; Copyright [2007] National Academy of Sciences, USA).

explicitly designed to test for introgression between *H. sapiens* and *H. neanderthalensis*, a more recent comparison of dental traits present in Neanderthals and extant humans detected traits characteristic of both species in a child buried *c.* 30,000 ybp (Bayle et al. 2010).

The evidence from fossil analyses within the genus *Homo* suggests a complex pattern of speciation and evolution that includes genetic exchange. As discussed, this is particularly apparent for Neanderthals and anatomically modern humans. Indeed, the evidence for introgression between these latter two lineages even led Wolpoff (2009) to recommend that Neanderthals be relegated to subspecific status within *H. sapiens*. In the following sections, I will review the genomic evidence that has provided additional tests for the model of divergence-with-introgression within the *Homo* species complex. I will discuss the whole-genome and population genomic data sets used to test for ancient admixture between taxa located within and outside the continent of Africa.

Genomics and evidence of admixture between African lineages of Homo

The advent of genome analyses, beginning with mitochondrial DNA, provided tests of alternative models of human evolution. As emphasized by Disotell (2012), these earliest studies led to the inference that *H. sapiens* evolved recently within Africa, migrated out of this region, and replaced its congeners without any significant introgression. This recent African origin (i.e. RAO) model held sway for two decades, in spite of the growing number of fossil and genomic data sets as well as theoretical treatments that were consistent with ancient admixture (e.g., Stringer and Andrews 1988; Hawks and Wolpoff 2001; Takahata et al. 2001; Templeton 2002; Garrigan et al. 2005; Ackermann et al. 2006; Lordkipanidze et al. 2007; DeGiorgio et al. 2009). It is now understood that one of the main reasons that human evolution, viewed from the earlier molecular studies, appeared to fit so well with the RAO model was due to the limitations of mtDNA data. In this regard, Takahata et al. (2001) concluded that "our inability to detect ancient admixture results largely from single-locus information, and many independent regions are needed to improve the power."

By analyzing DNA sequences from X-, Y-, and autosomal loci, Takahata et al. (2001) did indeed detect ancestry within extant *H. sapiens* from both ancient African and Asian lineages. In fact, even with the limitations of being a single, non-recombining molecular data source, some of the sequence variation from the mitochondrial genome was also consistent with an admixture model. Thus, mtDNA haplotypes identified as being of African origin were present in clades of non-African samples (Ingman et al. 2000) and estimates of the number of mtDNA introgression events were found to be low, but not zero (Currat and Excoffier 2004).

The application of genome sequencing and population genomics to samples of both extant and extinct species within *Homo*, in concert with new analytical techniques, has caused a paradigm shift with regard to the estimation of the significance of genetic exchange (both ancient and more recent) during the evolution of *H. sapiens* (Wall et al. 2009; Currat and Excoffier 2011; Lalueza-Fox and Gilbert 2011; Alves et al. 2012; Schlebusch et al. 2012; Meyer et al. 2014; Yang et al. 2014). Though many of the studies testing for reticulate evolution in *Homo* have focused on lineages outside of Africa (e.g., Yotova et al. 2011), a number of analyses have also allowed an assessment of whether or not introgressive hybridization occurred between different species within the geographic point of origin for *H. sapiens* (Veeramah and Hammer 2014). For example, Plagnol and Wall (2006) detected genomic variation in extant European and West African populations consistent with introgression between archaic species and *H. sapiens*. These authors estimated that the degree of introgression from the archaic species into anatomically modern humans was at least 5% (Plagnol and Wall 2006). Similarly, Hammer et al. (2011) utilized sequence data from 61 intergenic, autosomal loci (*c.* 20 kb/locus) for three sub-Saharan African populations, to test a null hypothesis of no admixture between ancient *H. sapiens* and African congeners. Simulations based on these genomic data led to the rejection of the no-introgression model. Specifically, the analyses of Hammer et al. (2011) supported the occurrence of introgressive hybridization between *H. sapiens* and a species that diverged *c.* 0.7 mya from a common ancestor. The admixture event was estimated to have occurred approximately 35,000 ybp, with 2% of the genomes of the sub-Saharan African populations being of archaic origin (Hammer et al. 2011). Likewise, whole-genome sequencing of individuals from three hunter-gatherer populations, also of sub-Saharan origin, detected introgressed regions reflecting ancient hybridization between either members of each hunter-gatherer class, or the shared ancestor of all three classes (Lachance et al. 2012).

Genomics and evidence of admixture between non-African lineages of Homo

The genome sequences (both nuclear and mitochondrial) for individuals from archaic *Homo* taxa as well as from fossils of anatomically modern humans from previous millennia (e.g., Green et al. 2010; Reich et al. 2010; Lazaridis et al. 2014; Meyer et al. 2014; Prüfer et al. 2014) have provided the most rigorous tests to-date for divergence-with-introgression within this genus. Because of the availability of such "ancient" genomic data sets, Pickrell and Reich (2014) could thus argue that "accessing the genetic make-up of populations living at archaeologically known times and places . . . makes it possible to directly track migrations and responses to natural selection." Furthermore, the analysis of genomic sequences from past populations of extinct *Homo* species and *H. sapiens* allows the testing for not only genetic exchange, but also the spatial and taxonomic diversity involved in introgression events. Therefore, when Green et al. (2010) and Reich et al. (2010) reported the transfer of genomic sequences into *H. sapiens*, their observations indicated events involving hybridization with two separate archaic taxa in very different geographic regions— Neanderthals (Europe) and Denisovans (Siberia), respectively. More recent analyses have identified an even more complex contribution of different archaic lineages in producing the mosaic genomes of ancient and present-day humans. For example, Neanderthal genomic material has been identified in all non-African and even some African populations of *H. sapiens* (Fu et al. 2013; Wall et al. 2013), while Denisovan introgression was detected primarily within Oceanic and Asian *H. sapiens* populations (Abi-Rached et al. 2011; Rasmussen et al. 2011; Skogland and Jakobsson 2011; Mendez et al. 2012; Meyer et al. 2012). Finally, Prüfer et al. (2014) not

only detected similar patterns of introgression into *H. sapiens* by *H. neanderthalensis* and Denisovans, they also inferred possible introgression between Neanderthals and Denisovans as well as from an unidentified hominin into the Denisovan lineage. Thus, the inferences emerging from these genomic analyses reflect the observations for other primate assemblages; introgression occurs when different taxa overlap in space and time (Arnold and Meyer 2006; Arnold 2009).

Analyses of ancient samples (i.e., 35,000—45,000 ybp) of *H. sapiens* that lived in various sites throughout present-day Europe, Siberia, and China have also helped to refine the understanding of the temporal and spatial components of archaic introgression (Fu et al. 2013, 2014; Seguin-Orlando et al. 2014). In addition, the results from studies of ancient *H. sapiens* samples have uncovered mechanisms of genomic evolution. Thus, in contrast to the regions of Neanderthal and Denisovan introgressed material in contemporaneous *H. sapiens*, those found in the human genomes from nearer the time period of introgression are of much longer length (Fu et al. 2014; Seguin-Orlando et al. 2014). This pattern fits with the prediction of continued recombination within the introgressed blocks during the subsequent matings by the hybrids formed between *H. sapiens* and the archaic species (Fu et al. 2014; Seguin-Orlando et al. 2014). Therefore, samples closer to the initial hybridization event between anatomically modern humans and now-extinct lineages had longer stretches of introgressed material that were subsequently fragmented by meiotic recombination.

Genomics and evidence of adaptive introgression from archaic species into H. sapiens

Garrigan and Kingan (2007) reflected that, given admixture between archaic species of *Homo* and *H. sapiens*, an "intriguing hypothesis is that the expanding anatomically modern human population acquired locally adapted genetic variants from endemic archaic populations." A portion of the data supporting adaptive introgression into *H. sapiens* came from analyses of allelic variation at the *microcephalin* locus, *MCPH1*, a gene contributing to brain size (Jackson et al. 2002). The observation of increased levels of both allelic variation and rates of change at *MCPH1*, especially within humans, led to the inference of strong positive selection affecting this locus (Wang and Su 2004; Evans et al. 2005). The putative adaptive evolutionary change demonstrated by *microcephalin* provided a motivation for Evans et al. (2006) to determine the age and source of the alleles present in extant *H. sapiens* populations. Figure 8.11 reflects a model constructed to explain the following findings (derived from the results of Evans et al. 2004, 2005, 2006): (1) the non-D (*H. sapiens'*) and the D (*H. neanderthalensis'*) *MCPH1* alleles diverged from a common hominin ancestor *c.* 1.7 mya; (2) *c.* 37,000 ybp asymmetric introgression transferred the D allele from *H. neanderthalensis* into *H. sapiens*; and (3) strong, directional selection favored the D allele leading to its current worldwide distribution and ~70% frequency in human populations. If the introduced allele was indeed favored, the evolutionary process reflects adaptive trait introgression (Evans et al. 2006).

As with estimates of the taxonomic and geographic components of archaic × anatomically modern human admixture, tests for adaptive introgression have benefited greatly from the production of whole-genome sequences for both extinct and extant lineages (Hawks 2013). The comparisons of the genomes of the different *Homo* taxa have thus identified numerous introgressed alleles in *H. sapiens* that affect fitness. Though some of the reported introgression events resulted in a reduction of

fitness in at least present-day humans (e.g., loci contributing to diseases in *H. sapiens*—prostate cancer, Ding et al. 2014a; type 2 diabetes, Sankararaman et al. 2014), the majority of the effects reported to-date have reflected the action of positive selection for the introgressed alleles. Thus, a wide range of phenotypes expressed in humans, including lipid metabolism, skin color, response to UV radiation, response to limited O_2, hair characteristics, and immunity (Abi-Rached et al. 2011; Mendez et al. 2013; Ding et al. 2014a, b; Huerta-Sánchez et al. 2014; Khrameeva et al. 2014; Sankararaman et al. 2014; Vernot and Akey 2014) suggest not only the involvement of introgressive hybridization, but also the action of directional [positive] selection on the introduced alleles.

Two cases of apparent adaptive trait introgression into *H. sapiens* from Denisovans and *H. neanderthalensis* are reflected in responses to hypoxia and ultraviolet-B irradiation, respectively. With regard to adaptation to the hypoxic environments associated with extreme altitudes, Huerta-Sánchez et al. (2014) resequenced the region around the hypoxia pathway gene, *EPAS1*, in 40 Tibetan and 40 Han Chinese individuals. The motivation for the choice of this transcription factor (induced under hypoxic conditions) derived from previous work identifying *EPAS1* as having the strongest signal of positive selection in Tibetan populations (e.g., Yi et al. 2010).

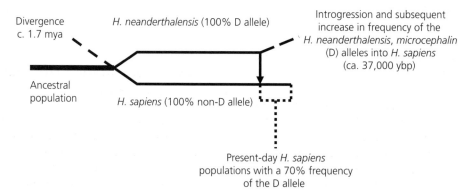

Figure 8.11 Model of the evolutionary history of alleles found at the microcephalin locus, *MCPH1*, in extant *H. sapiens* populations. The model reflects the ancient origin of the non-D and D alleles from a common [hominin] ancestor and the introgression of the D allele from *H. neanderthalensis* into *H. sapiens c.* 37,000 ybp. The increase of the D allele to a frequency of *c.* 70% in extant human populations suggests positive directional selection and thus adaptive trait introgression (from Evans et al. 2006).

The resequencing analysis, along with the examination of the genotypes at this locus across worldwide samples of *H. sapiens*, resulted in the following inferences: (1) the unique genotypic structure at this locus was due to introgression of *EPAS1* alleles from the Denisovan lineage; and (2) the *EPAS1* alleles shared at high frequency by Denisovans and Tibetans do not occur in high frequency in other human populations. The observation of strong positive selection at the *EPAS1* locus in Tibetan populations and its origin from an introgression event between *H. sapiens* and Denisovans suggested "that admixture with other hominin species has provided genetic variation that helped humans to adapt to new environments" (Huerta-Sánchez et al. 2014). In a similar vein, Ding et al. (2014a) provided evidence of adaptive introgression involving genes affecting the cellular response to ultraviolet-B exposure.

Like the *EPAS1* locus, the chromosome region known as *HYAL*, containing 18 genes, demonstrated sequence variation indicating introgression from an archaic donor (in this instance, from *H. neanderthalensis*) into *H. sapiens*. Furthermore, and also like the case of *EPAS1*, there were signatures indicative of positive selection favoring the introduced allelic variation in the *H. sapiens* genomic background, particularly affecting genes associated with protection against UVB irradiation. Ding et al. (2014a) thus detected a significantly higher frequency of the introgressed alleles in East Asians than Europeans and evidence for a latitude-dependent selective gradient. These findings implicated positive selection resulting in adaptive trait transfer with the catalyst for the selective response being differential fitness of the introgressed and non-introgressed genotypes in the presence of different degrees of UV exposure (Ding et al. 2014a).

These studies of fossil and genomic traits of *H. sapiens* and its now extinct congeners provide the evidence needed to once-and-for-all falsify the null hypothesis that the genus *Homo* is somehow different from all other primates in terms of being a non-introgressing clade. Instead, this assemblage resembles baboons, or howler monkeys, or gorillas, or chimpanzees, or indeed any primate clade considered; its evolutionary history is characterized by divergence-with-gene-flow, some leading to new adaptations in anatomically modern humans.

8.6 Genetic exchange and the ecological setting of *Homo sapiens*

In Sections 8.7 to 8.9, I will address the role of divergence-with-genetic-exchange in the evolutionary history of a few of the organisms that provide the ecological setting for humans. Necessarily, this discussion will touch upon a fraction of the available data for organisms upon which *H. sapiens* depends for survival and improved quality of life (e.g., clothing, food, companions, drugs, ornamentation). Furthermore, I will not discuss the many organisms that pose a threat to individuals and populations (i.e., disease vectors and pathogens; but see discussions in previous chapters). That this discussion will be incredibly abbreviated—in light of possible subjects and examples—can be illustrated by the observation that this topic was covered in *c.* 100 pp. in *Reticulate Evolution and Humans—Origins and Ecology*. Furthermore, since this book appeared in 2009, research concerning the genomic mosaicism of cotton, pigs, grapes, horses, schistosomes, black flies, and so much more has gone through what can be accurately typified as an explosive radiation (e.g., Rezaei et al. 2010; Besnard et al. 2013; Pagès et al. 2013; Rogers et al. 2014; Triplett et al. 2014; Willis et al. 2014). In the following discussion, I will thus reflect a very small portion of the breadth of categories of organisms that are components of the *H. sapiens* ecological backdrop. Specifically, I will describe genetic exchange-mediated evolution among plants and animals from which we derive the benefits of food, companionship, and relaxation.

8.7 Genetic exchange and the evolution of human food sources

Humans utilize a myriad of plants, animals, and microorganisms as food sources. Furthermore, it is difficult to identify a source of nourishment that our species has adopted that does *not* reflect the web-of-life. Animal protein sources ranging from sea urchins, eels, kangaroos, ibex, pheasants, turtles, mussels, and even extinct species such as mammoths, illustrate the process of divergence-with-gene-flow (Arnold 2004; Addison and Pogson

2009; Neaves et al. 2010; Enk et al. 2011; Vilaça et al. 2012; Dong et al. 2013; Gosset and Bierne 2013; Grossen et al. 2014; Pujolar et al. 2014). Likewise, the vast majority, if not all, of the plant species providing food from flowers, fruits, leaves, seeds, stems, and roots also contain mosaic genomes (e.g., Meyer et al. 2012; Roullier et al. 2013; Bock et al. 2014; Chin et al. 2014; Kaur et al. 2014; Mandáková et al. 2014; Morrell et al. 2014). Thus, domesticated plants and animals reflect the various processes associated with reticulate evolution, including hybrid speciation (both homoploid and allopolyploid), introgressive hybridization, and horizontal gene transfer (Heiser 1979; Brown et al. 2009; Soltis and Soltis 2009; Larson and Fuller 2014). To illustrate the role played by reticulate evolutionary processes in the origin and diversification of domesticated plants and animals, I will consider studies of apples, maize, pigs, and chickens. These taxa thus reflect common evolutionary processes in dicots, monocots, mammals, and birds.

8.7.1 Genetic exchange and the evolution of human food sources: Apples

The angiosperm family Rosaceae contains a variety of fleshy-fruited cultivars (Potter et al. 2007) including members of the genera *Aronia* (i.e., chokeberry), *Cydonia* (i.e., quince), *Pyrus* (i.e., pear), and *Malus* (i.e., apples). One of the common inferences drawn for clades within this family is that evolutionary history has been reticulate rather than simply bifurcating. For example, Lo and Donahue

(2012) reported molecular data reflecting the likely hybrid origin of the genera *Micromeles* and *Pseudocydonia*. Likewise, the wild species from which domesticated flowering cherry was derived, *Prunus yedoensis*, has recently been shown to be a hybrid species (Cho et al. 2014); many members of the *Prunus* assemblage reflect hybrid derivations (Chin et al. 2014). Finally, the discordance/lack of resolution found for members of the apple genus, *Malus* (Lo and Donahue 2012; Nikiforova et al. 2013), could well reflect past reticulate evolution between and within species (Cornille et al. 2013, 2014).

Divergence-with-gene-flow has been of particular significance during the evolutionary history of the domesticated apple, *Malus domestica* (Figure 8.12; Harris et al. 2002; Gross et al. 2014). The types of genetic exchange events during the development and spread of cultivated apples into their current range have included introgression from wild forms into the domesticated lineage and from the cultivar into nearby populations of wild *Malus* species (Figure 8.12; Coart et al. 2006; Cornille et al. 2012, 2015). For example, "Bidirectional gene flow between the domesticated apple and the European crabapple [i.e., *M. sylvestris*] resulted in the current *M. domestica* being genetically more closely related to this species than to its Central Asian progenitor, *M. sieversii*" (Figure 8.12; Cornille et al. 2012). However, though there is an overall, greater similarity of *M. domestica* with *M. sylvestris*, genomic analyses have detected contributions from a number of species to the evolutionary history of the domesticated apple.

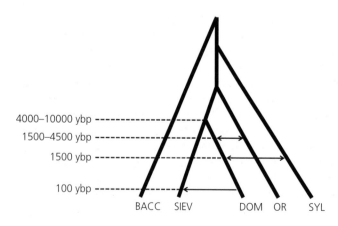

Figure 8.12 Evolutionary relationships among wild (*M. sylvestris* = SYL; *M. sieversii* = SIEV; *M. orientalis* = OR; *M. baccata* = BACC) and cultivated apples (*M. domestica* = DOM). Approximate dates of the domestication and hybridization events between wild and cultivated species are indicated with horizontal lines. Double-headed arrows indicate bidirectional introgression and the single-headed arrow unidirectional introgression (from Cornille et al. 2014).

As illustrated in Figure 8.12, *M. domestica* arose from the Central Asian species *M. sieversii*, and subsequently was transported westward along the Silk Route during which introgression from the wild species *M. baccata* (Siberia), *M. orientalis* (Caucasus), and finally *M. sylvestris* (Europe; Cornille et al. 2012, 2014) occurred. As expected, Cornille et al. (2012) detected a greater relative contribution of *ancient* introgression from the original progenitor *M. sieversii* into *M. domestica*. The amount of *recent* introgression was estimated to be greatest from *M. sylvestris*, with 26% of the cultivars possessing material from the European crabapple, but with relatively a small percentage from other species (i.e., *M. sieversii*, *M. orientalis*, and *M. baccata* contributed *c.* 2%, 3%, and 0.02% to the *M. domestica* genotypes, respectively; Cornille et al. 2012). The evolutionary history of the genus *Malus*, including the species complex that gave rise to domesticated apples, thus reflects the web-of-life both within and between the various species.

8.7.2 Genetic exchange and the evolution of human food sources: Maize

The evolutionary history of the clade containing domesticated maize (i.e., *Zea mays* ssp. *mays*) has included hybrid speciation and both ancient and recent introgressive hybridization. With regard to the hybrid speciation event, molecular analyses have provided evidence of the formation of the *Zea* lineage via either allotetraploidization or homoploidy (Gaut and Doebley 1997; Wei et al. 2007). Thus, the uncertainty of defining maize evolutionary history, due to the apparent multifaceted processes leading to its origin, has been apparent from hypotheses concerning the first step in its formation. For example, different data sets have led to the inference of chromosome numbers in the two progenitors ranging from 5–10 (Molina and Naranjo 1987; Wilson et al. 1999; Wei et al. 2007). Though the detection of duplicated chromosomal segments suggested the occurrence of allopolyploid formation (Gaut and Doebley 1997), it was hypothesized that the patterns of genomic variation could also be explained by divergent progenitors with a haploid number of ten chromosomes (Wei et al. 2007). Given that the progenitors possessed n = 10 (as does maize), the production of this clade might have been initiated through homoploid hybrid speciation rather than allotetraploidy (Wei et al. 2007).

Regardless of whether the formation of the *Zea* lineage/clade involved WGD or homoploidy, the first step was reticulate (Gaut and Doebley 1997; Wei et al. 2007). Likewise, the evolutionary history of the wild and cultivated species and subspecies in the *Zea* assemblage reflects divergence-with-introgression (Figure 8.13). In particular, the work of Doebley, Ross-Ibarra, and their colleagues (Doebley et al. 1984; Doebley 1989; Matsuoka et al. 2002; Fukunaga et al. 2005; Vigouroux et al. 2008; Bitocchi et al. 2009; Ross-Ibarra et al. 2009; van Heerwaarden et al. 2011; Hufford et al. 2013; Mir et al. 2013; da Fonseca et al. 2015) has revealed multiple instances of reticulate processes affecting the origin and adaptive evolution of *Zea* lineages (Figure 8.13). For example, Ross-Ibarra et al. (2009) tested four models predicting the presence/absence and timing of introgression during the derivation of three wild taxa, *Z. luxurians*, *Z. m.* ssp. *parviglumis* (the progenitor of maize), *Z. m.* ssp. *mexicana*, and domesticated *Z. m.* ssp. *mays*. Ross-Ibarra et al. (2009) defined patterns of genomic variability that supported some level of introgression (ancient and/or recent) in a majority of the intertaxonomic comparisons (Figure 8.13).

One of the paradoxical observations associated with maize relates to the adaptation of some landraces that are genomically similar to the lowland-occurring progenitor, *Z. m.* ssp. *parviglumis*, to high elevation environments (Hufford et al. 2012). Likewise, there has also been apparent derivation of additional landraces in the Southwestern United States from highland landraces (da Fonseca et al. 2015). The studies of van Heerwaarden et al. (2011), Hufford et al. (2013), and da Fonseca et al. (2015) each provided evidence that genetic exchange catalyzed the spread of *Z. m.* ssp. *mays* first from lowland into highland habitats and subsequently into the divergent habitats in the US Southwest (Figure 8.13). Specifically, these population genomic analyses detected evidence of adaptive introgression from both *Z. m.* ssp. *mexicana* into maize as well as among different maize landraces. Thus, the derivation of maize landraces adapted to highland environments, from landraces developed from *Z. m.* ssp. *parviglumis*, was at least partially

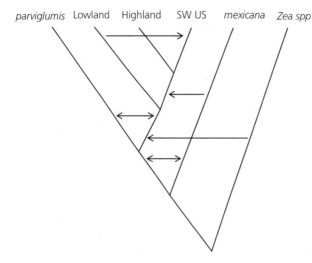

parviglumis Lowland Highland SW US *mexicana* *Zea spp.*

Figure 8.13 Evolutionary relationships among wild ("*Zea spp.*" = species other than *Z. mays*; "*mexicana*" = *Z. mays* ssp. *mexicana*; "*parviglumis*" = *Z. mays* ssp. *parviglumis*) and *Z. mays* ssp. *mays* cultivars ("Lowland" = low-elevation, Mexican populations; "Highland" = high-elevation populations in central Mexico; "SW US" = Southwestern United States populations). Arrows between the various lineages indicate introgression events. Double-headed arrows indicate bidirectional introgression and single-headed arrows indicate unidirectional introgression (from Doebley et al. 1984; Ross-Ibarra et al. 2009; van Heerwaarden et al. 2011; Hufford et al. 2013; da Fonseca et al. 2015).

facilitated by the introduction of loci from the high-elevation adapted *Z. m.* ssp. *mexicana* (Figure 8.13; van Heerwaarden et al. 2011; Hufford et al. 2013). Likewise, the adaptation of the Mexican high elevation landraces to the divergent habitats characterizing the Southwestern United States was promoted by introgression from resident landraces (Figure 8.13; da Fonseca et al. 2015).

Each of the studies discussed confirms the significant role played by web-of-life processes during the evolution of the *Zea* species complex. From the derivation of the entire clade through hybrid speciation (either homoploid or allopolyploid) to introgression among maize cultivars, the entire assemblage reflects not only widespread reticulations, but genetic exchange-mediated adaptive evolution as well.

8.7.3 Genetic exchange and the evolution of human food sources: Pigs

The genus *Sus* has been used as a model system to test for the components of evolutionary diversification. Furthermore, because of the economic importance of *Sus scrofa* (i.e., wild boars and domesticated pig breeds), this assemblage has been investigated in terms of the origin and post-domestication evolutionary processes associated with this species (Ramos-Onsins et al. 2014). The various stages leading to the domesticated pig (reviewed by Bosse

et al. 2014) included: (1) the divergence of *S. scrofa* from the common ancestor of several other *Sus* species in the islands of Southeast Asia, at the beginning of the Pliocene, 3–6 mya (Larson et al. 2005; LAF Frantz et al. 2013); (2) the spread of the wild boar throughout Eurasia *c.* 1.2 mya; (3) the divergence of European and Asian lineages of *S. scrofa* near the time of its spread across Eurasia (Groenen et al. 2012; LAF Frantz et al. 2013); and (4) the independent domestication from the European and Asian wild boars, beginning possibly as early as 10,000 ybp (Bosse et al. 2014).

As with other species assemblages that include domesticated forms, the pig genus provides numerous examples of molecular divergence consistent with divergence-with-gene-flow (Kijas and Andersson 2001; Larson et al. 2005; Mona et al. 2007; Ramírez et al. 2009; Scandura et al. 2011; Groenen et al. 2012; Ottoni et al. 2013; Frantz et al. 2014). These reticulate events occurred both pre- and post-domestication, and involved both domestic pigs and wild taxa. Indeed, Darwin (1868; p. 67) concluded that "the breeds of the *Sus scrofa* type have either descended from, or been modified by crossing with, forms . . . which are, according to some naturalists, distinct species."

In terms of interspecific introgressive hybridization among wild species of *Sus*, LAF Frantz et al. (2013, 2014) documented multiple cases of genetic exchange among various species inhabiting island

and mainland regions of Southeast Asia. For example, they reported asymmetric introgression from *S. verrucosus* into wild populations of *S. scrofa* (LAF Frantz et al. 2014). Likewise, their earlier analysis of whole-genome sequences from several wild *Sus* species (LAF Frantz et al. 2013) detected introgression from *S. verrucosus* and *S. cebifrons* into *S. celebensis*. It appears that in the case of this assemblage of wild *Sus* species (including wild boars), variations in sea level gave rise to alternating corridors and barriers for genetic exchange (LAF Frantz et al. 2013, 2014).

The post-domestication evolution of *S. scrofa* also involved numerous instances of introgressive hybridization. Introgression between divergent domesticated lineages was facilitated by the independent origin of pigs from the highly differentiated European and Asian wild boar populations (Groenen et al. 2012; LAF Frantz et al. 2013). Furthermore, during the domestication process and following the establishment of pigs, genetic exchange (i.e., bidirectional) occurred with the local wild boars (G Larson et al. 2005, 2007; Scandura et al. 2008; AC Frantz et al. 2013; Ottoni et al. 2013; Ramírez et al. 2015). Finally, human-mediated introgression between Asian and European lineages, as well as between breeds developed from a single center of domestication, resulted in extensive admixture (Ramírez et al. 2009; Alves et al. 2010; White 2011; Groenen et al. 2012; Bosse et al. 2014), providing the genetic basis for the present-day industrial swine-production (White 2011). Thus, from interspecific hybridization due to natural environmental perturbations to human-mediated crosses among lineages of wild boar and domestic pig breeds, divergence-with-gene-flow has occurred throughout the evolutionary history and geographic distribution of the genus *Sus*.

8.7.4 Genetic exchange and the evolution of human food sources: Chickens

The domesticated chicken, *Gallus gallus domesticus*, represents a significant protein source and economic factor across human populations. As one example, broiler chicken (i.e., a young animal utilized for meat; United States Department of Agriculture 2012) production in the United

States in 2013 accounted for *c*. 37 billion pounds of ready-to-cook meat (United States Department of Agriculture 2015).

As with domestic pigs, Darwin also contributed to hypotheses concerning the origin of the domestic chicken (Darwin 1868, pp. 233–275), positing that the derivation of *G. g. domesticus* occurred in the area of South or Southeast Asia from the red junglefowl, *G. g. gallus* (i.e., referred to as "*G. bankiva*" by Darwin 1868, pp. 233–275). Darwin's conclusion that the red junglefowl was an ancestor of the domesticated form has been repeatedly supported (e.g., Fumihito et al. 1994, 1996), however, recent genomic analyses have falsified the hypothesis of a monophyletic origin of *G. g. domesticus* from a single geographic region (Wong et al. 2004; Eriksson et al. 2008; Wright et al. 2010; Miao et al. 2013; Xiang et al. 2014). Indeed, evidence has now been obtained indicating that the genus *Gallus* in general has been impacted by ancient and recent genetic exchange events (Nishibori et al. 2005). Likewise, post-domestication introgression has also been detected between wild species and domesticated populations (Berthouly et al. 2009).

With regard to evolution within the genus *Gallus*, Nishibori et al. (2005) utilized both mitochondrial and nuclear sequences to test for the genomic constitution of *G. g. domesticus* relative to its wild congeners. The molecular data also provided a means to test for reticulate evolution among the domesticated chicken and wild taxa, including the red junglefowl, grey junglefowl (*G. sonneratii*), green junglefowl (*G. varius*), and the Ceylon junglefowl (*G. lafayetii*). Figure 8.14 illustrates the results from a phylogenetic analysis based upon sequence data from the nuclear locus, *ornithine carbamoyltransferase*. The discordance, as reflected by the non-monophyly of samples of domesticated chicken samples, as well as the red, grey, and Ceylon junglefowls (Figure 8.14) was also found in phylogenies based upon other nuclear and mitochondrial loci (Nishibori et al. 2005). These phylogenetic patterns (Figure 8.14) led to the conclusion that introgressive hybridization had occurred between *G. sonneratii* and *G. g. domesticus*/*G. g. gallus* as well as between *G. sonneratii* and *G. lafayetii* (Nishibori et al. 2005).

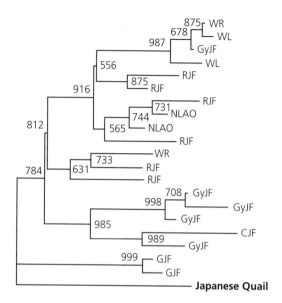

Figure 8.14 Phylogeny for wild species and the domesticated chicken from the genus *Gallus*. The phylogeny was constructed using sequences from the nuclear gene, *ornithine carbamoyltransferase*. The numbers on each node indicate the bootstrap values out of 1000 replicates. The outgroup taxon was the Japanese quail. Designations for *Gallus* samples are as follows: chicken = WR, WL, NLAO; red junglefowl = RJF; grey junglefowl = GyJF; green junglefowl = GJF; Ceylon junglefowl = CJF (from Nishibori et al. 2005).

Hybridization among wild species of *Gallus* and between domesticated and wild populations has also been implicated in the origin and continued evolution of *G. g. domesticus*. Specifically, Eriksson et al. (2008) and Xiang et al. (2014) detected evidence suggesting that the chicken originated from hybridization between the red and grey junglefowls. The earlier of these two studies demonstrated that the *yellow skin* phenotype, found in most chickens from populations in the Western world, derived not from the red junglefowl, but rather from another species. The most likely source for the alleles underlying this phenotype were the grey junglefowl, leading to the hypothesis of a hybrid origin for *G. g. domesticus* (Eriksson et al. 2008). Similarly, Xiang et al. (2014) detected the presence of mitochondrial haplotypes of both *G. g. gallus* and *G. sonneratii* in both ancient (up to 10,000 ybp) and modern *G. g. domesticus* populations.

Finally, genetic exchange has also played a significant role during the period following the origin of the domesticated chicken. In particular, a number of studies have documented the likelihood of bidirectional introgression involving the wild and domesticated lineages (Wong et al. 2004; Berthouly et al. 2009; Miao et al. 2013; Xiang et al. 2014). Thus, ancient and recent divergence-with-gene-flow has apparently helped shape the genomic and phenotypic constitution of both native and domesticated forms belonging to the genus *Gallus*.

8.8 Genetic exchange and the evolution of human companions

The utilization of animals, plants, and microorganisms that provide humans with some measure of protection, relaxation, and companionship has grown, at least for Western cultures, into a major lifestyle and economic factor. For example, during 2013 United States pet owners spent 55.72 billion dollars on their domesticated animals, with *c.* $23 billion paid for pet food alone (American Pet Products Association 2015). I recognize that placing the following two animal clades (i.e., dogs and horses) in the category of "companions," and indeed even having such a category, demonstrates my cultural bias. For the vast majority of *Homo* populations, both ancient and contemporary, these animals have not primarily been companions, but instead protein sources and beasts of burden. Old and New World human populations (including modern-day) have thus utilized *Canis familiaris* and *Equus caballus* (i.e., domestic dogs and horses, respectively) as both beasts of burden and an important portion of their diet (see <http://coombs.anu.edu.au/~vern/wild-trade/eats/eats.html>). For example, *C. familiaris* was utilized extensively during some great gatherings of Native Americans. Thus, at the "Laramie Council" of 1851, "Father De Smet, the famous Catholic missionary, who was there, declared that 'no epoch in Indian annals probably shows a greater massacre of the canine race'" (Ambrose 1996, p. 54). Yet, it is also accurate to typify the, largely Western, outlook of companionship as being as significant for human cultural and economic development as that of domestication as a source of food (e.g., Barnes 2004).

8.8.1 Genetic exchange and the evolution of human companions: Dogs

Reticulate evolutionary processes have been detected in many canid taxa, both under natural conditions and due to human-mediated introductions (e.g., Lehman et al. 1991; Mercure et al. 1993; Roy et al. 1994; Norén et al. 2005; Sacks et al. 2011; Witt et al. 2015). An example of human-caused introgressive hybridization between canid species can be illustrated by the recent introduction of the red wolf (i.e., *Canis rufus*) into the Southeast United States, which subsequently underwent genetic exchange with resident populations of the coyote, *C. latrans* (Miller et al. 2003; Adams et al. 2007; Bohling et al. 2013). Ironically, the effort to minimize introgression between these two species (e.g., through the removal of hybrid animals) in order to conserve the "red wolf" genotypic and phenotypic characteristics might seem ill-advised given the evidence for a history of extensive natural introgressive hybridization between *C. rufus* and both wolves (i.e., *C. lupus*) and coyotes (Wayne and Jenks 1991; Gray et al. 2009).

Conservation directives notwithstanding, introgression among numerous New and Old World wolf-like species has been reported, often involving the domestic dog and/or the coyote (e.g., Figure 8.15; Gottelli et al. 1994; Randi et al. 2000; Andersone et al. 2002; Randi and Lucchini 2002; Adams et al. 2003; Vilà et al. 2003; Verardi et al. 2006; Anderson et al. 2009; Godinho et al. 2011, 2015; Kopaliani et al. 2014; Monzón et al. 2014), and sometimes involving the exchange of adaptive traits (Anderson et al. 2009; Monzón et al. 2014; see Section 1.5.1). For example, an analysis of *C. familiaris* and *C. lupus* from the Caucasus detected extensive admixture with *c.* 10% and 13% of the samples being from introgressed individuals, respectively (Kopaliani et al. 2014). A similar inference of extensive introgression was also reached in an analysis of North American canids, including western and eastern lineages of coyotes and wolves, as well as domestic dogs. Thus, Monzón et al. (2014) concluded that every *C. latrans* sampled possessed genomic material from other North American canids (i.e., western wolves, eastern wolves, and/or domestic dogs). Significantly, as with the example

of coat color introgression from domestic dogs into wolves (Anderson et al. 2009; Section 1.5.1), Monzón et al. (2014) also concluded that at least a portion of the introgression from wolves into eastern coyote populations might have involved adaptive trait introgression. In this regard, they observed that *C. latrans* "living in areas of high deer density are more wolf-like genetically . . . supporting the idea that introgressive hybridization with wolves . . . introduced adaptive genetic variation that allowed coyotes to exploit a prey base rich with ungulates."

The details of the origin and evolution of domestic dogs has been highly contentious (Thalmann et al. 2013). This is likely due to the fact that not only during post-domestication (Sundqvist et al. 2006), but also prior to and during domestication of *C. familiaris* from *C. lupus*, introgressive hybridization occurred between the progenitor lineages and the domesticated form (Figure 8.15; Freedman et al. 2014; Matsumura et al. 2014). For example, Wayne and Ostrander (2007), based upon a number of molecular data sets, inferred "that dogs have a diverse origin in East Asia that subsequently involved multiple contributions from several wolf populations through backcrossing." Similarly, Pang et al. (2009), using mtDNA sequences, concluded that *C. familiaris* arose less than 16,300 ybp from a limited number of southern Chinese *C. lupus* (Pang et al. 2009). In contrast, the conclusions of both Wayne and Ostrander (2007) and Pang et al. (2009) were brought into question by a subsequent analysis of variation in dogs and wolves using *c.* 50,000 SNP loci. Specifically, vonHoldt et al. (2010) uncovered genomic variation indicative of a larger contribution from Middle Eastern wolves than from *C. lupus* from any other region, including East Asia. In addition, they detected evidence of admixture between both domesticated breeds and between *C. familiaris* and local populations of *C. lupus* from different geographic regions (vonHoldt et al. 2010; see also Niskanen et al. 2013). However, a more recent analysis of *C. familiaris* and *C. lupus*, once again, provided data that seemingly falsified previous inferences concerning the evolutionary history of the domesticated dog. Thus, Thalmann et al. (2013) used mitochondrial sequences from both prehistoric and modern dogs and wolves to estimate the temporal and geographical parameters

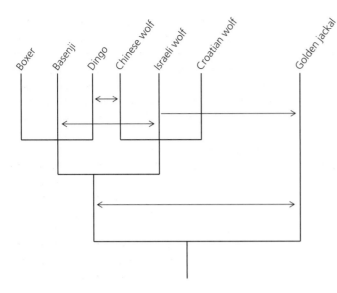

Figure 8.15 Evolutionary relationships among various wolf (i.e., *C. lupus* = Chinese, Israeli, and Croatian wolves) and domesticated/wild dogs (i.e., *C. familiaris* = Boxer and Basenji; *C. dingo* = Dingo) from whole-genome sequence data. Arrows indicate introgression events among the various lineages, with single- and double-headed arrows reflecting uni- and bidirectional genetic exchange, respectively (from Freedman et al. 2014).

of domestication. They concluded that domestication took place as early as 32,000 ybp from European wolf lineages (Thalmann et al. 2013).

The newest hypotheses concerning the evolutionary history of dogs suggested a primary role for introgressive hybridization (Figure 8.15) in causing the alternative inferences concerning when, where, and how *C. familiaris* arose and evolved in the context of other canids. Thus, the analysis of whole genome sequence data for wolves and dogs failed to identify any wolf lineage, from any putative center of domestication, that was more closely related to *C. familiaris*; *C. lupus* lineages from across Eurasia formed a sister clade to domestic dogs (Figure 8.15; Freedman et al. 2014). Like previous analyses, Freedman et al. (2014) detected admixed genotypes indicative of post-domestic introgression. In addition, the whole-genome sequences provided strong evidence of ancient genetic exchange between the various wolf-like canids that would have confounded the analyses designed to test for the geographic and demographic parameters of domestication (Figure 8.15; Freedman et al. 2014). Therefore, the lack of resolution of many of the details of canid evolutionary history, and in particular the origin of the human companion, *C. familiaris*, likely resulted from ancient and contemporaneous admixture.

8.8.2 Genetic exchange and the evolution of human companions: Horses

The original domestication of horses as a food source has been well demonstrated, with this utilization continuing in present-day human populations (Anthony 1986; Chrisafis 2013). However, in addition to its usage as a source of protein, Diamond (1991) has pointed out how the domestication of horses revolutionized human travel, warfare, economics, and social organization. Indeed, he argued that the derivation of *E. caballus* (i.e., from the wild *E. ferus*; see McCue et al. 2012 for a discussion and additional references) provided the catalyst for the origin and spread of Indo-European languages spoken, by the time of the Roman Empire, from Britain to India (Diamond 1991; see also Anthony 1986). Given its important role in human interactions, an understanding of the history and factors associated with the origin and spread of *E. caballus* can provide a mechanism for understanding aspects of an animal with far-reaching effects on human cultural evolution.

Vilà et al. (2001) examined sequence variation in the non-genic control region of mitochondrial DNA from contemporary, domestic horse breeds and from fossil horse remains. One surprising conclusion was that, unlike many other domesticated animals, numerous, ancient matrilineal lineages had been utilized in the domestication of *E. caballus* (Vilà et al. 2001). In fact, the divergences of the various matrilineal lineages detected in present-day breeds ranged from 100,000 to 600,000 ybp (Vilà et al. 2001), reflecting orders of magnitude earlier time periods relative to the estimated domestication at *c.* 6000 ybp (Bennett and Hoffmann 1999). Furthermore, the presence of many, highly divergent mtDNA haplotypes argued against the formation of the domesticated form from a geographically restricted founder. Instead, it was concluded that horse domestication involved the capture of wild *Equus* from an extensive geographic region (Vilà et al. 2001).

Since the domestication of *E. caballus*, and as modern breeds were developed, both inbreeding and admixture occurred (McCue et al. 2012; Schubert et al. 2014). Initially, admixture within *E. caballus* was likely due to repeated introgressive hybridization with *E. ferus* individuals as humans spread the domesticated horse from its origin in Eurasia (Warmuth et al. 2012). More recently, significant admixture has also been pursued in horse husbandry resulting in highly variable breeds such as the Quarter Horse, Hanoverian, and Swiss Warmblood (McCue et al. 2012).

The human-mediated introgression between *E. caballus* and its progenitor is analogous to the type of speciation-accompanied-by-genetic-exchange inferred for many of the other members of the *Equus* clade (e.g., *E. asinus asinus*, *E. asinus somaliensis*, *E. zebra hartmannae*, *E. grevyi*, *E. quagga boehmi*; Jónsson et al. 2014). Therefore, like canid lineages, the clade containing the domesticated horse reflects effects before and after human intervention resulting in extensive web-of-life processes for the entire *Equus* assemblage.

8.9 Genetic exchange and the evolution of human drugs

As the Psalmist proclaimed, God makes "wine to gladden the heart of man" (Psalm 104, vs. 15; English Standard Version). In this last example of organisms from which *H. sapiens* derives benefits I will thus discuss the evolutionary history of *Vitis vinifera* (i.e., the grapevine), a history marked by numerous reticulate processes (This et al. 2006). In this regard, the genome of *V. vinifera*, like other dicots, possesses a genome with contributions from three lineages (Jaillon et al. 2007). This genomic constitution reflects ancient allopolyploidization (i.e., hexaploidization) in the ancestral lineage of all dicots (Jaillon et al. 2007) and represents the first of many reticulate evolutionary events leading to the production of wine grapes.

McGovern et al. (1997) have inferred that viniculture was established during the Neolithic period, between 6000 and 10,500 ybp. There have been alternative inferences concerning the initial stages of the domestication of *V. vinifera*, specifically with regard to the number of regions of domestication and the amount of variation in the progenitor populations. Aradhya et al. (2003) detected high levels of polymorphism at microsatellite loci in cultivated (i.e., *V. vinifera* ssp. *vinifera*) and wild (*V. vinifera* ssp. *sylvestris*) accessions. The overall pattern of variation led to the inference of a single, highly heterogeneous founder population from which all the current cultivars were developed (Aradhya et al. 2003). In contrast, Arroyo-García et al. (2006) posited the existence of at least two separate areas of domestication. Specifically, based upon chloroplast DNA sequence diversity (i.e., representing matrilineal variation), they detected a western Mediterranean and Near Eastern center of domestication (Arroyo-García et al. 2006). Support for this model has come from an investigation of the domestication history of Portuguese grapevines that suggested the contribution of several geographic regions to present-day cultivars (Lopes et al. 2009).

Regardless of the number of regions that contributed to domestication, introgressive hybridization between both the cultivated form and native *V. vinifera* ssp. *sylvestris* and different cultivar lineages has provided genomic heterogeneity within *V. vinifera* ssp. *vinifera* (Bowers et al. 1999; Aradhya et al. 2003; Lopes et al. 2009; De Andrés et al. 2012). One example of human-mediated genomic enrichment was reported for 16 types of Northeastern France wine grapes, including such varieties as Beaunoir, Chardonnay, Dameron, and Peurion; the genomic variation indicated the formation of all 16 forms from a cross between plants of the Pinot and Gouais blanc varieties (Bowers et al. 1999). Likewise, possibly the most famous of the red wine varieties, Cabernet Sauvignon, was produced through hybridization between Cabernet franc and Sauvignon blanc (Bowers and Meredith 1997). In addition to the role of cultivar × cultivar reticulate evolution, post-domestication introgression between the wild progenitor and cultivated grapevines has been detected in many regions. For example, such introgression has been documented in France, Italy, Spain, and Portugal (Di Vecchi-Staraz et al. 2009; Lopes et al. 2009; De Andrés et al. 2012).

These studies point once again to the significant role of both ancient and recent divergence-with-gene-flow in the formation of an organism that benefits *H. sapiens* individuals. Thus, ancient allopolyploidy and introgression between wild and cultivated subspecies and human-mediated hybridization between cultivars has provided the genomic and phenotypic variability that forms the basis of a product that can gladden the human heart.

8.10 Conclusions

As indicated by the title of this chapter and all the examples I have chosen to discuss, this section of the book is intentionally human-centric. However, I believe that there are many unique lessons to be garnered from this penultimate chapter. The first of these is that the *H. sapiens* genome is a mosaic of genetic material from many primate lineages. For example, proto-*Homo* individuals appear to have participated in introgressive hybridization with proto-*Pan* and proto-*Gorilla*. However, this introgression did not end with the origin of the *Homo* clade. In fact, it now appears that multiple members of the *Homo* assemblage contributed to the genomic and phenotypic characteristics of anatomically modern humans. What the material in this chapter also indicates, in a very abbreviated form, is that the evolutionary trajectory of organisms upon which we depend for food, companionship, etc. has been modified significantly by the introduction of foreign genomic material. In short, we are fed, entertained, sheltered, and as discussed in previous chapters, attacked and killed, by organisms that possess genomes compiled from events of introgressive hybridization, viral recombination, and/or horizontal gene transfer. Thus, the evolutionary history of *H. sapiens* and of those organisms making up our species' ecological surroundings was reticulate.

CHAPTER 9

Epilogue

"Hybridization between populations having very different genetic systems of adaptation may lead to several different results." **(Anderson and Stebbins 1954)**

"The introduction of genes from another species can serve as the raw material for an adaptive evolutionary advance." **(Lewontin and Birch 1966)**

"As far as animals are concerned, there can be no doubt that even in certain groups where premating isolating mechanisms . . . have played a major part in speciation, occasional hybridization occurs and may lead to a significant level of genic 'introgression' from one species into another." **(White 1978, p. 346)**

"These ancient HGT events include a gene replacement during the early evolution of the fungi, which could be a defining trait for the kingdom Fungi" **(McDonald et al. 2012)**

"Phylogenetic analysis reveals important discrepancies with the phenotype-based taxonomy. We find extensive evidence for interspecific gene flow throughout the radiation. Hybridization has given rise to species of mixed ancestry." **(Lamichhaney et al. 2015)**

9.1 Genetic exchange is [still] pervasive

I began the final chapters of my two most recent books by considering first a question I often encountered from others—"If hybrids are relatively fit, why don't natural populations consist of hybrids rather than well-defined species?" (Arnold 2006, p. 187). Second, I reflected on the many data sets for viruses, prokaryotes, and eukaryotes resulting in the paradoxical observation (in light of the first question), "the majority (if not all) of organisms have an evolutionary history . . . that includes genetic exchange events" (Arnold 2009, p. 183). To address these seemingly contrary observations, I argued first that genetic exchange had historically, especially by zoologists, been grossly underestimated. In fact, I pointed to my own inadequate understanding (Arnold 1997) to reflect on the universality

of web of life processes. Second, I observed that, if evolutionary lineages across all major clades had diverged with some measure of episodic genetic exchange, then species were not "well-defined" in the typological, philosophical sense of the biological species concept.

These conclusions have significant implications for an appreciation of the roles played by geographic distribution, mutation, genetic drift, and natural selection during evolutionary diversification. Specifically, given the apparent common occurrence of divergence-with-introgression, the separation of populations into non-overlapping geographic distributions, until reproduction is no longer possible, is now seen as a naïve viewpoint for the process of speciation. In fact, the data presented in the previous eight chapters illustrate the pervasiveness of non-allopatric divergence. Given

Divergence with Genetic Exchange. Michael L. Arnold.
© Michael L. Arnold 2016. Published 2016 by Oxford University Press.

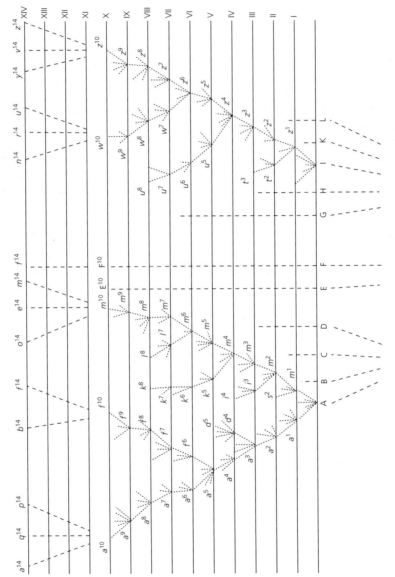

Figure 9.1 Darwin's depiction of evolutionary diversification as a bifurcating tree of life (from Darwin 1859, pp. 514–515).

parapatric or sympatric diversification as a major mechanism by which new lineages evolve, it becomes necessary to address how much greater a role natural selection must play in evolutionary radiations. This paradigm reflects a major transition from previous models constructed during the neo-Darwinian synthesis (Mayr 1942, 1963), but is illustrated by more recently proposed frameworks (e.g., Schluter 2000; Seehausen 2004).

Another application arising from the studies reviewed in this text can be a greater respect for the degree to which genetic exchange has, and continues to, structure the biological world. In this regard, an accurate appreciation for the taxonomic diversity encompassed by genetic exchange should mitigate against the practice of relegating "hybrid" lineages to a category of less value than that occupied by "pure" lineages. In other words, if all organisms possess mosaic genomes, the concept of "purity" becomes nonsensical. I pointed to the danger of such an emphasis in the conservation arena, given the widespread argument that any population of hybrid origin should not be protected from extirpation/extinction.

As mentioned, I too have underestimated the degree to which genetic exchange has affected organismic evolution. Furthermore, the prevalent role for web of life processes reflects the diversity of modes providing bridges for the transfer of genetic material between closely or distantly related organisms. Though I have a great respect and fondness for Darwin's illustration of the evolutionary process (Figure 9.1) as a simple bifurcating, tree-like structure, such a model has been falsified by a wealth of data. Instead, the genotypic, phenotypic, ecological,

and evolutionary relationships among organisms are much more complex. Indeed, because of the degree to which organisms reflect diverse "connections", I have continued to emphasize the web-of-life metaphor to replace the longstanding tree of life. Thus, genetic exchange cannot now be accurately typified as "anomalous" or "rare" or "unimportant." In this light, I hope that a consideration of the concepts and examples reviewed in the preceding chapters will catalyze studies that approach evolutionary diversification as a dynamic process impacted repeatedly by the exchange of genetic material between organisms on different strands of the web-of-life. Given my penchant for neat, clean answers, such a viewpoint leads me to the uncomfortable position of accepting nature as even more messy than I formerly believed it to be.

Although I have concluded that Darwin's tree-of-life metaphor is too simplistic as a representation of the evolutionary process, I concluded *Natural Hybridization and Evolution* (Arnold 1997, p. 185) and *Evolution Through Genetic Exchange* (Arnold 2006, p. 190) with a quote from *On the Origin of Species*. I believe this quote is a wonderful analogy for the web-of-life metaphor. "It is interesting to contemplate an entangled bank, clothed with many plants of many kinds, with birds singing on the bushes, with various insects flitting about, and with worms crawling through the damp earth, and to reflect that these elaborately constructed forms, so different from each other, and dependent on each other in so complex a manner, have all been produced by laws acting around us" (Darwin 1859, p. 489).

It remains accurate to recognize genetic exchange as one of those laws.

Glossary

Adaptive radiation The evolution of ecological and phenotypic diversity within a rapidly multiplying lineage (Seehausen 2004).

Adaptive trait transfer (or adaptive trait introgression) The transfer of genes and thus the phenotype of an adaptive trait through viral recombination, horizontal gene transfer, or introgressive hybridization (see Anderson 1949 and Arnold 2006 for discussion and references).

Allopatric speciation Divergence into separate species when "populations are separated by uninhabited space (even if it is only a very short distance), across which migration (movement) occurs at very low frequency" (Futuyma and Mayer 1980).

Allopolyploid speciation Species formation through hybridization between members of evolutionary lineages with "strongly differentiated genomes" followed by chromosomal doubling, trebling, etc. (Stebbins 1947).

Autopolyploidy A type of polyploidy in which the derivatives originate from crosses within a single evolutionary lineage (Stebbins 1947).

Biological species concept "Species are groups of actually or potentially interbreeding natural populations, which are reproductively isolated from other such groups" (Mayr 1942).

Cohesion species concept "An evolutionary lineage or set of lineages with genetic exchangeability and/or ecological interchangeability" (Templeton 2001).

Conjugation The transfer of DNA through physical contact between donor and recipient cells. This process can mediate the transfer of genetic material between such divergent evolutionary lineages as bacteria and plants (Ochman et al. 2000).

Concerted evolution Molecular processes that facilitate changes in a single genetic element to be incorporated into all genes belonging to a multigene family (Zimmer et al. 1980).

Cyclic parthenogenesis Occurs when "unisexual propagation via eggs is periodically interrupted by a bisexual phase" (Lynch 1984). In the cladoceran *Daphnia*, "cyclic parthenogens produce sexual resting eggs; that is, the eggs are haploid and require fertilization by sperm to develop" (Crease et al. 1989).

Donor Organism that acts as the source for the DNA sequences transferred through horizontal gene transfer.

Gene conversion "a process where a gene converts another homologous gene into its own kind, and it is thought to involve heteroduplex formation" (Ohta and Dover 1983).

Genetic rescue "An increase in population fitness (growth) owing to immigration of new alleles" (Whiteley et al. 2015).

Genic species concept Species are defined on the basis of an integrated set of genes that determine a species' unique set of adaptations, with the process of speciation thus involving "differential adaptation to different natural or sexual environments" (Wu 2001).

Gynogenesis Mode of asexual reproduction in which the presence of sperm from a related bisexual species is required to stimulate egg development in an asexually reproducing female. In concert with the stimulation, "the female's nuclear genome is transmitted intact to the egg, which then develops into an offspring genetically identical to the mother" (Avise et al. 1992).

Homoploid hybrid speciation Species formation through hybridization between members of divergent lineages at the same ploidal level, giving rise to a new, hybrid species at the same ploidal level as the parental lineages (e.g. see Grant 1958; Rieseberg 1997).

Horizontal gene transfer The transfer of genetic material between individuals from two populations, or groups of populations, that are distinguishable on the basis of one or more heritable characters through the processes of transformation, transduction, conjugation, or vector-mediated transfer.

Hybridogenesis Mode of asexual reproduction in which "an ancestral genome from the maternal line is transmitted to the egg without recombination, while paternally-derived chromosomes are discarded only to be replaced in each generation through fertilization by sperm from a related sexual species" (Avise et al. 1992).

Hybrid speciation Process in which natural hybridization results in the production of an evolutionary lineage that is at least partially reproductively isolated from both parental lineages, and which demonstrates a distinct evolutionary and ecological trajectory (Arnold and Burke 2006).

Hybrid species At least partially reproductively isolated lineages arising as a result of natural hybridization. These lineages demonstrate distinct evolutionary and ecological trajectories as defined by distinguishable (and heritable) morphological, ecological, and/or reproductive differences relative to their progenitors (Arnold and Burke 2006).

Hybrid zone Geographical region in which natural hybridization occurs (Arnold 1997 as adapted from Harrison 1990).

Incomplete lineage sorting "could also be called deep coalescence, the failure of ancestral copies to coalesce (looking backwards in time) into a common ancestral copy until deeper than previous speciation events" (Maddison 1997).

Introgressive hybridization or introgression The transfer of DNA between individuals from two populations, or groups of populations, that are distinguishable on the basis of one or more heritable characters via hybridization followed by repeated backcrossing between hybrid and parental individuals (Anderson and Hubricht 1938).

Mosaic genomes As with mosaic tile-work or a quilt, mosaic genomes are mixtures of components from different sources (i.e. parental genomes). Like the different fabrics in a quilt, the sources of the different genomic segments in a mosaic genome are recognizable.

Natural hybrid Offspring resulting from a cross in nature between individuals from two populations, or groups of populations, that are distinguishable on the basis of one or more heritable characters (Arnold 1997 as adapted from Harrison 1990).

Natural hybridization Successful matings in nature between individuals from two populations, or groups of populations, that are distinguishable on the basis of one or more heritable characters (Arnold 1997 as adapted from Harrison 1990).

Neo-functionalization "When one of two duplicate genes acquires a mutation in coding or regulatory sequences that allows the gene to take on a new and useful function" (Prince and Pickett 2002).

Next-generation DNA sequencing "These newer technologies constitute various strategies that rely on a combination of template preparation, sequencing, and imaging, and genome alignment and assembly methods . . . The major advance offered by NGS is the ability to produce an enormous volume of data cheaply in some cases in excess of one billion short reads per instrument run" (Metzker 2010).

Non-functionalization (silencing) "When one of two duplicate genes acquires a mutation in coding or regulatory sequences that ultimately renders the gene nonfunctional" (Prince and Pickett 2002).

Numts mtDNA sequences translocated into nuclear genomes of humans and other primates. These nuclear insertions of mtDNA represent a potential source for artifactual data.

Parthenogenesis Reproduction "in which the female's nuclear genome is transmitted intact to the egg, which then develops into an offspring genetically identical to the mother" (Avise et al. 1992).

Phylogenetic species concept Two concepts can be grouped under this general heading, character-based and history-based (Baum and Donoghue 1995). The [character-based] phylogenetic species is "the smallest aggregation of populations (sexual) or lineages (asexual) diagnosable by a unique combination of character states in comparable individuals" (Nixon and Wheeler 1990). The [history-based] phylogenetic species is "an exclusive group of organisms whose members are more closely related to each other than to any organisms outside the group" (Olmstead 1995).

Prokaryotic species concept Two concepts can be placed under this category. A [phylo-phenetic] species is "a monophyletic and genomically coherent cluster of individual organisms that show a high degree of overall similarity in many independent characteristics, and is diagnosable by a discriminative phenotypic property" (Rosselló-Móra and Amann 2001). A [phylogenetic and ecological divergence] species is "a group of organisms whose divergence is capped by a force of cohesion; divergence between different species is irreversible; and different species are ecologically distinct" (Cohan 2002).

Recipient Organism that receives the DNA from the donor through horizontal gene transfer.

Reticulate evolution Web-like phylogenetic relationships reflecting genetic exchange (through horizontal gene transfer, viral recombination, introgressive hybridization) between diverging lineages.

Sub-functionalization "after duplication, the two gene copies acquire complementary loss-of-function mutations in independent subfunctions, such that both genes are required to produce the full complement of functions of the single ancestral gene" (Prince and Pickett 2002).

Sympatric speciation (1) The origin of reproductive isolation between two lineages in the absence of geographic separation (Mayr 1963, p. 449) or (2) "The emergence of new species from a population where mating is random with respect to the place of birth of the mating partners" (Gavrilets 2003).

Syngameon "an habitually interbreeding community" (Lotsy 1931) or "the most inclusive unit of interbreeding in a hybridizing species group" (Grant 1981).

Transcriptomic shock "Hybrids and allopolyploids typically exhibit radically altered gene expression patterns relative to their parents" (Xu et al. 2014).

Transduction The transfer of DNA through a bacteriophage intermediate (Ochman et al. 2000).

Transformation The uptake of naked DNA from the environment (Ochman et al. 2000).

Transmission ratio distortion "occurs when one of the two alleles from either parent is preferentially transmitted to the offspring. This leads to a statistical departure from the Mendelian law of inheritance, which states that each of the two parental alleles is transmitted to offspring with a probability of 0.5" (Huang et al. 2013).

Vector-mediated transfer The transfer of DNA from a donor to a recipient by a vector intermediate (e.g. between insect species through a shared parasite).

References

Abbott R, Albach D, Ansell S, Arntzen JW, Baird SJE, Bierne N, *et al.* (2013). Hybridization and speciation. *Journal of Evolutionary Biology* **26**, 229–246.

Abbott RJ and Brennan AC (2014). Altitudinal gradients, plant hybrid zones and evolutionary novelty. *Philosophical Transactions of the Royal Society of London B* **369**, 20130346.

Abi-Rached L, Jobin MJ, Kulkami S, McWhinnie A, Dalva K, Gragert L, *et al.* (2011). The shaping of modern human immune systems by multiregional admixture with archaic humans. *Science* **334**, 89–94.

Acevedo P, Jiménez-Valverde A, Melo-Ferreira J, Real R, and Alves PC (2012a). Parapatric species and the implications for climate change studies: A case study on hares in Europe. *Global Change Biology* doi: 10.1111/j.1365-2486.2012.02655.x.

Acevedo P, Melo-Ferreira J, Real R, and Alves PC (2012b). Past, present and future distributions of an Iberian endemic, *Lepus granatensis*: Ecological and evolutionary clues from species distribution models. *PLoS ONE* **7**, e51529.

Achilli A, Olivieri A, Pellecchia M, Uboldi C, Colli L, Al-Zahery N, *et al.* (2008). Mitochondrial genomes of extinct aurochs survive in domestic cattle. *Current Biology* **18**, R157–R158.

Achtman M, Morelli G, Zhu P, Wirth T, Diehl I Kusecek B, *et al.* (2004). Microevolution and history of the plague bacillus, *Yersinia pestis. Proceedings of the National Academy of Sciences USA* **101**, 17837–17842.

Achtman M, Zurth K, Morelli G, Torrea G, Guiyoule A, and Carniel E (1999). *Yersinia pestis*, the cause of plague, is a recently emerged clone of *Yersinia pseudotuberculosis. Proceedings of the National Academy of Sciences USA* **96**, 14043–14048.

Ackermann R (2007). Craniofacial variation and developmental divergence in primate and human evolution, pp. 262–279. In Novartis Foundation Symposium 284. Tinkering: The micro-evolution of development. Wiley, Chichester, U.K.

Ackermann RR (2010). Phenotypic traits of primate hybrids: Recognizing admixture in the fossil record. *Evolutionary Anthropology* **19**, 258–270.

Ackermann RR and Bishop JM (2010). Morphological and molecular evidence reveals recent hybridization between gorilla taxa. *Evolution* **64**, 271–290.

Ackermann RR, Brink JS, Vrahimis S, and de Klerk B. (2010). Hybrid wildebeest (Artiodactyla: Bovidae) provide further evidence for shared signatures of admixture in mammalian crania. *South African Journal of Science* **106**, doi: 10.4102/sajs. v106i11/12.423.

Ackermann RR, Rogers J, and Cheverud JM (2006). Identifying the morphological signatures of hybridization in primate and human evolution. *Journal of Human Evolution* **51**, 632–645.

Ackermann RR, Schroeder L, Rogers J, and Cheverud JM (2014). Further evidence for phenotypic signatures of hybridization in descendent baboon populations. *Journal of Human Evolution* **76**, 54–62.

Adamowicz SJ, Gregory TR, Marinone MC, and Hebert PDN (2002). New insights into the distribution of polyploid *Daphnia*: The Holarctic revisited and Argentina explored. *Molecular Ecology* **11**, 1209–1217.

Adamowicz SJ, Hebert PDN, and Marinone MC (2004). Species diversity and endemism in the *Daphnia* of Argentina: A genetic investigation. *Zoological Journal of the Linnean Society* **140**, 171–205.

Adams RI, Goldberry S, Whitham TG, Zinkgraf MS, and Dirzo R (2011). Hybridization among dominant tree species correlates positively with understory plant diversity. *American Journal of Botany* **98**, 1623–1632.

Adams JR, Leonard JA, and Waits LP (2003). Widespread occurrence of a domestic dog mitochondrial DNA haplotype in southeastern US coyotes. *Molecular Ecology* **12**, 541–546.

Adams JR, Lucash C, Schutte L, and Waits LP (2007). Locating hybrid individuals in the red wolf (*Canis rufus*) experimental population area using a spatially targeted sampling strategy and faecal DNA genotyping. *Molecular Ecology* **16**, 1823–1834.

Addison JA and Pogson GH (2009). Multiple gene genealogies reveal asymmetrical hybridization and introgression among strongylocentrotid sea urchins. *Molecular Ecology* **18**, 1239–1251.

Adu F, Iber J, Bukbuk D, Gumede N, Yang S-J, Jorba J, *et al.* (2007). Isolation of recombinant type 2 vaccine-derived poliovirus (VDPV) from a Nigerian child. *Virus Research* **127**, 17–25.

Agol VI (2006). Vaccine-derived polioviruses. *Biologicals* **34**, 103–108.

Agostini I, Holzmann I, and Di Bitetti MS (2008). Infant hybrids in a newly formed mixed-species group of howler monkeys (*Alouatta guariba clamitans* and *Alouatta caraya*) in northeastern Argentina. *Primates* **49**, 304–307.

Aguiar LM, Mellek DM, Abreu KC, Boscarato TG, Bernardi IP, Miranda JMD, and Passos FC (2007). Sympatry between *Alouatta caraya* and *Alouatta clamitans* and the rediscovery of free-ranging potential hybrids in Southern Brazil. *Primates* **48**, 245–248.

Aguiar LM, Pie MR, and Passos FC (2008). Wild mixed groups of howler species (*Alouatta caraya* and *Alouatta clamitans*) and new evidence for their hybridization. *Primates* **49**, 149–152.

Ainouche ML, Baumel A, Salmon A, and Yannic G (2003). Hybridization, polyploidy and speciation in *Spartina* (Poaceae). *New Phytologist* **161**, 165–172.

Alberici da Barbiano L, Rangel L, Aspbury AS, and Gabor CR (2012). Male permissiveness in a unisexual-bisexual mating complex promotes maintenance of a vertebrate unisexual sperm-dependent species. *Behaviour* **149**, 869–879.

Alberts SC and Altmann J (2001). Immigration and hybridization patterns of yellow and anubis baboons in and around Amboseli, Kenya. *American Journal of Primatology* **53**, 139–154.

Alcaide M, Scordato ESC, Price TD, and Irwin DE (2014). Genomic divergence in a ring species complex. *Nature* **511**, 83–85.

Alcázar R, von Reth M, Bautor J, Chae E, Weigel D, Koornneef M, and Parker JE (2014). Analysis of a plant complex resistance gene locus underlying immune-related hybrid incompatibility and its occurrence in nature. *PLoS Genetics* **10**, e1004848.

Aldridge G (2005). Variation in frequency of hybrids and spatial structure among *Ipomopsis* (Polemoniaceae) contact sites. *New Phytologist* **167**, 279–288.

Aldridge G and Campbell DR (2006). Asymmetrical pollen success in *Ipomopsis* (Polemoniaceae) contact sites. *American Journal of Botany* **93**, 903–909.

Aldridge G and Campbell DR (2007). Variation in pollinator preference between two *Ipomopsis* contact sites that differ in hybridization rate. *Evolution* **61**, 99–110.

Aldridge G and Campbell DR (2009). Genetic and morphological patterns show variation in frequency of hybrids between *Ipomopsis* (Polemoniaceae) zones of sympatry. *Heredity* **102**, 257–265.

Allendorf FW, Leary RF, Hitt NP, Knudsen KL, Boyer MC, and Spruell P (2005). Cutthroat trout hybridization and the U.S. Endangered Species Act: One species, two policies. *Conservation Biology* **19**, 1326–1328.

Allendorf FW, Leary RF, Hitt NP, Knudsen KL, Lundquist LL, and Spruell P (2004). Intercrosses and the U.S. Endangered Species Act: Should hybridized populations be included as westslope cutthroat trout? *Conservation Biology* **18**, 1203–1213.

Allendorf FW, Leary RF, Spruell P, and Wenburg JK (2001). The problems with hybrids: Setting conservation guidelines. *Trends in Ecology and Evolution* **16**, 613–622.

Alves PC, Ferrand N, Suchentrunk F, and Harris DJ (2003). Ancient introgression of *Lepus* timidus mtDNA into *L. granatensis* and *L. europaeus* in the Iberian Peninsula. *Molecular Phylogenetics and Evolution* **27**, 70–80.

Alves PC, Harris DJ, Melo-Ferreira J Branco M, and Ferrand N (2006). Hares on thin ice: Introgression of mitochondrial DNA in hares and its implications for recent phylogenetic analyses. *Molecular Phylogenetics and Evolution* **40**, 640–641.

Alves PC, Melo-Ferreira J, Freitas H, and Boursot P (2008). The ubiquitous mountain hare mitochondria: multiple introgressive hybridization in hares, genus *Lepus*. *Philosophical Transactions of the Royal Society of London B* **363**, 2831–2839.

Alves PC, Pinheiro I, Godinho R, Vicente J, Gortázar C, and Scandura M (2010). Genetic diversity of wild boar populations and domestic pig breeds (*Sus scrofa*) in South-western Europe. *Biological Journal of the Linnean Society* **101**, 797–822.

Alves I, Šrámková Hanulová A, Foll M, and Excoffier L (2012) Genomic data reveal a complex making of humans. *PLoS Genetics* **8**, e1002837.

Amaral AR, Lovewell G, Coelho MM, Amato G, and Rosenbaum HC (2014). Hybrid speciation in a marine mammal: The clymene dolphin (*Stenella clymene*). *PLoS ONE* **9**, e83645.

Ambrose SE (1996). *Crazy Horse and Custer—The Parallel Lives of Two American Warriors*. Random House Inc, New York.

American Pet Products Association (2015). Pet Industry Market Size & Ownership Statistics. <http://american-petproducts.org/press_industrytrends.asp>

Anderson E (1936). Hybridization in American Tradescantias. *Annals of the Missouri Botanical Garden* **23**, 511–525.

Anderson E (1948). Hybridization of the habitat. *Evolution* **2**, 1–9.

Anderson E (1949). *Introgressive Hybridization*. John Wiley and Sons, Inc, New York.

Anderson DW and Evans BJ (2009). Regulatory evolution of a duplicated heterodimer across species and tissues of allopolyploid clawed frogs (*Xenopus*). *Journal of Molecular Evolution* **68**, 236–247.

Anderson E and Hubricht L (1938). Hybridization in *Tradescantia*. III. The evidence for introgressive hybridization. *American Journal of Botany* **25**, 396–402.

Anderson E and Stebbins GL, Jr (1954). Hybridization as an evolutionary stimulus. *Evolution* **8**, 378–388.

Anderson EC and Thompson EA (2002). A model-based method for identifying species hybrids using multilocus genetic data. *Genetics* **160**, 1217–1229.

Anderson TM, vonHoldt BM, Candille SI, Musiani M, Greco C, Stahler DR, et al. (2009). Molecular and evolutionary history of melanism in North American Gray Wolves. *Science* **323**, 1339–1343.

Andersone Z, Lucchini V, Randi E, and Ozolins J (2002). Hybridisation between wolves and dogs in Latvia as documented using mitochondrial and microsatellite DNA markers. *Mammalian Biology* **67**, 79–90.

Andújar C, Arribas P, Ruiz C, Serrano J, and Gómez-Zurita J (2014). Integration of conflict into integrative taxonomy: Fitting hybridization in species delimitation of *Mesocarabus* (Coleoptera: Carabidae). *Molecular Ecology* **23**, 4344–4361.

Anthony DW (1986). The "Kurgan Culture," Indo-European Origins, and the domestication of the horse: A Reconsideration. *Current Anthropology* **27**, 291–304.

Anthony NM, Johnson-Bawe M, Jeffery K, Clifford SL, Abernethy KA, Tutin CE, et al. (2007). The role of Pleistocene refugia and rivers in shaping gorilla genetic diversity in central Africa. *Proceedings of the National Academy of Sciences USA* **104**, 20432–20436.

Antón SC and Swisher CC III (2004). Early dispersals of *Homo* from Africa. *Annual Review of Anthropology* **33**, 271–296.

Aradhya MK, Dangl GS, Prins BH, Boursiquot J-M, Walker MA, Meredith CP, and Simon CJ (2003). Genetic structure and differentiation in cultivated grape, *Vitis vinifera* L. *Genetical Research* **81**, 179–192.

Ari E, Ittzés P, Podani J, Thi QCL, and Jakó E (2012). Comparison of Boolean analysis and standard phylogenetic methods using artificially evolved and natural mt-tRNA sequences from great apes. *Molecular Phylogenetics and Evolution* **63**, 193–202.

Arias CF, Rosales C, Salazar C, Castaño J, Bermingham E, Linares M, and McMillan WO (2012). Sharp genetic discontinuity across a unimodal *Heliconius* hybrid zone. *Molecular Ecology* **21**, 5778–5794.

Arias CF, Salazar C, Rosales C, Kronforst MR, Linares M, Bermingham E, and McMillan WO (2014). Phylogeography of *Heliconius cydno* and its closest relatives: Disentangling their origin and diversification. *Molecular Ecology* **23**, 4137–4152.

Arioli M, Jakob C, and Reyer H-U (2010). Genetic diversity in water frog hybrids (*Pelophylax esculentus*) varies with population structure and geographic location. *Molecular Ecology* **19**, 1814–1828.

Arita I, Nakane M, and Fenner F (2006). Is polio eradication realistic? *Science* **312**, 852–854.

Arnegard ME, McGee MD, Matthews B, Marchinko KB, Conte GL, Kabir S, et al. (2014). Genetics of ecological divergence during speciation. *Nature* **511**, 307–311.

Arnold ML (1986). The heterochromatin of grasshoppers from the *Caledia captiva* species complex III. Cytological organization and sequence evolution in a dispersed highly repeated DNA family. *Chromosoma* **94**, 183–188.

Arnold ML (1992). Natural hybridization as an evolutionary process. *Annual Review of Ecology and Systematics* **23**, 237–261.

Arnold ML (1993). *Iris nelsonii*: origin and genetic composition of a homoploid hybrid species. *American Journal of Botany* **80**, 577–583.

Arnold ML (1994). Natural hybridization and Louisiana Irises. *BioScience* **44**, 141–147.

Arnold ML (1997). *Natural Hybridization and Evolution*. Oxford University Press, Oxford.

Arnold ML (2000). Anderson's paradigm: Louisiana Irises and the study of evolutionary phenomena. *Molecular Ecology* **9**, 1687–1698.

Arnold ML (2004). Natural hybridization and the evolution of domesticated, pest, and disease organisms. *Molecular Ecology* **13**, 997–1007.

Arnold ML (2006). *Evolution Through Genetic Exchange*. Oxford University Press, Oxford.

Arnold ML (2009). *Reticulate Evolution and Humans—Origins and Ecology*. Oxford University Press, Oxford.

Arnold ML, Appels R, and Shaw DD (1986). The heterochromatin of grasshoppers from the *Caledia captiva* species complex I. Sequence evolution and conservation in a highly repeated DNA family. *Molecular Biology and Evolution* **3**, 29–43.

Arnold ML, Ballerini ES, and Brothers AN (2012). Hybrid fitness, adaptation and evolutionary diversification: lessons learned from Louisiana Irises. *Heredity* **108**, 159–166.

Arnold ML and Bennett BD (1993). Natural hybridization in Louisiana irises: genetic variation and ecological determinants. In RG Harrison, ed., *Hybrid Zones and the Evolutionary Process*, pp. 115–139. Oxford University Press, Oxford.

Arnold ML, Bennett BD, and Zimmer EA (1990a). Natural hybridization between *Iris fulva* and *I. hexagona*: Pattern of ribosomal DNA variation. *Evolution* **44**, 1512–1521.

Arnold ML, Brothers AN, Hamlin JAP, Taylor SJ, and Martin NH (2015). Divergence-with-gene-flow—What humans and other mammals got up to. In N. Gontier, ed., *Reticulate Evolution: Symbiosis, Lateral Gene Transfer and Hybridization*, **In press**. Springer International Publishing, Switzerland.

Arnold ML, Buckner CM, and Robinson JJ (1991). Pollen mediated introgression and hybrid speciation in Louisiana irises. *Proceedings of the National Academy of Sciences USA* **88**, 1398–1402.

Arnold ML, Bulger MR, Burke JM, Hempel AL, and Williams JH (1999). Natural hybridization—How low can you go? (and still be important). *Ecology* **80**, 371–381.

Arnold ML and Burke JM (2006). Natural hybridization, pp. 399–413. In CW Fox and JB Wolf, eds., *Evolutionary Genetics: Concepts and Case Studies*, Oxford University Press, Oxford.

Arnold ML, Contreras N, and Shaw DD (1988). Biased gene conversion and asymmetrical introgression between subspecies. *Chromosoma* **96**, 368–371.

Arnold ML and Fogarty ND (2009). Reticulate evolution and marine organisms: The final frontier? *International Journal of Molecular Sciences* **10**, 3836–3860.

Arnold ML, Hamrick JL, and Bennett BD (1990b). Allozyme variation in Louisiana Irises: a test for introgression and hybrid speciation. *Heredity* **65**, 297–306.

Arnold ML, Hamrick JL, and Bennett BD (1993). Interspecific pollen competition and reproductive isolation in *Iris*. *Journal of Heredity* **84**, 13–16.

Arnold ML and Hodges SA (1995). Are natural hybrids fit or unfit relative to their parents? *Trends in Ecology and Evolution* **10**, 67–71.

Arnold ML and Jackson RC (1978). Biochemical, cytogenetic and morphological relationships of a new species of *Machaeranthera* section *Arida* (Compositae). *Systematic Botany* **3**, 208–217.

Arnold ML and Larson EJ (2004). Evolution's new look. *The Wilson Quarterly* (Autumn) pp. 60–72.

Arnold ML and Martin NH (2010). Hybrid fitness across time and habitats. *Trends in Ecology and Evolution* **25**, 530–536.

Arnold ML and Meyer A (2006). Natural hybridization in primates: One evolutionary mechanism. *Zoology* **109**, 261–276.

Arnold ML and Shaw DD (1985). The heterochromatin of grasshoppers from the *Caledia captiva* species complex II. Cytological organization of tandemly repeated DNA sequences. *Chromosoma* **93**, 183–190.

Arnold ML, Shaw DD, and Contreras N (1987a). Ribosomal RNA encoding DNA introgression across a narrow hybrid zone between two subspecies of grasshopper. *Proceedings of the National Academy of Sciences USA* **84**, 3946–3950.

Arnold ML, Wilkinson P, Shaw DD, Marchant AD, and Contreras N (1987b). Highly repeated DNA and allozyme variation between sibling species: Evidence for introgression. *Genome* **29**, 272–279.

Arroyo-García R, Ruiz-García L, Bolling L, Ocete R, López MA, Arnold C, *et al.* (2006). Multiple origins of cultivated grapevine (*Vitis vinifera* L. ssp. *sativa*) based on chloroplast DNA polymorphisms. *Molecular Ecology* **15**, 3707–3714.

Ashalakshmi NC, Nag KSC, and Karanth KP (2015). Molecules support morphology: Species status of South Indian populations of the widely distributed Hanuman langur. *Conservation Genetics* **16**, 43–58.

Attard CRM, Beheregaray LB, Jenner KCS, Gill PC, Jenner M-N, Morrice MG, Robertson KM, and Möller LM (2012). Hybridization of Southern Hemisphere blue whale subspecies and a sympatric area off Antarctica: Impacts of whaling or climate change? *Molecular Ecology* **21**, 5715–5727.

Aubert J and Solignac M (1990). Experimental evidence for mitochondrial DNA introgression between *Drosophila* species. *Evolution* **44**, 1272–1282.

Ausich WI and Meyer DL (1994). Hybrid crinoids in the fossil record (Early Mississipian, Phylum Echinodermata). *Paleobiology* **20**, 362–367.

Avise JC (1994). *Molecular Markers, Natural History and Evolution*. Chapman & Hall Inc, New York.

Avise JC (2000a). Cladists in wonderland. *Evolution* **54**, 1828–1832.

Avise JC (2000b). *Phylogeography*. Harvard University Press, Cambridge, Massachusetts.

Avise JC, Ankney CD, and Nelson WS (1990). Mitochondrial gene trees and the evolutionary relationship of mallard and black ducks. *Evolution* **44**, 1109–1119.

Avise JC, Quattro JM, and Vrijenhoek RC (1992). Molecular clones within organismal clones: Mitochondrial DNA phylogenies and the evolutionary histories of unisexual vertebrates. *Evolutionary Biology* **26**, 225–246.

Avise JC, Trexler JC, Travis J, and Nelson WS (1991). *Poecilia mexicana* is the recent female parent of the unisexual fish *P. formosa*. *Evolution* **45**, 1530–1533.

Avrani S, Wurtzel O, Sharon I, Sorek R, and Lindell D (2011). Genomic island variability facilitates *Prochlorococcus*-virus coexistence. *Nature* **474**, 604–608.

Ayres DR, Grotkopp E, Zaremba K, Sloop CM, Blum MJ, Bailey JP, *et al.* (2008). Hybridization between invasive *Spartina densiflora* (Poaceae) and native *S. foliosa* in San Francisco Bay, California, USA. *American Journal of Botany* **95**, 713–719.

Babik W, Butlin RK, Baker WJ, Papadopulus AST, Boulesteix M, Anstett M-C, *et al.* (2009). How sympatric is speciation in the *Howea* palms of Lord Howe Island? *Molecular Ecology* **18**, 3629–3638.

Baird AB, Hillis DM, Patton JC, and Bickham JW (2009). Speciation by monobrachial centric fusions: A test of the model using nuclear DNA sequences from the bat genus *Rhogeessa*. *Molecular Phylogenetics and Evolution* **50**, 256–267.

Baird SJE, Ribas A, Macholán M, Albrecht T, Piálek J, and de Bellocq JG (2012). Where are all the wormy mice?

A reexamination of hybrid parasitism in the European house mouse hybrid zone. *Evolution* **66**, 2757–2772.

Baker RJ and Bickham JW (1986). Speciation by monobrachial centric fusions. *Proceedings of the National Academy of Sciences USA* **83**, 8245–8248.

Balao F, Casimiro-Soriguer R, García-Castaño JL, Terrab A, and Talavera S (2015). Big thistle eats the little thistle: does unidirectional introgressive hybridization endanger the conservation of *Onopordum hinojense*? *New Phytologist* **206**, 448–458.

Baldassarre DT, White TA, Karubian J, and Webster MS (2014). Genomic and morphological analysis of a semipermeable avian hybrid zone suggests asymmetrical introgression of a sexual signal. *Evolution* **68**, 2644–2657.

Baldwin BG, Kyhos DW, and Dvorak J (1990). Chloroplast DNA evolution and adaptive radiation in the Hawaiian silversword alliance (Asteraceae-Madiinae). *Annals of the Missouri Botanical Garden* **77**, 96–109.

Baldwin BG, Kyhos DW, Dvorak J, and Carr GD (1991). Chloroplast DNA evidence for a North American origin of the Hawaiian silversword alliance (Asteraceae). *Proceedings of the National Academy of Sciences USA* **88**, 1840–1843.

Baldwin BG and Sanderson MJ (1998). Age and rate of diversification of the Hawaiian silversword alliance (Compositae). *Proceedings of the National Academy of Sciences USA* **95**, 9402–9406.

Ballard JWO (2000). When one is not enough: Introgression of mitochondrial DNA in *Drosophila*. *Molecular Biology and Evolution* **17**, 1126–1130.

Ballerini ES, Brothers AN, Tang S, Knapp SJ, Bouck A, Taylor SJ, et al. (2012). QTL mapping reveals the genetic architecture of loci affecting pre- and post-zygotic isolating barriers in Louisiana Iris. *BMC Plant Biology* **12**, 91.

Ballerini ES, Mockaitis K, and Arnold ML (2013). Transcriptome sequencing and phylogenetic analysis of Type II MIKC^C MADS-box and R2R3 MYB transcription factors in reproductive tissue from the non-grass monocot *Iris fulva*. *Gene* **531**, 337–346.

Baltrus DA (2013). Exploring the costs of horizontal gene transfer. *Trends in Ecology and Evolution* **28**, 489–495.

Bapteste E, Lopez P, Bouchard F, Baquero F, McInerney JO, and Burian RM (2012). Evolutionary analyses of non-genealogical bonds produced by introgressive descent. *Proceedings of the National Academy of Sciences USA* **109**, 18266–18272.

Bapteste E, van Iersel L, Janke A, Kelchner S, Kelk S, McInerney JO, et al. (2013). Networks: Expanding evolutionary thinking. *Trends in Genetics* **29**, 439–441.

Barbash DA (2010). Genetic testing of the hypothesis that hybrid male lethality results from a failure in dosage compensation. *Genetics* **184**, 313–316.

Barluenga M and Meyer A (2004). The Midas cichlid species complex: incipient sympatric speciation in Nicaraguan cichlid fishes? *Molecular Ecology* **13**, 2061–2076.

Barluenga M, Stolting KN, Salzburger W, Muschick M, and Meyer A (2006). Sympatric speciation in Nicaraguan crater lake cichlid fish. *Nature* **439**, 719–723.

Barnabas J, Goodman M and Moore GW (1972). Descent of mammalian alpha globin chain sequences investigated by the maximum parsimony method. *Journal of Molecular Biology* **69**, 249–278.

Barnes NG (2004). A market analysis of the US pet food industry to determine new opportunities for the cranberry industry, pp. 1–192. Report for the Center for Business Research, University of Massachusetts, Dartmouth, Massachusetts.

Barr CM and Fishman L (2010). The nuclear component of a cytonuclear hybrid incompatibility in *Mimulus* maps to a cluster of pentatricopeptide repeat genes. *Genetics* **184**, 455–465.

Barreto FS and Burton RS (2013). Evidence for compensatory evolution of ribosomal proteins in response to rapid divergence of mitochondrial rRNA. *Molecular Biology and Evolution* **30**, 310–314.

Barreto FS, Pereira RJ, and Burton RS (2015). Hybrid dysfunction and physiological compensation in gene expression. *Molecular Biology and Evolution* **32**, 613–622.

Barrier M, Baldwin BG, Robichaux RH, and Purugganan MD (1999). Interspecific hybrid ancestry of a plant adaptive radiation: allopolyploidy of the Hawaiian silversword alliance (Asteraceae) inferred from floral homeotic gene duplications. *Molecular Biology and Evolution* **16**, 1105–1113.

Barrier M, Robichaux RH, and Purugganan MD (2001). Accelerated regulatory gene evolution in an adaptive radiation. *Proceedings of the National Academy of Sciences USA* **98**, 10208–10213.

Barton NH (1979a). Gene flow past a cline. *Heredity* **43**, 333–339.

Barton NH (1979b). The dynamics of hybrid zones. *Heredity* **43**, 341–359.

Barton NH (1980). The hybrid sink effect. *Heredity* **44**, 277–278.

Barton NH and Hewitt GM (1985). Analysis of hybrid zones. *Annual Review of Ecology and Systematics* **16**, 113–148.

Baum D (2007). Concordance trees, concordance factors, and the exploration of reticulate genealogy. *Taxon* **56**, 417–426.

Baum DA and Donoghue MJ (1995). Choosing among alternative "phylogenetic" species concepts. *Systematic Botany* **20**, 560–573.

Baumel A, Ainouche ML, Bayer RJ, Ainouche AK, and Misset MT (2002). Molecular phylogeny of hybridizing species from the genus *Spartina* Schreb (Poaceae). *Molecular Phylogenetics and Evolution* **22**, 303–314.

Bayes JJ and Malik HS (2009). Altered heterochromatin binding by a hybrid sterility protein in *Drosophila* sibling species. *Science* **326**, 1538–1541.

Bayle P, Macchiarelli R, Trinkaus E, Duarte C, Mazurier A, and Zilhão J (2010). Dental maturational sequence and dental tissue proportions in the early Upper Paleolithic child from Abrigo do Lagar Velho, Portugal. *Proceedings of the National Academy of Sciences USA* **107**, 1338–1342.

Beardsley PM, Yen A, and Olmstead RG (2003). AFLP phylogeny of *Mimulus* section Erythranthe and the evolution of hummingbird pollination. *Evolution* **57**, 1397–1410.

Beaton MJ and Hebert PDN (1988). Geographical parthenogenesis and polyploidy in *Daphnia pulex*. *American Naturalist* **132**, 837–845.

Beaumont M, Barratt EM, Gottelli D, Kitchener AC, Daniels MJ, Pritchard JK, and Bruford MW (2001). Genetic diversity and introgression in the Scottish wildcat. *Molecular Ecology* **10**, 319–336.

Beck JB, Alexander PJ, Allphin L, Al-Shehbaz IA, Rushworth C, Bailey CD, and Windham MD (2011). Does hybridization drive the transition to asexuality in diploid *Boechera*? *Evolution* **66**, 985–995.

Becquet C, Patterson N, Stone AC, Przeworski M, and Reich D (2007). Genetic structure of chimpanzee populations. *PLoS Genetics* **3**, 0617–0626.

Beebee TJC (2005). Conservation genetics of amphibians. *Heredity* **95**, 423–427.

Beehner JC, Phillips-Conroy JE, and Whitten PL (2005). Female testosterone, dominance rank, and aggression in an Ethiopian population of hybrid baboons. *American Journal of Primatology* **67**, 101–119.

Bell MA and Travis MP (2005). Hybridization, transgressive segregation, genetic covariation, and adaptive radiation. *Trends in Ecology and Evolution* **20**, 358–361.

Belle EMS, Benazzo A, Ghirotto S, Colonna V, and Barbujani G (2009). Comparing models on the genealogical relationships among Neandertal, Cro-Magnoid and modern Europeans by serial coalescent simulations. *Heredity* **102**, 218–225.

Belser JA, Gustin KM, Pearce MB, Maines TR, Zeng H, Pappas C, *et al.* (2013). Pathogenesis and transmission of avian influenza A (H7N9) virus in ferrets and mice. *Nature* **501**, 556–559.

Bendiksby M, Tribsch A, Borgen L, Trávníček P, and Brysting AK (2011). Allopolyploid origins of the *Galeopsis* tetraploids—revisiting Müntzing's classical textbook example using molecular tools. *New Phytologist* **191**, 1150–1167.

Bennett BD and Grace JB (1990). Shade tolerance and its effect on the segregation of two species of Louisiana iris and their hybrids. *American Journal of Botany* **77**, 100–107.

Bennett D and Hoffmann RS (1999). *Equus caballus*. *Mammalian Species* **628**, 1–14.

Benson WW (1972). Natural selection for Müllerian mimicry in *Heliconius erato* in Costa Rica. *Science* **176**, 936–939.

Benzie JAH (1986). Phenetic and cladistic analyses of the phylogenetic relationships within the genus *Daphnia* worldwide. *Hydrobiologia* **140**, 105–124.

Bergman TJ and Beehner JC (2004). Social system of a hybrid baboon group (*Papio anubis* × *P. hamadryas*). *International Journal of Primatology* **25**, 1313–1330.

Bergman TJ, Phillips-Conroy JE, and Jolly CJ (2008). Behavioral variation and reproductive success of male baboons (*Papio anubis* x *Papio hamadryas*) in a hybrid social group. *American Journal of Primatology* **70**, 136–147.

Bergthorsson U, Adams KL, Thomason B, and Palmer JD (2003). Widespread horizontal transfer of mitochondrial genes in flowering plants. *Nature* **424**, 197–201.

Berthouly C, Leroy G, Nhu Van T, Hoang Thanh H, Bed'Hom B, Trong Nguyen T, *et al.* (2009). Genetic analysis of local Vietnamese chickens provides evidence of gene flow from wild to domestic populations. *BMC Genetics* **10**, 1.

Besansky NJ, Krzywinski J, Lehmann T, Simard F, Kern M, Mukabayire O, *et al.* (2003). Semipermeable species boundaries between *Anopheles gambiae* and *Anopheles arabiensis*: Evidence from multilocus DNA sequence variation. *Proceedings of the National Academy of Sciences USA* **100**, 10818–10823.

Besansky NJ, Lehmann T, Fahey GT, Fontenille D, Braack LEO, Hawley WA, and Collins FH (1997). Patterns of mitochondrial variation within and between African malaria vectors, *Anopheles gambiae* and *An. arabiensis*, suggest extensive gene flow. *Genetics* **147**, 1817–1828.

Besnard G, Khadari B, Navascués M, Fernández-Mazuecos M, El Bakkali A, Arrigo N, *et al.* (2013). The complex history of the olive tree: From Late Quaternary diversification of Mediterranean lineages to primary domestication in the northern Levant. *Proceedings of the Royal Society of London B* **280**, 20122833.

Beysard M and Heckel G (2014). Structure and dynamics of hybrid zones at different stages of speciation in the common vole (*Microtus arvalis*). *Molecular Ecology* **23**, 673–687.

Bezault E, Mwaiko S, and Seehausen O (2011). Population genomic tests of models of adaptive radiation in Lake Victoria region cichlid fish. *Evolution* **65**, 3381–3397.

Bhaya D, Dufresne A, Vaulot D, and Grossman A (2002). Analysis of the hli gene family in marine and freshwater cyanobacteria. *FEMS Microbiology Letters* **215**, 209–219.

Bi Y, Ren X, Yan C, Shao J, Xie D, and Zhao Z (2015). A genome-wide hybrid incompatibility landscape between *Caenorhabditis briggsae* and *C. nigoni*. *PLoS Genetics* **11**, e1004993.

Bicca-Marques JC, Mattjie Prates H, de Aguiar FRC, and Jones CB (2008). Survey of *Alouatta caraya*, the black-and-gold howler monkey, and *Alouatta guariba clamitans*, the brown howler monkey, in a contact zone, State of Rio Grande do Sul, Brazil: Evidence for hybridization. *Primates* **49**, 246–252.

Bidon T, Janke A, Fain SR, Geir Eiken H, Hagen SB, Saarma U, et al. (2014). Brown and polar bear Y chromosomes reveal extensive male-biased gene flow within brother lineages. *Molecular Biology and Evolution* **31**, 1353–1363.

Biju-Duval C, Ennafaa H, Dennebouy N, Monnerot M, Mignotte F, Soriguer RC, et al. (1991). Mitochondrial DNA evolution in lagomorphs: Origin of systematic heteroplasmy and organization of diversity in European rabbits. *Journal of Molecular Evolution* **33**, 92–102.

Bing J, Han P-J, Liu W-Q, Wang Q-M, and Bai F-Y (2014). Evidence for a Far East Asian origin of lager beer yeast. *Current Biology* **24**, R380–R381.

Bischoff M, Jürgens A, and Campbell DR (2014). Floral scent in natural hybrids of *Ipomopsis* (Polemoniaceae) and their parental species. *Annals of Botany* **113**, 533–544.

Bischoff M, Raguso RA, Jürgens A, and Campbell DR (2015). Context-dependent reproductive isolation mediated by floral scent and color. *Evolution* **69**, 1–13.

Bitocchi E, Nanni L, Rossi M, Rau D, Bellucci E, Giardini A, et al. (2009). Introgression from modern hybrid varieties into landrace populations of maize (*Zea mays* ssp. *mays* L.) in central Italy. *Molecular Ecology* **18**, 603–621.

Blacker HP, Kirkpatrick NC, Rubite A, O'Rourke D, and Noormohammadi AH (2011). Epidemiology of recent outbreaks of infectious laryngotracheitis in poultry in Australia. *Australian Veterinary Journal* **89**, 89–94.

Blair WF (1955). Mating call and stage of speciation in the *Microhyla olivacea-M. carolinensis* complex. *Evolution* **9**, 469–480.

Blakeslee AF (1945). Removing some of the barriers to crossability in plants. *Proceedings of the American Philosophical Society* **89**, 561–574.

Blum MJ (2008). Ecological and genetic associations across a *Heliconius* hybrid zone. *Journal of Evolutionary Biology* **21**, 330–341.

Blum MGB and Jakobsson M (2011). Deep divergences of human gene trees and models of human origins. *Molecular Biology and Evolution* **28**, 889–898.

Bock DG, Kane NC, Ebert DP, and Rieseberg LH (2014). Genome skimming reveals the origin of the Jerusalem Artichoke tuber crop species: neither from Jerusalem nor an artichoke. *New Phytologist* **201**, 1021–1030.

Bogart JP and Bi K (2013). Genetic and genomic interactions of animals with different ploidy levels. *Cytogenetic and Genome Research* **140**, 117–136.

Bohling JH, Adams JR, and Waits LP (2013). Evaluating the ability of Bayesian clustering methods to detect hybridization and introgression using an empirical red wolf data set. *Molecular Ecology* **22**, 74–86.

Böhne A, Sengstag T, and Salzburger W (2014). Comparative Transcriptomics in East African cichlids reveals sex- and species-specific expression and new candidates for sex differentiation in fishes. *Genome Biology and Evolution* **6**, 2567–2585.

Boissinot S, Alvarez L, Giraldo-Ramirez J, and Tollis M (2014). Neutral nuclear variation in baboons (genus *Papio*) provides insights into their evolutionary and demographic histories. *American Journal of Physical Anthropology* **155**, 621–634.

Bolnick DI and Fitzpatrick BM (2007). Sympatric speciation: Models and empirical evidence. *Annual Review of Ecology, Evolution and Systematics* **38**, 459–487.

Bombarely A, Coate JE, and Doyle JJ (2014). Mining transcriptomic data to study the origins and evolution of a plant allopolyploid complex. *PeerJ* **2**, e391.

Boratyński Z, Melo-Ferreira J, Alves PC, Berto S, Koskela E, Pentikäinen OT, et al. (2014). Molecular and ecological signs of mitochondrial adaptation: Consequences for introgression? *Heredity* **113**, 277–286.

Bosse M, Megens H-J, Madsen O, Frantz LAF, Paudel Y, Crooijmans RPMA, and Groenen MAM (2014). Untangling the hybrid nature of modern pig genomes: A mosaic derived from biogeographically distinct and highly divergent *Sus scrofa* populations. *Molecular Ecology* **23**, 4089–4102.

Boto L (2014). Horizontal gene transfer in the acquisition of novel traits by metazoans. *Proceedings of the Royal Society of London B* **281**, 20132450.

Boucher Y, Douady CJ, Papke RT, Walsh DA, Boudreau MER, Nesbø CL, et al. (2003). Lateral gene transfer and the origins of prokaryotic groups. *Annual Review of Genetics* **37**, 283–328.

Bouck AC (2004). *The Genetic Architecture of Reproductive Isolation in Louisiana Irises*. PhD thesis, University of Georgia, Athens, GA.

Bouck AC, Peeler R, Arnold ML, and Wessler SR (2005). Genetic mapping of species boundaries in Louisiana Irises using IRRE retrotransposon display markers. *Genetics* **171**, 1289–1303.

Bouck AC, Wessler SR, and Arnold ML (2007). QTL analysis of floral traits in Louisiana Iris hybrids. *Evolution* **61**, 2308–2319.

Bowers J, Boursiquot J-M, This P, Chu K, Johansson H, and Meredith C (1999). Historical genetics: The parentage of Chardonnay, Gamay, and other wine grapes of northeastern France. *Science* **285**, 1562–1565.

Bowers JE, Chapman BA, Rong J, and Paterson AH (2003). Unraveling angiosperm genome evolution by phylogenetic analysis of chromosomal duplication events. *Nature* **422**, 433–438.

Bowers JE and Meredith CP (1997). The parentage of a classic wine grape, Cabernet Sauvignon *Nature Genetics* **16**, 84–87.

Bradshaw HD Jr and Schemske DW (2003). Allele substitution at a flower colour locus produces a pollinator shift in monkeyflowers. *Nature* **426**, 176–178.

Branco M, Ferrand N, and Monnerot M (2000). Phylogeography of the European rabbit (*Oryctolagus cuniculus*) in the Iberian Peninsula inferred from RFLP analysis of the cytochrome b gene. *Heredity* **85**, 307–317.

Branco M, Monnerot M, Ferrand N, and Templeton AR (2002). Postglacial dispersal of the European rabbit (*Oryctolagus cuniculus*) on the Iberian Peninsula reconstructed from nested clade and mismatch analyses of mitochondrial DNA genetic variation. *Evolution* **56**, 792–803.

Brand CL, Kingan SB, Wu L, and Garrigan D (2013). A selective sweep across species boundaries in *Drosophila*. *Molecular Biology and Evolution* **30**, 2177–2186.

Brandvain Y, Kenney AM, Flagel L, Coop G, and Sweigart AL (2014). Speciation and Introgression between *Mimulus nasutus* and *Mimulus guttatus*. *PLoS Genetics* **10**, e1004410.

Brawand D, Wagner CE, Li YI, Malinsky M, Keller I, Fan S, *et al.* (2014). The genomic substrate for adaptive radiation in African cichlid fish. *Nature* **513**, 375–381.

Brede N, Sandrock C, Straile D, Spaak P, Jankowski T, Streit B, and Schwenk K (2009). The impact of human-made ecological changes on the genetic architecture of *Daphnia* species. *Proceedings of the National Academy of Sciences USA* **106**, 4758–4763.

Brennan AC, Barker D, Hiscock SJ, and Abbott RJ (2012). Molecular genetic and quantitative trait divergence associated with recent homoploid hybrid speciation: a study of *Senecio squalidus* (Asteraceae). *Heredity* **108**, 87–95.

Brennan AC, Hiscock SJ, and Abbott RJ (2014).Interspecific crossing and genetic mapping reveal intrinsic genomic incompatibility between two *Senecio* species that form a hybrid zone on Mount Etna, Sicily. *Heredity* **113**, 195–204.

Brenner DJ, Steigerwalt AG, and McDade JE (1979). Classification of the Legionnaires' disease bacterium: *Legionella pneumophila*, genus novum, species nova, of the family Legionellaceae, familia nova. *Annals of Internal Medicine* **90**, 656–658.

Britch SC, Cain ML, and Howard DJ (2001). Spatio-temporal dynamics of the *Allonemobius fasciatus—A. socius* mosaic hybrid zone: A 14-year perspective. *Molecular Ecology* **10**, 627–638.

Britten RJ (2006). Transposable elements have contributed to thousands of human proteins. *Proceedings of the National Academy of Sciences USA* **103**, 1798–1803.

Britten RJ and Kohne DE (1968). Repeated sequences in DNA. *Science* **161**, 529–540.

Bronson CL, Grubb TC Jr, and Braun MJ (2003a). A test of the endogenous and exogenous selection hypotheses for the maintenance of a narrow avian hybrid zone. *Evolution* **57**, 630–637.

Bronson CL, Grubb TC Jr, Sattler GD, and Braun MJ (2003b). Mate preference: a possible causal mechanism for a moving hybrid zone. *Animal Behaviour* **65**, 489–500.

Brothers AN, Barb JG, Ballerini ES, Drury DW, Knapp SJ, and Arnold ML (2013). The genetic architecture of floral traits in *Iris hexagona* and *Iris fulva*. *Journal of Heredity* **104**, 853–861.

Brothers AN and Delph LF (2010). Haldane's Rule is extended to plants with sex chromosomes. *Evolution* **64**, 3643–3648.

Brown TA, Jones MK, Powell W, and Allaby RG (2009). The complex origins of domesticated crops in the Fertile Crescent. *Trends in Ecology and Evolution* **24**, 103–109.

Brumfield RT, Jernigan RW, McDonald DB, and Braun MJ (2001). Evolutionary implications of divergent clines in an avian (*Manacus*: Aves) hybrid zone. *Evolution* **55**, 2070–2087.

Buchholz JT, Williams LF, and Blakeslee AF (1935). Pollen-tube growth of ten species of *Datura* in interspecific pollinations. *Proceedings of the National Academy of Sciences USA* **21**, 651–656.

Budd AF and Pandolfi JM (2004). Overlapping species boundaries and hybridization within the *Montastraea "annularis"* reef coral complex in the Pleistocene of the Bahama Islands. *Paleobiology* **30**, 396–425.

Budd AF and Pandolfi JM (2010). Evolutionary novelty is concentrated at the edge of coral species distributions. *Science* **328**, 1558–1561.

Burgess KS, Morgan M, and Husband BC (2008). Interspecific seed discounting and the fertility cost of hybridization in an endangered species. *New Phytologist* **177**, 276–284.

Burkart-Waco D, Ngo K, Lieberman M, and Comai L (2015). Perturbation of parentally biased gene expression during interspecific hybridization. *PLoS ONE* **10**, e0117293.

Burke JM, Bulger MR, Wesselingh RA, and Arnold ML (2000a). Frequency and spatial patterning of clonal reproduction in Louisiana Iris hybrid populations. *Evolution* **54**, 137–144.

Burke JM, Voss TJ, and Arnold ML (1998). Genetic interactions and natural selection in Louisiana Iris hybrids. *Evolution* **52**, 1304–1310.

Burke GR, Walden KKO, Whitfield JB, Robertson HM, and Strand MR (2014). Widespread genome reorganization of an obligate virus mutualist. *PLoS Genetics* **10**, e1004660.

Burke JM, Wyatt R, DePamphilis CW, and Arnold ML (2000b). Nectar characteristics of interspecific hybrids and their parents in *Aesculus* and *Iris*. *Journal of the Torrey Botanical Society* **127**, 200–206.

Burns CC, Shaw J, Jorba J, Bukbuk D, Adu F, Gumede N, *et al.* (2013). Multiple independent emergences of type 2 vaccine-derived polioviruses during a large outbreak in northern Nigeria. *Journal of Virology* **87**, 4907–4922.

Burton RS (1990). Hybrid breakdown in developmental time in the copepod *Tigriopus californicus*. *Evolution* **44**, 1814–1822.

Burton RS, Pereira RJ, and Barreto FS (2013). Cytonuclear genomic interactions and hybrid breakdown. *Annual Review of Ecology, Evolution and Systematics* **44**, 281–302.

Burton RS, Rawson PD, and Edmands S (1999). Genetic architecture of physiological phenotypes: Empirical evidence for coadapted gene complexes. *American Zoologist* **39**, 451–462.

Bush GL (1969). Sympatric host race formation and speciation in frugivorous flies of the genus *Rhagoletis* (Diptera, Tephritidae). *Evolution* **23**, 237–251.

Bush GL (1998). The conceptual radicalization of an evolutionary biologist. In DJ Howard and SH Berlocher, eds., *Endless Forms Species and Speciation*, pp. 425–438. Oxford University Press, New York.

Butcher PA, Skinner AK, and Gardiner CA (2005). Increased inbreeding and inter-species gene flow in remnant populations of the rare *Eucalyptus benthamii*. *Conservation Genetics* **6**, 213–226.

Cahill JA, Green RE, Fulton TL, Stiller M, Jay F, Ovsyanikov N, *et al.* (2013). Genomic evidence for island population conversion resolves conflicting theories of polar bear evolution. *PLoS Genetics* **9**, e1003345.

Campbell DR (2003). Natural selection in *Ipomopsis* hybrid zones: Implications for ecological speciation. *New Phytologist* **161**, 83–90.

Campbell DR, Crawford M, Brody AK, and Forbis TA (2002a). Resistance to pre-dispersal seed predators in a natural hybrid zone. *Oecologia* **131**, 436–443.

Campbell DR and Dooley JL (1992). The spatial scale of genetic differentiation in a hummingbird-pollinated plant: Comparison with models of isolation by distance. *American Naturalist* **139**, 735–748.

Campbell DR, Galen C, and Wu CA (2005). Ecophysiology of first and second generation hybrids in a natural plant hybrid zone. *Oecologia* **144**, 214–225.

Campbell DR and Waser NM (1989). Variation in pollen flow within and among populations of *Ipomopsis aggregata*. *Evolution* **43**, 1444–1455.

Campbell DR and Waser NM (2001). Genotype-by-environment interaction and the fitness of plant hybrids in the wild. *Evolution* **55**, 669–676.

Campbell DR and Waser NM (2007). Evolutionary dynamics of an *Ipomopsis* hybrid zone: Confronting models with lifetime fitness data. *American Naturalist* **169**, 298–310.

Campbell DR, Waser NM, Aldridge G, and Wu CA (2008). Lifetime fitness in two generations of *Ipomopsis* hybrids. *Evolution* **62**, 2616–2627.

Campbell DR, Waser NM, and Meléndez-Ackerman EJ (1997). Analyzing pollinator-mediated selection in a plant hybrid zone: Hummingbird visitation patterns on three spatial scales. *American Naturalist* **149**, 295–315.

Campbell DR, Waser NM, and Pederson GT (2002b). Predicting patterns of mating and potential hybridization from pollinator behavior. *American Naturalist* **159**, 438–450.

Campbell DR, Waser NM, and Price MV (1994). Indirect selection of stigma position in *Ipomopsis aggregata* via a genetically correlated trait. *Evolution* **48**, 55–68.

Campbell DR, Waser NM, Price MV, Lynch EA, and Mitchell RJ (1991). Components of phenotypic selection: pollen export and flower corolla width in *Ipomopsis aggregata*. *Evolution* **45**, 1458–1467.

Campbell DR and Wendlandt C (2013). Altered precipitation affects plant hybrids differently than their parental species. *American Journal of Botany* **100**, 1322–1331.

Campbell DR, Wu CA, and Travers SE (2010). Photosynthetic and growth responses of reciprocal hybrids to variation in water and nitrogen availability. *American Journal of Botany* **97**, 925–933.

Cann RL, Stoneking M, and Wilson AC (1987). Mitochondrial DNA and human evolution. *Nature* **325**, 31–36.

Cannon SB, McKain MR, Harkess A, Nelson MN, Dash S, Deyholos MK, *et al.* (2015). Multiple polyploidy events in the early radiation of nodulating and nonnodulating legumes. *Molecular Biology and Evolution* **32**, 193–210.

Carleton KL, Parry JWL, Bowmaker JK, Hunt DM, and Seehausen O (2005). Colour vision and speciation in Lake Victoria cichlids of the genus *Pundamilia*. *Molecular Ecology* **14**, 4341–4353.

Carneiro M, Afonso S, Geraldes A, Garreau H, Bolet G, Boucher S, *et al.* (2011). The genetic structure of domestic rabbits. *Molecular Biology and Evolution* **28**, 1801–1816.

Carneiro M, Albert FW, Afonso S, Pereira RJ, Burbano H, Campos R, *et al.* (2014a). The genomic architecture of population divergence between subspecies of the European rabbit. *PLoS Genetics* **10**, e1003519.

Carneiro M, Baird SJE, Afonso S, Ramirez E, Tarroso P, Teotónio H, *et al.* (2013). Steep clines within a highly permeable genome across a hybrid zone between two subspecies of the European rabbit. *Molecular Ecology* **22**, 2511–2525.

Carneiro M, Blanco-Aguiar JA, Villafuerte R, Ferrand N, and Nachman MW (2010). Speciation in the European rabbit (*Oryctolagus cuniculus*): Islands of differentiation on the X chromosome and autosomes. *Evolution* **64**, 3443–3460.

Carneiro M, Ferrand N, and Nachman MW (2009). Recombination and speciation: Loci near centromeres are more differentiated than loci near telomeres between subspecies of the European rabbit (*Oryctolagus cuniculus*). *Genetics* **181**, 593–606.

Carneiro M, Rubin C-J, Di Palma F, Albert FW, Alfoldi J, Martinez Barrio A, *et al.* (2014b). Rabbit genome analysis reveals a polygenic basis for phenotypic change during domestication. *Science* **345**, 1074–1079.

Carney SE, Cruzan MB, and Arnold ML (1994). Reproductive interactions between hybridizing irises: Analyses of pollen tube growth and fertilization success. *American Journal of Botany* **81**, 1169–1175.

Carney SE, Gardner KA, and Rieseberg LH (2000). Evolutionary changes over the fifty-year history of a hybrid population of sunflowers (*Helianthus*). *Evolution* **54**, 462–474.

Carney SE, Hodges SA, and Arnold ML (1996). Effects of differential pollen-tube growth on hybridization in the Louisiana Irises. *Evolution* **50**, 1871–1878.

Carr SM, Brothers AJ, and Wilson AC (1987). Evolutionary inferences from restriction maps of mitochondrial DNA from nine taxa of *Xenopus* frogs. *Evolution* **41**, 176–188.

Carr GD and Kyhos DW (1981). Adaptive radiation in the Hawaiian silversword alliance (Compositae-Madiinae) I. Cytogenetics of spontaneous hybrids. *Evolution* **35**, 543–556.

Carr GD, Robichaux RH, Witter MS, and Kyhos DW (1989). Adaptive radiation of the Hawaiian silversword alliance (Compositae-Madiinae): A comparison with Hawaiian picture-winged *Drosophila*. In LV Giddings, KY Kaneshiro, and WW Anderson, eds., *Genetics, Speciation, and the Founder Principle*, pp. 79–97. Oxford University Press, Oxford.

Carvajal-Rodríguez A, Crandall KA, and Posada D (2007). Recombination favors the evolution of drug resistance in HIV-1 during antiretroviral therapy. *Infections, Genetics and Evolution* **7**, 476–483.

Casadevall A and Pirofski L-a (2001). Host-pathogen interactions: The attributes of virulence. *Journal of Infectious Diseases* **184**, 337–344.

Casjens S (2003). Prophages and bacterial genomics: What have we learned so far? *Molecular Microbiology* **49**, 277–300.

Cassone BJ, Kamdem C, Cheng C, Tan JC, Hahn MW, Costantini C, and Besansky NJ (2014). Gene expression divergence between malaria vector sibling species *Anopheles gambiae* and *An. coluzzii* from rural and urban Yaoundé Cameroon. *Molecular Ecology* **23**, 2242–2259.

Castro-Nallar E, Pérez-Losada M, Burton GF, and Crandall KA (2012). The evolution of HIV: Inferences using phylogenetics. *Molecular Phylogenetics and Evolution* **62**, 777–792.

Caswell H (2001). *Matrix Population Models*, 2nd edition. Sinauer, Sunderland, Massachusetts.

Caswell JL, Mallick S, Richter DJ, Neubauer J, Schirmer C, Gnerre S, and Reich D (2008). Analysis of chimpanzee history based on genome sequence alignments. *PLoS Genetics* **4**, e1000057.

Cazalet C, Gomez-Valero L, Rusniok C, Lomma M, Dervins-Ravault D, Newton HJ, *et al.* (2010). Analysis of the *Legionella longbeachae* genome and transcriptome uncovers unique strategies to cause Legionnaires' disease. *PLoS Genetics* **6**, e1000851.

Centers for Disease Control and Prevention (2002). Ebola hemorrhagic fever information packet.

Centers for Disease Control and Prevention (2014). Questions and answers on experimental treatments and vaccines for Ebola.

Chain PSG, Carniel E, Larimer FW, Lamerdin J, Stoutland PO, Regala WM, *et al.* (2004). Insights into the evolution of *Yersinia pestis* through whole-genome comparison with *Yersinia pseudotuberculosis*. *Proceedings of the National Academy of Sciences USA* **101**, 13826–13831.

Chain FJJ, Ilieva D, and Evans BJ (2008). Duplicate gene evolution and expression in the wake of vertebrate allopolyploidization. *BMC Evolutionary Biology* **8**, 43.

Chamberlain NL, Hill RI, Baxter SW, Jiggins CD, and Kronforst MR (2011). Comparative population genetics of a mimicry locus among hybridizing *Heliconius* butterfly species. *Heredity* **107**, 200–204.

Champigneulle A and Cachera S (2003). Evaluation of large-scale stocking of early stages of brown trout, *Salmo trutta*, to angler catches in the French-Swiss part of the River Doubs. *Fisheries Management and Ecology* **10**, 79–85.

Chan JM, Carlsson G, and Rabadan R (2013). Topology of viral evolution. *Proceedings of the National Academy of Sciences USA* **110**, 18566–18571.

Chang AS and Noor MAF (2007). The genetics of hybrid male sterility between the allopatric species pair *Drosophila persimilis* and *D. pseudoobscura bogotana*: Dominant sterility alleles in collinear autosomal regions. *Genetics* **176**, 343–349.

Charney ND (2012). Relating hybrid advantage and genome replacement in unisexual salamanders. *Evolution* **66**, 1387–1397.

Charpentier MJE, Fontaine MC, Cherel E, Renoult JP, Jenkins T, Benoit L, *et al.* (2012). Genetic structure in a dynamic baboon hybrid zone corroborates behavioural observations in a hybrid population. *Molecular Ecology* **21**, 715–731.

Charpentier MJE, Tung J, Altmann J, and Alberts SC (2008). Age at maturity in wild baboons: genetic, environmental and demographic influences. *Molecular Ecology* **17**, 2026–2040.

Charron G, Leducq J-B, and Landry CR (2014). Chromosomal variation segregates within incipient species and correlates with reproductive isolation. *Molecular Ecology* **23**, 4362–4372.

Chavez AS, Saltzberg CJ, and Kenagy GJ (2011). Genetic and phenotypic variation across a hybrid zone between ecologically divergent tree squirrels (*Tamiasciurus*). *Molecular Ecology* **20**, 3350–3366.

Chen F and Lu J (2002). Genomic sequence and evolution of marine cyanophage P60: A new insight on lytic and lysogenic phages. *Applied and Environmental Microbiology* **68**, 2589–2594.

Chen J and Novick RP (2009). Phage-mediated intergeneric transfer of toxin genes. *Science* **323**, 139–141.

Cheng E, Hodges KE, Melo-Ferreira J, Alves PC, and Mills LS (2014). Conservation implications of the evolutionary history and genetic diversity hotspots of the snowshoe hare. *Molecular Ecology* **23**, 2929–2942.

Chien M, Morozova I, Shi S, Sheng H, Chen J, Gomez SM, *et al.* (2004). The genomic sequence of the accidental pathogen *Legionella pneumophila*. *Science* **305**, 1966–1968.

Chin MPS, Rhodes TD, Chen J, Fu W, and Hu W-S (2005). Identification of a major restriction in HIV-1 intersubtype recombination. *Proceedings of the National Academy of Sciences USA* **102**, 9002–9007.

Chin S-W, Shaw J, Haberle R, Wen J, and Potter D (2014). Diversification of almonds, peaches, plums and cherries—Molecular systematics and biogeographic history of *Prunus* (Rosaceae). *Molecular Phylogenetics and Evolution* **76**, 34–48.

Chistoserdova L, Vorholt JA, Thauer RK, and Lidstrom ME (1998). C1 transfer enzymes and coenzymes linking methylotrophic bacteria and methanogenic Archaea. *Science* **281**, 99–102.

Cho M-S, Kim C-S, Kim S-H, Kim TO, Heo K-I, Jun J, and Kim S-C (2014). Molecular and morphological data reveal hybrid origin of wild *Prunus yedoensis* (Rosaceae) from Jeju Island, Korea: Implications for the origin of the flowering cherry. *American Journal of Botany* **101**, 1976–1986.

Choi SC and Hey J (2011). Joint inference of population assignment and demographic history. *Genetics* **189**, 561–577.

Chou H-H, Chiu H-C, Delaney NF, Segrè D, and Marx CJ (2011). Diminishing returns epistasis among beneficial mutations decelerates adaptation. *Science* **332**, 1190–1192.

Chrisafis A (2013). Horsemeat scandal triggers 15% rise in sales for France's equine butchers. <http://www. theguardian.com/uk/2013/feb/21/horsemeat-scandal-rise-sales-france-butcher>

Christiansen DG, Fog K, Pedersen BV, and Boomsma JJ (2005). Reproduction and hybrid load in all-hybrid populations of *Rana esculenta* water frogs in Denmark. *Evolution* **59**, 1348–1361.

Christiansen DG and Reyer H-U (2009). From clonal to sexual hybrids: Genetic recombination via triploids in all-hybrid populations of water frogs. *Evolution* **63**, 1754–1768.

Christiansen DG and Reyer H-U (2011). Effects of geographic distance, sea barriers and habitat on the genetic structure and diversity of all-hybrid water frog populations. *Heredity* **106**, 25–36.

Christin P-A, Edwards EJ, Besnard G, Boxall SF, Gregory R, Kellogg EA, *et al.* (2012). Adaptive evolution of C4 photosynthesis through recurrent lateral gene transfer. *Current Biology* **22**, 445–449.

Chung Y and Ané C (2011). Comparing two Bayesian methods for gene tree/species tree reconstruction: Simulations with incomplete lineage sorting and horizontal gene transfer. *Systematic Biology* **60**, 261–275.

Clark AG (1985). Natural selection with nuclear and cytoplasmic transmission. II. Tests with *Drosophila* from diverse populations. *Genetics* **111**, 97–112.

Clausen J, Keck DD, and Hiesey WM (1939). The concept of species based on experiment. *American Journal of Botany* **26**, 103–106.

Clausen J, Keck DD, and Hiesey WM (1945). Experimental studies on the nature of species. II. Plant evolution through amphiploidy and autoploidy, with examples from the Madiinae. *Carnegie Institution of Washington Publication* **564**, 1–163.

Clifford HT (1954). Analysis of suspected hybrid swarms in the genus *Eucalyptus*. *Heredity* **8**, 259–269.

Clifford SL, Anthony NM, Bawe-Johnson M, Abernethy KA, Tutin CEG, White LJT, *et al.* (2004). Mitochondrial DNA phylogeography of western lowland gorillas (*Gorilla gorilla gorilla*). *Molecular Ecology* **13**, 1551–1565.

Coart E, Van Glabeke S, De Loose M, Larsen AS, and Roldán-Ruiz I (2006). Chloroplast diversity in the genus *Malus*: New insights into the relationship between the European wild apple (*Malus sylvestris* (L.) Mill.) and the domesticated apple (*Malus domestica* Borkh.). *Molecular Ecology* **15**, 2171–2182.

Coate JE, Luciano AK, Seralathan V, Minchew KJ, Owens TG, and Doyle JJ (2012). Anatomical, biochemical, and photosynthetic responses to recent allopolyploidy in *Glycine dolichocarpa* (Fabaceae). *American Journal of Botany* **99**, 55–67.

Coate JE, Powell AF, Owens TG, and Doyle JJ (2013). Transgressive physiological and transcriptomic responses to light stress in allopolyploid *Glycine dolichocarpa* (Leguminosae). *Heredity* **110**, 160–170.

Coelho MA, Gonçalves C, Sampaio JP, and Gonçalves P (2013). Extensive intra-kingdom horizontal gene transfer converging on a fungal fructose transporter gene. *PLoS Genetics* **9**, e1003587.

Coetzee M, Hunt RH, Wilkerson R, della Torre A, Coulibaly MB, and Besansky NJ (2013). *Anopheles coluzzii* and *Anopheles amharicus*, new members of the *Anopheles gambiae* complex. *Zootaxa* **3619**, 246–274.

Coffin JM (1979). Structure, replication, and recombination of retrovirus genomes: Some unifying hypotheses. *Journal of General Virology* **42**, 1–26.

Cohan FM (2002). What are bacterial species? *Annual Review of Microbiology* **56**, 457–487.

Cohuet A, Dia I, Simard F, Raymond M, Rousset F, Antonio-Nkondjio C, *et al.* (2005). Gene flow between chromosomal forms of the malaria vector *Anopheles funestus* in Cameroon, Central Africa, and its relevance in malaria fighting. *Genetics* **169**, 301–311.

Colbourne JK, Crease TJ, Weider LJ, Hebert PDN, Dufresne F, and Hobæk A (1998). Phylogenetics and evolution of a circumarctic species complex (Cladocera: *Daphnia pulex*). *Biological Journal of the Linnean Society* **65**, 347–365.

Coleman RR, Gaither MR, Kimokeo B, Stanton FG, Bowen BW, and Toonen RJ (2014). Large-scale introduction of the Indo-Pacific damselfish *Abudefduf vaigiensis* into Hawai'i promotes genetic swamping of the endemic congener *A. abdominalis*. *Molecular Ecology* **23**, 5552–5565.

Colgan DJ (1985). Evidence for the evolutionary significance of developmental variation in an abundant protein of orthopteran muscle. *Genetica* **67**, 81–85.

Colgan DJ (1986). Developmental changes in the isoenzymes controlling glycolysis in the acridine grasshopper, *Caledia captiva*. *Development Genes and Evolution* **195**, 197–201.

Colmenares F, Esteban MM, and Zaragoza F (2006). One-male units and clans in a colony of hamadryas baboons (*Papio hamadryas hamadryas*): Effect of male number and clan cohesion on feeding success. *American Journal of Primatology* **68**, 21–37.

Coluzzi M, Sabatini A, della Torre A, Di Deco MA, and Petrarca V (2002). A polytene chromosome analysis of the *Anopheles gambiae* species complex. *Science* **298**, 1415–1418.

Combelas N, Holmblat B, Joffret M-L, Colbere-Garapin F, and Delpeyroux F (2011). Recombination between poliovirus and coxsackie A viruses of species C: A model of viral genetic plasticity and emergence. *Viruses* **3**, 1460–1484.

Combiescu M, Guillot S, Persu A, Baicus A, Pitigoi D, Balanant J, *et al.* (2007). Circulation of a type 1 recombin-

ant vaccine-derived poliovirus strain in a limited area in Romania. *Archives of Virology* **152**, 727–738.

Conesa MA, Mus M, and Rosselló JA (2010). Who threatens who? Natural hybridization between *Lotus dorycnium* and the island endemic *Lotus fulgurans* (Fabaceae). *Biological Journal of the Linnean Society* **101**, 1–12.

Cordaux R and Batzer MA (2009). The impact of retrotransposons on human genome evolution. *Nature Reviews Genetics* **10**, 691–703.

Cornelis G, Heidmann O, Bernard-Stoecklina S, Reynaud K, Vérond G, Mulote B, *et al.* (2012). Ancestral capture of *syncytin-Car1*, a fusogenic endogenous retroviral envelope gene involved in placentation and conserved in Carnivora. *Proceedings of the National Academy of Sciences USA* **109**, E432–E441.

Cornelis G, Vernochet C, Carradec Q, Souquere S, Mulot B, Catzeflis F, *et al.* (2015). Retroviral envelope gene captures and syncytin exaptation for placentation in marsupials. *Proceedings of the National Academy of Sciences USA* **112**, E487–E496.

Cornelis G, Vernochet C, Malicorne S, Souquere S, Tzika AC, Goodman SM, *et al.* (2014). Retroviral envelope *syncytin* capture in an ancestrally diverged mammalian clade for placentation in the primitive Afrotherian tenrecs. *Proceedings of the National Academy of Sciences USA* **111**, E4332–E4341.

Cornille A, Feurtey A, Gélin U, Ropars J, Misvanderbrugge K, Gladieux P, and Giraud T (2015). Anthropogenic and natural drivers of gene flow in a temperate wild fruit tree: a basis for conservation and breeding programs in apples. *Evolutionary Applications* 8(4), 373–384.

Cornille A, Giraud T, Bellard C, Tellier A, Le Cam B, Smulders MJM, *et al.* (2013). Postglacial recolonization history of the European crabapple (*Malus sylvestris* Mill.), a wild contributor to the domesticated apple. *Molecular Ecology* **22**, 2249–2263.

Cornille A, Giraud T, Smulders MJM, Roldán-Ruiz I, and Gladieux P (2014). The domestication and evolutionary ecology of apples. *Trends in Genetics* **30**, 57–65.

Cornille A, Gladieux P, Smulders MJM, Roldán-Ruiz I, Laurens F, Le Cam B, *et al.* (2012). New insight into the history of domesticated apple: Secondary contribution of the European wild apple to the genome of cultivated varieties. *PLoS Genetics* **8**, e1002703.

Cornman RS, Burke JM, Wesselingh RA, and Arnold ML (2004). Contrasting genetic structure of adults and progeny in a Louisiana Iris hybrid population. *Evolution* **58**, 2669–2681.

Cortés-Ortiz L, Bermingham E, Rico C, Rodríguez-Luna E, Sampaio I, and Ruiz-García M (2003). Molecular systematics and biogeography of the Neotropical monkey genus, *Alouatta*. *Molecular Phylogenetics and Evolution* **26**, 64–81.

Cortés-Ortiz L, Duda TF Jr, Canales-Espinosa D, García-Orduña F, Rodríquez-Luna E, and Bermingham E (2007). Hybridization in large-bodied New World primates. *Genetics* **176**, 2421–2425.

Coscollá M, Comas I, and González-Candelas F (2011). Quantifying nonvertical inheritance in the evolution of *Legionella pneumophila*. *Molecular Biology and Evolution* **28**, 985–1001.

Cox MP, Mendez FL, Karafet TM, Metni Pilkington M, Kingan SB, Destro-Bisol G, *et al.* (2008). Testing for archaic hominin admixture on the X chromosome: Model likelihoods for the modern human *RRM2P4* region from summaries of genealogical topology under the structured coalescent. *Genetics* **178**, 427–437.

Coyne JA, Kim SY, Chang AS, Lachaise D, and Elwyn S (2002). Sexual isolation between two sibling species with overlapping ranges: *Drosophila santomea* and *Drosophila yakuba*. *Evolution* **56**, 2424–2434.

Coyne JA and Orr HA (1989). Patterns of speciation in *Drosophila*. *Evolution* **43**, 362–381.

Coyne JA and Orr HA (2004). *Speciation*. Sinauer Associates, Inc. Sunderland, Massachusetts.

Cracraft J (1989). Speciation and its ontology: the empirical consequences of alternative species concepts for understanding patterns and processes of differentiation. In D Otte and JA Endler, eds., *Speciation and its Consequences*, pp. 28–59. Sinauer Associates Inc., Sunderland, Massachusetts.

Crease TJ, Stanton DJ, and Hebert PDN (1989). Polyphyletic origins of asexuality in *Daphnia pulex*. II. Mitochondrial-DNA variation. *Evolution* **43**, 1016–1026.

Cropp SJ, Larson A, and Cheverud JM (1999). Historical biogeography of tamarins, genus *Saguinus*: The molecular phylogenetic evidence. *American Journal of Physical Anthropology* **108**, 65–89.

Cruickshank TE and Hahn MW (2014). Reanalysis suggests that genomic islands of speciation are due to reduced diversity, not reduced gene flow. *Molecular Ecology* **23**, 3133–3157.

Cruzan MB and Arnold ML (1993). Ecological and genetic associations in an *Iris* hybrid zone. *Evolution* **47**, 1432–1445.

Cruzan MB and Arnold ML (1994). Assortative mating and natural selection in an *Iris* hybrid zone. *Evolution* **48**, 1946–1958.

Cuellar HS (1971). Levels of genetic compatibility of *Rana areolata* with southwestern members of the *Rana pipiens* Complex (Anura: Ranidae). *Evolution* **25**, 399–409.

Cunha C, Doadrio I, Abrantes J, and Coelho MM (2011). The evolutionary history of the allopolyploid *Squalius alburnoides* (Cyprinidae) complex in the northern Iberian Peninsula. *Heredity* **106**, 100–112.

Currat M and Excoffier L (2004). Modern humans did not admix with Neanderthals during their range expansion into Europe. *PLoS Biology* **2**, e421.

Currat M and Excoffier L (2011). Strong reproductive isolation between humans and Neanderthals inferred from observed patterns of introgression. *Proceedings of the National Academy of Sciences USA* **108**, 15129–15134.

Czypionka T, Cheng J, Pozhitkov A, and Nolte AW (2012). Transcriptome changes after genome-wide admixture in invasive sculpins (*Cottus*). *Molecular Ecology* **21**, 4797–4810.

da Cunha DB, Monteiro E, Vallinoto M, Sampaio I, Ferrari SF, and Schneider H (2011). A molecular phylogeny of the tamarins (genus *Saguinus*) based on five nuclear sequence data from regions containing *Alu* insertions. *American Journal of Physical Anthropology* **146**, 385–391.

da Fonseca RR, Smith BD, Wales N, Cappellini E, Skoglund P, Fumagalli M, *et al.* (2015). The origin and evolution of maize in the Southwestern United States. *Nature Plants* **1**, Article number: 14003 (2015) doi:10.1038/nplants.2014.3.

Daly JC, Wilkinson P, and Shaw DD (1981). Reproductive isolation in relation to allozymic and chromosomal differentiation in the grasshopper *Caledia captiva*. *Evolution* **35**, 1164–1179.

Darwin C (1845). *The Voyage of the Beagle*, 2nd edition, PF Collier and Son, New York.

Darwin C (1859). *On the Origin of Species by Means of Natural Selection or the Preservation of Favoured Races in the Struggle for Life*. John Murray, London.

Darwin C (1868). *The Variation of Animals and Plants Under Domestication*. Volume I. John Murray, London.

Das R, Hergenrother SD, Soto-Calderón ID, Dew JL, Anthony NM, and Jensen-Seaman MI (2014). Complete mitochondrial genome sequence of the Eastern gorilla (*Gorilla beringei*) and implications for African ape biogeography. *Journal of Heredity* **105**, 752–761.

Dasmahapatra KK, Walters JR, Briscoe AD, Davey JW, Whibley A, Nadeau NJ, *et al.* (2012). Butterfly genome reveals promiscuous exchange of mimicry adaptations among species. *Nature* **487**, 94–98.

Daubin V and Ochman H (2004). Bacterial genomes as new gene homes: The genealogy of ORFans in *E. coli*. *Genome Research* **14**, 1036–1042.

David J, Lemeunier F, Tsacas L, and Bocquet C (1974). Hybridation d'une nouvelle espèce, *Drosophila mauritiana* avec *D. melanogaster* et *D. simulans*. *Annales de Genetique* **17**, 235–241.

Davidson EH and Britten RJ (1979). Regulation of gene expression: Possible role of repetitive sequences. *Science* **204**, 1052–1059.

Davis WT (1892). Interesting oaks recently discovered on Staten Island. *Bulletin of the Torrey Botanical Club* **19**, 301–303.

Day JJ, Santini S, and Garcia-Moreno J (2007). Phylogenetic relationships of the Lake Tanganyika cichlid tribe Lamprologini: The story from mitochondrial DNA. *Molecular Phylogenetics and Evolution* **45**, 629–642.

De Andrés MT, Benito A, Pérez-Rivera G, Ocete R, Lopez MA, Gaforio L, *et al.* (2012). Genetic diversity of wild grapevine populations in Spain and their genetic relationships with cultivated grapevines. *Molecular Ecology* **21**, 800–816.

de Been M, Lanza VF, de Toro M, Scharringa J, Dohmen W, Du Y, *et al.* (2014). Dissemination of cephalosporin resistance genes between *Escherichia coli* strains from farm animals and humans by specific plasmid lineages. *PLoS Genetics* **10**, e1004776.

Decker JE, McKay SD, Rolf MM, Kim J, Molina Alcalá A, Sonstegard TS, *et al.* (2014). Worldwide patterns of ancestry, divergence, and admixture in domesticated cattle. *PLoS Genetics* **10**, e1004254.

DeGiorgio M, Jakobsson M, and Rosenberg NA (2009). Explaining worldwide patterns of human genetic variation using a coalescent-based serial founder model of migration outward from Africa. *Proceedings of the National Academy of Sciences USA* **106**, 16057–16062.

Degnan JH and Rosenberg NA (2009). Gene tree discordance, phylogenetic inference and the multispecies coalescent. *Trends in Ecology and Evolution* **24**, 332–340.

De Hert K, Jacquemyn H, van Glabeke S, Roldán-Ruiz I, Vandepitte K, Leus L, and Honnay O (2011). Patterns of hybridization between diploid and derived allotetraploid species of *Dactylorhiza* (Orchidaceae) co-occurring in Belgium. *American Journal of Botany* **98**, 946–955.

Deinard A and Kidd K (1999). Evolution of a HOXB6 intergenic region within the great apes and humans. *Journal of Human Evolution* **36**, 687–703.

Deinard A and Kidd K (2000). Identifying conservation units within captive chimpanzee populations. *American Journal of Physical Anthropology* **111**, 25–44.

De La Torre AR, Roberts DR, and Aitkin SN (2014a). Genome-wide admixture and ecological niche modelling reveal the maintenance of species boundaries despite long history of interspecific gene flow. *Molecular Ecology* **23**, 2046–2059.

De La Torre AR, Wang T, Jaquish B, and Aitkin SN (2014b). Adaptation and exogenous selection in a *Picea glauca* × *Picea engelmannii* hybrid zone: implications for forest management under climate change. *New Phytologist* **201**, 687–699.

De León LF, Raeymaekers JAM, Bermingham E, Podos J, Herrel A, and Hendry AP (2011). Exploring possible human influences on the evolution of Darwin's finches. *Evolution* **65**, 2258–2272.

della Torre A, Merzagora L, Powell JR, and Coluzzi M (1997). Selective introgression of paracentric inversions between two sibling species of the *Anopheles gambiae* complex. *Genetics* **146**, 239–244.

DeMarais BD, Dowling TE, Douglas ME, Minckley WL, and Marsh PC (1992). Origin of *Gila seminuda* (Teleostei: Cyprinidae) through introgressive hybridization: Implications for evolution and conservation. *Proceedings of the National Academy of Sciences USA* **89**, 2747–2751.

Demuth JP, Flanagan RJ, and Delph LF (2014). Genetic architecture of isolation between two species of *Silene* with sex chromosomes and Haldane's rule. *Evolution* **68**, 332–342.

Denef VJ, Kalnejaisa LH, Muellera RS, Wilmesa P, Baker BJ, Thomas BC, *et al.* (2010). Proteogenomic basis for ecological divergence of closely related bacteria in natural acidophilic microbial communities. *Proceedings of the National Academy of Sciences USA* **107**, 2383–2390.

Deng L, Ignacio-Espinoza JC, Gregory AC, Poulos BT, Weitz JS Hugenholtz P, and Sullivan MB (2014). Viral tagging reveals discrete populations in *Synechococcus* viral genome sequence space. *Nature* **513**, 242–245.

Dennis ES, Peacock WJ, White MJD, Appels R, and Contreras N (1981). Cytogenetics of the parthenogenetic grasshopper *Warramaba virgo* and its bisexual relatives. VII. Evidence from repeated DNA sequences for a dual origin of *W. virgo*. *Chromosoma* **82**, 453–469.

Department of Environment and Climate Change, New South Wales, Australia (2007). *Lord Howe Island Biodiversity Plan.* <http://www.environment.nsw.gov.au/resources/parks/LHI_bmp.pdf>

Derr JN, Hedrick PW, Halbert ND, Plough L, Dobson LK, King J, *et al.* (2012). Phenotypic effects of cattle mitochondrial DNA in American bison. *Conservation Biology* **26**, 1130–1136.

De Sá RO and Hillis DM (1990). Phylogenetic relationships of the pipid frogs *Xenopus* and *Silurana*: An integration of ribosomal DNA and morphology. *Molecular Biology and Evolution* **7**, 365–376.

de Souza FSJ, Bumaschny VF, Low MJ, and Rubinstein M (2005). Subfunctionalization of expression and peptide domains following the ancient duplication of the proopiomelanocortin gene in teleost fishes. *Molecular Biology and Evolution* **22**, 2417–2427.

de Souza Jesus A, Schunemann HE, Müller J, da Silva MA, and Bicca-Marques JC (2010). Hybridization between *Alouatta caraya* and *Alouatta guariba clamitans* in captivity. *Primates* **51**, 227–230.

Detwiler KM, Burrell AS, and Jolly CJ (2005). Conservation implications of hybridization in African cercopithecine monkeys. *International Journal of Primatology* **26**, 661–684.

Devlin JM, Hartley CA, Gilkerson JR, Coppo MJC, Vaz P, Noormohammadi AH, *et al.* (2011). Horizontal transmission dynamics of a glycoprotein G deficient candidate vaccine strain of infectious laryngotracheitis virus and the effect of vaccination on transmission of virulent virus. *Vaccine* **29**, 5699–5704.

Diabaté A, Dabiré RK, Kim EH, Dalton R, Millogo N, Baldet T, *et al.* (2005). Larval development of the molecular forms of *Anopheles gambiae* (Diptera: Culicidae) in different habitats: A transplantation experiment. *Journal of Medical Entomology* **42**, 548–553.

Diabaté A, Dabiré RK, Heidenberger K, Crawford J, Lamp WO, Culler LE, and Lehmann T (2008). Evidence for divergent selection between the molecular forms of *Anopheles gambiae*: role of predation. *BMC Evolutionary Biology* **8**, 5.

Diabaté A, Dao A, Yaro AS, Adamou A, Gonzalez R, Manoukis NC, *et al.* (2009). Spatial swarm segregation and reproductive isolation between the molecular forms of *Anopheles gambiae*. *Proceedings of the Royal Society of London B* **276**, 4215–4222.

Diamond JM (1991). The earliest horsemen. *Nature* **350**, 275–276.

Diaz A and Macnair MR (1999). Pollen tube competition as a mechanism of prezygotic reproductive isolation between *Mimulus nasutus* and its presumed progenitor *M. guttatus*. *New Phytologist* **144**, 471–478.

Di Candia MR and Routman EJ (2007). Cytonuclear discordance across a leopard frog contact zone. *Molecular Phylogenetics and Evolution* **45**, 564–575.

Dickman CTD and Moehring AJ (2014). Contribution of the X chromosome to a marked reduction in lifespan in interspecies female hybrids of *Drosophila simulans* and *D. mauritiana*. *Journal of Evolutionary Biology* **27**, 25–33.

Dijkstra PD, Wiegertjes GF, Forlenza M, van der Sluijs I, Hofman HA, Metcalfe NB, and Groothuis TGG (2011). The role of physiology in the divergence of two incipient cichlid species. *Journal of Evolutionary Biology* **24**, 2639–2652.

Ding Q, Hu Y, Xu S, Wang J, and Jin L (2014a). Neanderthal introgression at chromosome 3p21.31 was under positive natural selection in East Asians. *Molecular Biology and Evolution* **31**, 683–695.

Ding Q, Hu Y, Xu S, Wang C-C, Li H, Zhang R, *et al.* (2014b). Neanderthal origin of the haplotypes carrying the functional variant Val92Met in the *MC1R* in modern humans. *Molecular Biology and Evolution* **31**, 1994–2003.

Dirks W, Reid DJ, Jolly CJ, Phillips-Conroy JE, and Brett FL (2002). Out of the mouths of baboons: Stress, life history, and dental development in the Awash National Park hybrid zone, Ethiopia. *American Journal of Physical Anthropology* **118**, 239–252.

Disotell TR (2012). Archaic human genomics. *Yearbook of Physical Anthropology* **55**, 24–39.

Di Vecchi-Staraz M, Laucou V, Bruno G, Lacombe T, Gerber S, Bourse T, *et al.* (2009). Low level of pollen-mediated gene flow from cultivated to wild grapevine: Consequences for the evolution of the endangered subspecies *Vitis vinifera* L. subsp. *silvestris*. *Journal of Heredity* **100**, 66–75.

Dobson MC, Taylor SJ, Arnold ML, and Martin NH (2011). Patterns of herbivory and fungal infection in experimental Louisiana *Iris* hybrids. *Evolutionary Ecology Research* **13**, 543–552.

Dobzhansky Th (1935). A critique of the species concept in biology. *Philosophy of Science* **2**, 344–355.

Dobzhansky Th (1936). Studies on hybrid sterility. II. Localization of sterility factors in *Drosophila pseudoobscura* hybrids. *Genetics* **21**, 113–135.

Dobzhansky Th (1937) *Genetics and the Origin of Species*. Columbia University Press, New York.

Dobzhansky Th (1940). Speciation as a stage in evolutionary divergence. *American Naturalist* **74**, 312–321.

Dobzhansky Th (1946). Genetics of natural populations. XIII. Recombination and variability in populations of *Drosophila pseudoobscura*. *Genetics* **31**, 269–290.

Dobzhansky Th (1970). *Genetics of the Evolutionary Process*. Columbia University Press, New York.

Dobzhansky T (1973). Is there gene exchange between *Drosophila pseudoobscura* and *Drosophila persimilis* in their natural habitats? *American Naturalist* **107**, 312–314.

Doebley JF (1989). Molecular evidence for a missing wild relative of maize and the introgression of its chloroplast genome into *Zea perennis*. *Evolution* **43**, 1555–1559.

Doebley JF, Goodman MM, and Stuber CW (1984). Isoenzymatic variation in *Zea* (Gramineae). *Systematic Botany* **9**, 203–218.

Dong L, Heckel G, Liang W, and Zhang Y (2013). Phylogeography of Silver Pheasant (*Lophura nycthemera* L.) across China: Aggregate effects of refugia, introgression and riverine barriers. *Molecular Ecology* **22**, 3376–3390.

Donnelly MJ, Pinto J, Girod R, Besansky NJ, and Lehmann T (2004). Revisiting the role of introgression vs shared ancestral polymorphisms as key processes shaping genetic diversity in the recently separated sibling species of the *Anopheles gambiae* complex. *Heredity* **92**, 61–68.

Donoghue PCJ and Purnell MA (2005). Genome duplication, extinction and vertebrate evolution. *Trends in Ecology and Evolution* **20**, 312–319.

Donovan LA, Ludwig F, Rosenthal DM, Rieseberg LH, and Dudley SA (2009). Phenotypic selection on leaf ecophysiological traits in *Helianthus*. *New Phytologist* **183**, 868–879.

Doolittle WF (1999). Phylogenetic classification and the universal tree. *Science* **284**, 2124–2128.

Douglas GM, Gos G, Steige KA, Salcedo A, Holm K, Josephs EB, *et al.* (2015). Hybrid origins and the earliest stages of diploidization in the highly successful recent polyploid *Capsella bursa-pastoris*. *Proceedings of the National Academy of Sciences USA* **112**, 2806–2811.

Dowling TE and DeMarais BD (1993). Evolutionary significance of introgressive hybridization in cyprinid fishes. *Nature* **362**, 444–446.

Doyle JJ, Doyle JL, and Brown AHD (1999). Origins, colonization, and lineage recombination in a widespread perennial soybean polyploid complex. *Proceedings of the National Academy of Sciences USA* **96**, 10741–10745.

Doyle JJ, Doyle JL, Brown AHD, and Grace JP (1990a). Multiple origins of polyploids in the *Glycine tabacina* complex inferred from chloroplast DNA polymorphism. *Proceedings of the National Academy of Sciences USA* **87**, 714–717.

Doyle JJ, Doyle JL, Grace JP, and Brown AHD (1990b). Reproductively isolated polyploid races of *Glycine tabacina* (Leguminosae) had different chloroplast genome donors. *Systematic Botany* **15**, 173–181.

Doyle JJ, Doyle JL, Rauscher JT, and Brown AHD (2004). Evolution of the perennial soybean polyploid complex (*Glycine* subgenus *Glycine*): a study of contrasts. *Biological Journal of the Linnean Society* **82**, 583–597.

Doyle JJ, Flagel LE, Paterson AH, Rapp RA, Soltis DE, Soltis PS, and Wendel JF (2008). Evolutionary genetics of genome merger and doubling in plants. *Annual Review of Genetics* **42**, 443–461.

Driscoll CA, Menotti-Raymond M, Roca AL, Hupe K, Johnson WE, Geffen E, *et al.* (2007). The Near Eastern origin of cat domestication. *Science* **317**, 519–523.

Drummond AJ, Suchard MA, Xie D, and Rambaut A (2012). Bayesian phylogenetics with BEAUti and the BEAST 1.7. *Molecular Biology and Evolution* **29**, 1969–1973.

Du H, Ouyang Y, Zhang C, and Zhang Q (2011). Complex evolution of *S5*, a major reproductive barrier regulator, in the cultivated rice *Oryza sativa* and its wild relatives. *New Phytologist* **191**, 275–287.

Du FK, Peng XL, Liu JQ, Lascoux M, Hu FS, and Petit RJ (2011). Direction and extent of organelle DNA introgression between two spruce species in the Qinghai-Tibetan Plateau. *New Phytologist* **192**, 1024–1033.

Duarte JM, Cui L, Wall PK, Zhang Q, Zhang X, Leebens-Mack J, *et al.* (2006). Expression pattern shifts following duplication indicative of subfunctionalization and neofunctionalization in regulatory genes of *Arabidopsis*. *Molecular Biology and Evolution* **23**, 469–478.

Dudas G, Bedford T, Lycett S, and Rambaut A (2015). Reassortment between influenza B lineages and the emergence of a coadapted PB1-PB2-HA gene complex. *Molecular Biology and Evolution* **32**, 162–172.

Dufresne F and Hebert PDN (1994). Hybridization and origins of polyploidy. *Proceedings of the Royal Society of London Series B* **258**, 141–146.

Dufresne F and Hebert PDN (1997). Pleistocene glaciations and polyphyletic origins of polyploidy in an arctic cladoceran. *Proceedings of the Royal Society of London Series B* **264**, 201–206.

Dufresne A, Salanoubat M, Partensky F, Artiguenave F, Axmann IM, Barbe V, *et al.* (2003). Genome sequence of the cyanobacterium *Prochlorococcus marinus* SS120, a nearly minimal oxyphototrophic genome. *Proceedings of the National Academy of Sciences USA* **100**, 10020–10025.

Duintjer Tebbins RJ, Pallansch MA, Kim J-H, Burns CC, Kew OM, Oberste MS, *et al.* (2013). Oral poliovirus vaccine evolution and insights relevant to modeling the risks of circulating vaccine-derived polioviruses (cVDPVs). *Risk Analysis* **33**, 680–702.

Dumas D, Catalan J, and Britton-Davidian J (2015). Reduced recombination patterns in Robertsonian hybrids between chromosomal races of the house mouse: Chiasma analyses. *Heredity* **114**, 56–64.

Dunn B, Paulish T, Stanbery A, Piotrowski J, Koniges G, Kroll E, *et al.* (2013). Recurrent rearrangement during adaptive evolution in an interspecific yeast hybrid suggests a model for rapid introgression. *PLoS Genetics* **9**, e1003366.

Dunn B, Richter C, Kvitek DJ, Pugh T, and Sherlock G (2012). Analysis of the *Saccharomyces cerevisiae* pan-genome reveals a pool of copy number variants distributed in diverse yeast strains from differing industrial environments. *Genome Research* **22**, 908–924.

Durand EY, Patterson N, Reich D, and Slatkin M (2011). Testing for ancient admixture between closely related populations. *Molecular Biology and Evolution* 28, 2239–2252.

Dzur-Gejdošová M, Simecek P, Gregorova S, Bhattacharyya T, and Forejt J (2012). Dissecting the genetic architecture of F_1 hybrid sterility in house mice. *Evolution* **66**, 3321–3335.

Eberhard WG (1990). Evolution in bacterial plasmids and levels of selection. *Quarterly Review of Biology* **65**, 3–22.

Eckenwalder JE (1984). Natural intersectional hybridization between North American species of *Populus* (Salicaceae) in sections *Aigeiros* and *Tacamaha*ca. III. Paleobotany and evolution. *Canadian Journal of Botany* **62**, 336–342.

Eckert AJ and Carstens BC (2008). Does gene flow destroy phylogenetic signal? The performance of three methods for estimating species phylogenies in the presence of gene flow. *Molecular Phylogenetics and Evolution* **49**, 832–842.

Edmands S (2008). Recombination in interpopulation hybrids of the copepod *Tigriopus californicus*: Release of beneficial variation despite hybrid breakdown. *Journal of Heredity* 99, 316–318.

Edmands S and Burton RS (1999). Cytochrome C Oxidase activity in interpopulation hybrids of a marine copepod: A test for nuclear-nuclear or nuclear-cytoplasmic coadaptation. *Evolution* 53, 1972–1978.

Edmands S, Northrup SL, and Hwang AS (2009). Maladapted gene complexes within populations of the intertidal copepod *Tigriopus californicus*? *Evolution* 63, 2184–2192.

Edwards SV, Liu L, and Pearl DK (2007). High-resolution species trees without concatenation. *Proceedings of the National Academy of Sciences USA* 104, 5936–5941.

Edwards CJ, Suchard MA, Lemey P, Welch JJ, Barnes I, Fulton TL, *et al.* (2011). Ancient hybridization and an Irish origin for the modern polar bear matriline. *Current Biology* 21, 1251–1258.

Egger B, Sefc KM, Makasa L, Sturmbauer C, and Salzburger W (2012). Introgressive hybridization between color morphs in a population of cichlid fishes twelve years after human-induced secondary admixis. *Journal of Heredity* 103, 515–522.

Elgvin TO, Hermansen JS, Fijarczyk A, Bonnet TE, Borge T, Sæther SA, Voje KL, and Sætre G-P (2011). Hybrid speciation in sparrows II: A role for sex chromosomes? *Molecular Ecology* 20, 3823–3837.

Ellison CK and Burton RS (2006). Disruption of mitochondrial function in interpopulation hybrids of *Tigriopus californicus*. *Evolution* 60, 1382–1391.

Ellison CK and Burton RS (2008a). Genotype-dependent variation of mitochondrial transcriptional profiles in interpopulation hybrids. *Proceedings of the National Academy of Sciences USA* 105, 15831–15836.

Ellison CK and Burton RS (2008b). Interpopulation hybrid breakdown maps to the mitochondrial genome. *Evolution* 62, 631–638.

Ellstrand NC and Hoffman CA (1990). Hybridization as an avenue of escape for engineered genes. *BioScience* 40, 438–442.

Ellstrand NC, Meirmans P, Rong J, Bartsch D, Ghosh A, de Jong TJ, *et al.* (2013). Introgression of crop alleles into wild or weedy populations. *Annual Review of Ecology, Evolution and Systematics* 44, 325–345.

Ellstrand NC and Schierenbeck KA (2000). Hybridization as a stimulus for the evolution of invasiveness in plants? *Proceedings of the National Academy of Sciences USA* 97, 7043–7050.

Emms SK and Arnold ML (1997). The effect of habitat on parental and hybrid fitness: reciprocal transplant experiments with Louisiana Irises. *Evolution* 51, 1112–1119.

Emms SK and Arnold ML (2000). Site-to-site differences in pollinator visitation patterns in a Louisiana Iris hybrid zone. *Oikos* 91, 568–578.

Emms SK, Hodges SA, and Arnold ML (1996). Pollen-tube competition, siring success, and consistent asymmetric hybridization in Louisiana Irises. *Evolution* 50, 2201–2206.

Endler JA (1973). Gene flow and population differentiation. *Science* 179, 243–250.

Endler JA (1977). *Geographic Variation, Speciation, and Clines*. Princeton University Press, Princeton, New Jersey.

Enk J, Devault A, Debruyne R, King CE, Treangen T, O'Rourke D, *et al.* (2011). Complete Columbian mammoth mitogenome suggests interbreeding with woolly mammoths. *Genome Biology* 12, R51.

Eriksson J, Larson G, Gunnarsson U, Bed'hom B, Tixier-Boichard M, Strömstedt L, *et al.* (2008). Identification of the *yellow skin* gene reveals a hybrid origin of the domestic chicken. *PLoS Genetics* 4, e1000010.

Eriksson A and Manica A (2012). Effect of ancient population structure on the degree of polymorphism shared between modern human populations and ancient hominins. *Proceedings of the National Academy of Sciences USA* 109, 13956–13960.

Eroukhmanoff F, Elgvin TO, Gonzàlez Rojas MF, Haas F, Hermansen JS, and Sætre G-P (2014). Effect of species interaction on beak integration in an avian hybrid species complex. *Evolutionary Biology* 41, 452–458.

Eroukhmanoff F, Hermansen JS, Bailey RI, Sæther SA, and Sætre G-P (2013). Local adaptation within a hybrid species. *Heredity* 111, 286–292.

Escalante MA, Garcia-De-Leon FJ, Dillman CB, de los Santos Camarillo A, George A, de los A Barriga-Sosa I, *et al.* (2014). Genetic introgression of cultured rainbow trout in the Mexican native trout complex. *Conservation Genetics* 15, 1063–1071.

Estep MC, McKain MR, Vela Diaz D, Zhong J, Hodge JG, Hodkinson TR, *et al.* (2014). Allopolyploidy, diversification, and the Miocene grassland expansion. *Proceedings of the National Academy of Sciences USA* 111, 15149–15154.

Evans BJ (2007). Ancestry influences the fate of duplicated genes millions of years after polyploidization of clawed frogs (*Xenopus*). *Genetics* 176, 1119–1130.

Evans PD, Anderson JR, Vallender EJ, Choi SS, and Lahn BT (2004). Reconstructing the evolutionary history of *microcephalin*, a gene controlling human brain size. *Human Molecular Genetics* 13, 1139–1145.

Evans PD, Gilbert SL, Mekel-Bobrov N, Vallender EJ, Anderson JR, Vaez-Azizi LM, *et al.* (2005). *Microcephalin*, a gene regulating brain size, continues to evolve adaptively in humans. *Science* 309, 1717–1120.

Evans BJ, Kelley DB, Melnick DJ, and Cannatella DC (2005). Evolution of RAG-1 in polyploid clawed frogs. *Molecular Biology and Evolution* **22**, 1193–1207.

Evans BJ, Kelley DB, Tinsley RC, Melnick DJ, and Cannatella DC (2004). A mitochondrial DNA phylogeny of African clawed frogs: Phylogeography and implications for polyploid evolution. *Molecular Phylogenetics and Evolution* **33**, 197–213.

Evans PD, Mekel-Bobrov N, Vallender EJ, Hudson RR, and Lahn BT (2006). Evidence that the adaptive allele of the brain size gene *microcephalin* introgressed into *Homo sapiens* from an archaic *Homo* lineage. *Proceedings of the National Academy of Sciences USA* **103**, 18178–18183.

Evans BJ, Morales JC, Picker MD, Kelley DB, and Melnick DJ (1997). Comparative molecular phylogeography of two *Xenopus* species, *X. gilli* and *X. laevis*, in the southwestern Cape Province, South Africa. *Molecular Ecology* **6**, 333–343.

Fagundes NJR, Ray N, Beaumont M, Neuenschwander S, Salzano FM, Bonatto SL, and Excoffier L (2007). Statistical evaluation of alternative models of human evolution. *Proceedings of the National Academy of Sciences USA* **104**, 17614–17619.

Fan Z, Zhao G, Li P, Osada N, Xing J, Yi Y, *et al.* (2014). Whole-genome sequencing of Tibetan macaque (*Macaca thibetana*) provides new insight into the macaque evolutionary history. *Molecular Biology and Evolution* **31**, 1475–1489.

Farrer RA, Weinert LA, Bielby J, Garner TWJ, Balloux F, Clare F, *et al.* (2011). Multiple emergences of genetically diverse amphibian-infecting chytrids include a globalized hypervirulent recombinant lineage. *Proceedings of the National Academy of Sciences USA* **108**, 18732–18736.

Farrington HL, Lawson LP, Clark CM, and Petren K (2014). The evolutionary history of Darwin's finches: Speciation, gene flow, and introgression in a fragmented landscape. *Evolution* **68**, 2932–2944.

Farris JS, Källersjö M, Kluge AG, and Bult C (1994). Testing significance of incongruence. *Cladistics* **10**, 315–319.

Feil EJ, Enright MC, and Spratt BG (2000). Estimating the relative contributions of mutation and recombination to clonal diversification: A comparison between *Neisseria meningitidis* and *Streptococcus pneumonidae*. *Research in Microbiology* **151**, 465–469.

Fernandez-Manjarres JF, Gerard PR, Dufour J, Raquin C, and Frascaria-Lacoste N (2006). Differential patterns of morphological and molecular hybridization between *Fraxinus excelsior* L. and *Fraxinus angustifolia* Vahl (Oleaceae) in eastern and western France. *Molecular Ecology* **15**, 3245–3257.

Feulner PGD, Gratten J, Kijas JW, Visscher PM, Pemberton JM, and Slate J (2013). Introgression and the fate of domesticated genes in a wild mammal population. *Molecular Ecology* **22**, 4210–4221.

Field DL, Ayre DJ, Whelan RJ, and Young AG (2008). Relative frequency of sympatric species influences rates of interspecific hybridization, seed production and seedling performance in the uncommon *Eucalyptus aggregata*. *Journal of Ecology* **96**, 1198–1210.

Field DL, Ayre DJ, Whelan RJ, and Young AG (2009). Molecular and morphological evidence of natural interspecific hybridization between the uncommon *Eucalyptus aggregata* and the widespread *E. rubida* and *E. viminalis*. *Conservation Genetics* **10**, 881–896.

Field DL, Ayre DJ, Whelan RJ, and Young AG (2011a). Patterns of hybridization and asymmetrical gene flow in hybrid zones of the rare *Eucalyptus aggregata* and common *E. rubida*. *Heredity* **106**, 841–853.

Field DL, Ayre DJ, Whelan RJ, and Young AG (2011b). The importance of pre-mating barriers and the local demographic context for contemporary mating patterns in hybrid zones of *Eucalyptus aggregata* and *Eucalyptus rubida*. *Molecular Ecology* **20**, 2367–2379.

Fields BS, Benson RF, and Besser RE (2002). *Legionella* and Legionnaires' disease: 25 years of investigation. *Clinical Microbiology Reviews* **15**, 506–526.

Figueroa-Bossi N and Bossi L (1999). Inducible prophages contribute to *Salmonella* virulence in mice. *Molecular Microbiology* **33**, 167–176.

Filée J, Forterre P, and Laurent J (2003). The role played by viruses in the evolution of their hosts: A view based on informational protein phylogenies. *Research in Microbiology* **154**, 237–243.

Filée J, Tétart F, Suttle CA, and Krisch HM (2005). Marine T4-type bacteriophages, a ubiquitous component of the dark matter of the biosphere. *Proceedings of the National Academy of Sciences USA* **102**, 12471–12476.

Finlayson C (2005). Biogeography and evolution of the genus *Homo*. *Trends in Ecology and Evolution* **20**, 457–463.

Fischer WJ, Koch WA, and Elepfandt A (2000). Sympatry and hybridization between the clawed frogs *Xenopus laevis laevis* and *Xenopus muelleri* (Pipidae). *Journal of Zoology* **252**, 99–107.

Fishman L, Aagaard J, and Tuthill JC (2008). Toward the evolutionary genomics of gametophytic divergence: Patterns of transmission ratio distortion in monkeyflower (*Mimulus*) hybrids reveal a complex genetic basis for conspecific pollen precedence. *Evolution* **62**, 2958–2970.

Fishman L, Kelly AJ, Morgan E, and Willis JH (2001). A genetic map in the *Mimulus guttatus* species complex reveals transmission ratio distortion due to heterospecific interactions. *Genetics* **159**, 1701–1716.

Fishman L, Kelly AJ, and Willis JH (2002). Minor quantitative trait loci underlie floral traits associated with mating system divergence in *Mimulus*. *Evolution* **56**, 2138–2155.

Fishman L, Stathos A, Beardsley PM, Williams CF, and Hill JP (2013). Chromosomal rearrangements and the genetics of reproductive barriers in *Mimulus* (monkeyflowers). *Evolution* **67**, 2547–2560.

Fishman L and Willis JH (2005). A novel meiotic drive locus almost completely distorts segregation in *Mimulus* (monkeyflower) hybrids. *Genetics* **169**, 347–353.

Fitzpatrick BM (2004). Rates of evolution of hybrid inviability in birds and mammals. *Evolution* **58**, 1865–1870.

Fitzpatrick BM, Johnson JR, Kump DK, Shaffer HB, Smith JJ, and Voss SR (2009). Rapid fixation of non-native alleles revealed by genome-wide SNP analysis of hybrid tiger salamanders. *BMC Evolutionary Biology* **9**:176.

Fitzpatrick BM, Johnson JR, Kump DK, Smith JJ, Voss SR, and Shaffer HB (2010). Rapid spread of invasive genes into a threatened native species. *Proceedings of the National Academy of Sciences USA* **107**, 3606–3610.

Foley BR, Rose CG, Rundle DE, Leong W, and Edmands S (2013). Postzygotic isolation involves strong mitochondrial and sex-specific effects in *Tigriopus californicus*, a species lacking heteromorphic sex chromosomes. *Heredity* **111**, 391–401.

Fontaine MC, Pease JB, Steele A, Waterhouse RM, Neafsey DE, Sharakhov IV, et al. (2015). Extensive introgression in a malaria vector species complex revealed by phylogenomics. *Science* **347**, 1258524–1–1258524–6.

Forterre P (1999). Displacement of cellular proteins by functional analogues from plasmids or viruses could explain puzzling phylogenies of many DNA informational proteins. *Molecular Microbiology* **33**, 457–465.

Fortune PM, Schierenbeck K, Ayres D, Bortolus A, Catrice O, Brown S, and Ainouche ML (2008). The enigmatic invasive *Spartina densiflora*: A history of hybridizations in a polyploidy context. *Molecular Ecology* **17**, 4304–4316.

Foster RC (1937). A cyto-taxonomic survey of the North American species of *Iris*. *Contributions from the Gray Herbarium*, no. CXIX, pp. 3–80.

Fraïsse C, Roux C, Welch JJ, and Bierne N (2014). Geneflow in a mosaic hybrid zone: Is local introgression adaptive? *Genetics* **197**, 939–951.

Frantz LAF, Madsen O, Megens H-J, Groenen MAM, and Lohse K (2014). Testing models of speciation from genome sequences: Divergence and asymmetric admixture in Island South-East Asian *Sus* species during the Plio-Pleistocene climatic fluctuations. *Molecular Ecology* **23**, 5566–5574.

Frantz LAF, Schraiber JG, Madsen O, Megens H-J, Bosse M, Paudel Y, et al. (2013). Genome sequencing reveals fine scale diversification and reticulation history during speciation in *Sus*. *Genome Biology* **14**, R107.

Frantz AC, Zachos FE, Kirschning J, Cellina S, Bertouille S, Mamuris Z, et al. (2013). Genetic evidence for introgression between domestic pigs and wild boars (*Sus scrofa*) in Belgium and Luxembourg: A comparative approach with multiple marker systems. *Biological Journal of the Linnean Society* **110**, 104–115.

Fraser C, Donnelly CA, Cauchemez S, Hanage WP, Van Kerkhove MD, Hollingsworth TD, et al. (2009). Pandemic potential of a strain of influenza A (H1N1): Early findings. *Science* **324**, 1557–1561.

Fraser C, Hanage WP, and Spratt BG (2007). Recombination and the nature of bacterial speciation. *Science* **315**, 476–480.

Fraser DW, Tsai TR, Orenstein W, Parkin WE, Beecham HJ, Sharrar RG, et al. (1977). Legionnaires' disease—Description of an epidemic of pneumonia. *New England Journal of Medicine* **297**, 1189–1197.

Freedman AH, Gronau I, Schweizer RM, Ortega-Del Vecchyo D, Han E, Silva PM, et al. (2014). Genome sequencing highlights the dynamic early history of dogs. *PLoS Genetics* **10**, e1004016.

Friar EA, Prince LM, Cruse-Sanders JM, McGlaughlin ME, Butterworth CA, and Baldwin BG (2008). Hybrid origin and genomic mosaicism of *Dubautia scabra* (Hawaiian Silversword Alliance; Asteraceae, Madiinae). *Systematic Botany* **33**, 589–597.

Frost WH (1919). The epidemiology of influenza. *Journal of the American Medical Association* **73**, 313–318.

Frost DR, Grant T, Faivovich J, Bain RH, Haas A, Haddad CFB, et al. (2006). The amphibian tree of life. *Bulletin of the American Museum of Natural History* **297**, 1–370.

Frost LS, Ippen-Ihler K, and Skurray RA (1994). Analysis of the sequence and gene products of the transfer region of the F sex factor. *Microbiological Reviews* **58**, 162–210.

Frost JS and Platz JE (1983). Comparative assessment of modes of reproductive isolation among four species of leopard frogs (*Rana pipiens* complex). *Evolution* **37**, 66–78.

Fu Q, Li H, Moorjani P, Jay F, Slepchenko SM, Bondarev AA, et al. (2014). Genome sequence of a 45,000-year-old modern human from western Siberia. *Nature* **514**, 445–449.

Fu Q, Meyer M, Gao X, Stenzel U, Burbano HA, Kelso J, and Pääbo S (2013). DNA analysis of an early modern human from Tianyuan Cave, China. *Proceedings of the National Academy of Sciences USA* **110**, 2223–2227.

Fuentes I, Stegemann S, Golczyk H, Karcher D, and Bock R (2014). Horizontal genome transfer as an asexual path to the formation of new species. *Nature* **511**, 232–235.

Fukunaga K, Hill J, Vigouroux Y, Matsuoka Y, Sanchez J, Liu K, et al. (2005). Genetic diversity and population structure of teosinte. *Genetics* **169**, 2241–2254.

Fumihito A, Miyake T, Sumi S-I, Takada M, Ohno S, and Kondo N (1994). One subspecies of the red junglefowl (*Gallus gallus gallus*) suffices as the matriarchic ancestor of all domestic breeds. *Proceedings of the National Academy of Sciences USA* **91**, 12505–12509.

Fumihito A, Miyake T, Takada M, Shingu R, Endo T, Gojobori T, *et al.* (1996). Monophyletic origin and unique dispersal patterns of domestic fowls. *Proceedings of the National Academy of Sciences USA* **93**, 6792–6795.

Furlong RF and Holland PWH (2002). Were vertebrates octoploid? *Philosophical Transactions of the Royal Society of London* **357**, 531–544.

Furman BLS, Bewick AJ, Harrison TL, Greenbaum E, Gvoždík V, Kusamba C, and Evans BJ (2015). Pan-African phylogeography of a model organism, the African clawed frog "*Xenopus laevis.*" *Molecular Ecology* **24**, 909–925.

Futuyma DJ (2005). *Evolution.* Sinauer Associates, Inc. Sunderland, Massachusetts.

Futuyma DJ and Mayer GC (1980). Non-allopatric speciation in animals. *Systematic Zoology* **29**, 254–271.

Fuzessy LF, de Oliveira Silva I, Malukiewicz J, Silva FFR, do Carmo Ponzio M, Boere V, and Ackermann RR (2014). Morphological variation in wild marmosets (*Callithrix penicillata* and *C. geoffroyi*) and their hybrids. *Evolutionary Biology,* **41**, 480–493.

Gabor CR and Aspbury AS (2008). Non-repeatable mate choice by male sailfin mollies, *Poecilia latipinna*, in a unisexual-bisexual mating complex. *Behavioral Ecology* **19**, 871–878.

Gagnaire P-A, Pavey SA, Normandeau E, and Bernatchez L (2013). The genetic architecture of reproductive isolation during speciation-with-gene-flow in lake whitefish species pairs assessed by RAD sequencing. *Evolution* **67**, 2483–2497.

Gaither MR, Schultz JK, Bellwood DR, Pyle RL, DiBattista JD, Rocha LA, and Bowen BW (2014). Evolution of pygmy angelfishes: Recent divergences, introgression, and the usefulness of color in taxonomy. *Molecular Phylogenetics and Evolution* **74**, 38–47.

Gallardo MH, Bickham JW, Honeycutt RL, Ojeda RA, and Köhler N (1999). Discovery of tetraploidy in a mammal. *Nature* **401**, 341.

Gallardo MH, Suárez-Villota EY, Nuñez JJ, Vargas RA, Haro R, and Köhler N (2013). Phylogenetic analysis and phylogeography of the tetraploid rodent *Tympanoctomys barrerae* (Octodontidae): Insights on its origin and the impact of Quaternary climate changes on population dynamics. *Biological Journal of the Linnean Society* **108**, 453–469.

Galligan TH, Donnellan SC, Sulloway FJ, Fitch AJ, Bertozzi T, and Kleindorfer S (2012). Panmixia supports divergence with gene flow in Darwin's small ground finch, *Geospiza fuliginosa*, on Santa Cruz, Galápagos Islands. *Molecular Ecology* **21**, 2106–2115.

Gao F, Bailes E, Robertson DL, Chen Y, Rodenburg CM, Michael SF, *et al.* (1999). Origin of HIV-1 in the chimpanzee *Pan troglodytes troglodytes. Nature* **397**, 436–441.

Gao J, Wang B, Mao J-F, Ingvarsson P, Zeng Q-Y, and Wang X-R (2012). Demography and speciation history of the homoploid hybrid pine *Pinus densata* on the Tibetan Plateau. *Molecular Ecology* **21**, 4811–4827.

García M and Riblet SM (2001). Characterization of infectious Laryngotracheitis virus isolates: Demonstration of viral subpopulations within vaccine preparations. *Avian Disease* **45**, 558–566.

Garrigan D and Kingan SB (2007). Archaic human admixture. *Current Anthropology* **48**, 895–902.

Garrigan D, Kingan SB, Geneva AJ, Andolfatto P, Clark AG, Thornton KR, and Presgraves DC (2012). Genome sequencing reveals complex speciation in the *Drosophila simulans* clade. *Genome Research* **22**, 1499–1511.

Garrigan D, Mobasher Z, Severson T, Wilder JA, and Hammer MF (2005). Evidence for archaic Asian ancestry on the human X chromosome. *Molecular Biology and Evolution* **22**, 189–192.

Garten RJ, Davis CT, Russell CA, Shu B, Lindstrom S, Balish A, *et al.* (2009). Antigenic and genetic characteristics of swine-origin 2009 A(H1N1) influenza viruses circulating in humans. *Science* **325**, 197–201.

Gartside DF, Littlejohn MJ, and Watson GF (1979). Structure and dynamics of a narrow hybrid zone between *Geocrinia laevis* and *G. victoriana* (Anura: Leptodactylidae) in south-eastern Australia. *Heredity* **43**, 165–177.

Gaut BS and Doebley JF (1997). DNA sequence evidence for the segmental allotetraploid origin of maize. *Proceedings of the National Academy of Sciences USA* **94**, 6809–6814.

Gauthier O and Lapointe F-J (2007). Hybrids and phylogenetics revisited: A statistical test of hybridization using quartets. *Systematic Botany* **32**, 8–15.

Gavrilets S (2003). Models of speciation: What have we learned in 40 years? *Evolution* **57**, 2197–2215.

Geiger MF, McCrary JK, and Schliewen UK (2010). Not a simple case—A first comprehensive phylogenetic hypothesis for the Midas cichlid complex in Nicaragua (Teleostei: Cichlidae: *Amphilophus*). *Molecular Phylogenetics and Evolution* **56**, 1011–1024.

Geneva A and Garrigan D (2010). Population genomics of secondary contact. *Genes* **1**, 124–142.

Genner MJ and Turner GF (2012). Ancient hybridization and phenotypic novelty within Lake Malawi's cichlid fish radiation. *Molecular Biology and Evolution* **29**, 195–206.

Gentile G, della Torre A, Maegga B, Powell JR, and Caccone A (2002). Genetic differentiation in the African malaria vector, *Anopheles gambiae* s.s., and the problem of taxonomic status. *Genetics* **161**, 1561–1578.

Geraldes A, Carneiro M, Delibes-Mateos M, Villafuerte R, Nachman MW, and Ferrand N (2008). Reduced introgression of the Y chromosome between subspecies of the European rabbit (*Oryctolagus cuniculus*) in the Iberian Peninsula. *Molecular Ecology* **17**, 4489–4499.

Geraldes A, Ferrand N, and Nachman MW (2006). Contrasting patterns of introgression at X-linked loci across the hybrid zone between subspecies of the European rabbit (*Oryctolagus cuniculus*). *Genetics* **173**, 919–933.

Geraldes A, Rogel-Gaillard C, and Ferrand N (2005). High levels of nucleotide diversity in the European rabbit (*Oryctolagus cuniculus*) SRY gene. *Animal Genetics* **36**, 349–351.

Gerard D, Gibbs HL, and Kubatko L (2011). Estimating hybridization in the presence of coalescence using phylogenetic intraspecific sampling. *BMC Evolutionary Biology* **11**, 291.

Gibbs MJ, Armstrong JS, and Gibbs AJ (2002). Questioning the evidence for genetic recombination in the 1918 "Spanish flu" virus. *Science* **296**, 211a.

Gifford RJ (2012). Viral evolution in deep time: lentiviruses and mammals. *Trends in Genetics* **28**, 89–100.

Giménez MD, White TA, Hauffe HC, Panithanarak T, and Searle JB (2013). Understanding the basis of diminished gene flow between hybridizing chromosome races of the house mouse. *Evolution* **67**, 1446–1462.

Gire SK, Goba A, Andersen KG, Sealfon RSG, Park DJ, Kanneh L, *et al.* (2014). Genomic surveillance elucidates Ebola virus origin and transmission during the 2014 outbreak. *Science* **345**, 1369–1372.

Gladieux P, Ropars J, Badouin H, Branca A, Aguileta G, de Vienne DM, *et al.* (2014). Fungal evolutionary genomics provides insight into the mechanisms of adaptive divergence in eukaryotes. *Molecular Ecology* **23**, 753–773.

Glöckner G, Hülsmann N, Schleicher M, Noegel AA, Eichinger L, Gallinger C, *et al.* (2014). The genome of the foraminiferan *Reticulomyxa filosa*. *Current Biology* **24**, 11–18.

Godinho R, Llaneza L, Blanco JC, Lopes S, Álvares F, García EJ, *et al.* (2011). Genetic evidence for multiple events of hybridization between wolves and domestic dogs in the Iberian Peninsula. *Molecular Ecology* **20**, 5154–5166.

Godinho R, López-Bao JV, Castro D, Llaneza L, Lopes S, Silva P, and Ferrand N (2015). Real-time assessment of hybridization between wolves and dogs: Combining noninvasive samples with ancestry informative markers. *Molecular Ecology Resources* **15**, 317–328.

Gomes S and Civetta A (2014). Misregulation of spermatogenesis genes in *Drosophila* hybrids is lineage-specific and driven by the combined effects of sterility and fast

male regulatory divergence. *Journal of Evolutionary Biology* **27**, 1775–1783.

Gompert Z and Buerkle CA (2009). INTROGRESS: A software package for mapping components of isolation in hybrids. *Molecular Ecology Resources* **10**, 378–384.

Gompert Z and Buerkle CA (2011). Bayesian estimation of genomic clines. *Molecular Ecology* **20**, 2111–2127.

Gompert Z, Fordyce JA, Forister ML, Shapiro AM, and Nice CC (2006). Homoploid hybrid speciation in an extreme habitat. *Science* **314**, 1923–1925.

Gompert Z, Lucas LK, Buerkle CA, Forister ML, Fordyce JA, and Nice CC (2014). Admixture and the organization of genetic diversity in a butterfly species complex revealed through common and rare genetic variants. *Molecular Ecology* **23**, 4555–4573.

Gonder MK, Locatelli S, Ghobrial L, Mitchell MW, Kujawski JT, Lankester FJ, *et al.* (2011). Evidence from Cameroon reveals differences in the genetic structure and histories of chimpanzee populations. *Proceedings of the National Academy of Sciences USA* **108**, 4766–4771.

Gong L, Olson M, and Wendel JF (2014). Cytonuclear evolution of rubisco in four allopolyploid lineages. *Molecular Biology and Evolution* **31**, 2624–2636.

González-Orozco CE, Brown AHD, Knerr N, Miller JT, and Doyle JJ (2012). Hotspots of diversity of wild Australian soybean relatives and their conservation in situ. *Conservation Genetics* **13**, 1269–1281.

Good JM, Hird S, Reid N, Demboski JR, Steppan SJ, Martin-Nims TR, *et al.* (2008). Ancient hybridization and mitochondrial capture between two species of chipmunks. *Molecular Ecology* **17**, 1313–1327.

Goodfriend GA and Gould SJ (1996). Paleontology and chronology of two evolutionary transitions by hybridization in the Bahamian land snail *Cerion*. *Science* **274**, 1894–1897.

Gosset CC and Bierne N (2013). Differential introgression from a sister species explains high FST outlier loci within a mussel species. *Journal of Evolutionary Biology* **26**, 14–26.

Gottelli D, Sillero-Zubiri C, Applebaum GD, Roy MS, Girman DJ, Garcia-Moreno J, *et al.* (1994). Molecular genetics of the most endangered canid: The Ethiopian wolf *Canis simensis*. *Molecular Ecology* **3**, 301–312.

Grant V (1958). The regulation of recombination in plants. *Cold Spring Harbor Symposium on Quantitative Biology* **23**, 337–363.

Grant V (1981). *Plant Speciation*. Columbia University Press, New York.

Grant PR and Grant BR (1992). Hybridization of bird species. *Science* **256**, 193–197.

Grant BR and Grant PR (2008). Fission and fusion of Darwin's finches populations. *Philosophical Transactions of the Royal Society of London B* **363**, 2821–2829.

Grant PR and Grant BR (2009). The secondary contact phase of allopatric speciation in Darwin's finches. *Proceedings of the National Academy of Sciences USA* **106**, 20141–20148.

Grant PR and Grant BR (2010). Natural selection, speciation and Darwin's finches. *Proceedings of the California Academy of Sciences* **61(Supplement II)**, 245–260.

Grant PR and Grant BR (2014a). *40 Years of Evolution Darwin's Finches on Daphne Major Island*. Princeton University Press, Princeton.

Grant PR and Grant BR (2014b). Synergism of natural selection and introgression in the origin of a new species. *American Naturalist* **183**, 671–681.

Grant PR, Grant BR, Markert JA, Keller LF, and Petren K (2004). Convergent evolution of Darwin's finches caused by introgressive hybridization and selection. *Evolution* **58**, 1588–1599.

Gravina B, Mellars P, and Ramsey B (2005). Radiocarbon dating of interstratified Neanderthal and early modern human occupations at the Chatelperronian type-site. *Nature* **438**, 51–56.

Gray MM, Granka JM, Bustamante CD, Sutter NB, Boyko AR, Zhu L, *et al.* (2009). Linkage disequilibrium and demographic history of wild and domestic canids. *Genetics* **181**, 1493–1505.

Greaves IK, Groszmann M, Wang A, Peacock WJ, and Dennis ES (2014). Inheritance of trans chromosomal methylation patterns from *Arabidopsis* F$_1$ hybrids. *Proceedings of the National Academy of Sciences USA* **111**, 2017–2022.

Green RE, Krause J, Briggs AW, Maricic T, Stenzel U, Kircher M, *et al.* (2010). A draft sequence of the Neandertal genome. *Science* **328**, 710–722.

Green RE, Malaspinas A-S, Krause J, Briggs AW, Johnson PLF, Uhler C, *et al.* (2008). A complete Neandertal mitochondrial genome sequence determined by high-throughput sequencing. *Cell* **134**, 416–426.

Green DM, Sharbel TF, Kearsley J, and Kaiser H (1996). Postglacial range fluctuation, genetic subdivision and speciation in the western North American spotted frog complex, *Rana pretiosa*. *Evolution* **50**, 374–390.

Grismer JL, Bauer AM, Grismer LL, Thirakhupt K, Aowphol A, Oaks JR, *et al.* (2014). Multiple origins of parthenogenesis, and a revised species phylogeny for the Southeast Asian butterfly lizards, *Leiolepis*. *Biological Journal of the Linnean Society* **113**, 1080–1093.

Groenen MAM, Archibald AL, Uenishi H, Tuggle CK, Takeuchi Y, Rothschild MF, *et al.* (2012). Analyses of pig genomes provide insight into porcine demography and evolution. *Nature* **491**, 393–398.

Groeters FR (1994). The adaptive role of facultative embryonic diapause in the grasshopper *Caledia captiva* (Orthoptera: Acrididae) in southeastern Australia. *Ecogeography* **17**, 221–228.

Groeters FR and Shaw DD (1992). Association between latitudinal variation for embryonic development time and chromosome structure in the grasshopper *Caledia captiva* (Orthoptera: Acrididae). *Evolution* **46**, 245–257.

Groeters FR and Shaw DD (1996). Evidence for association of chromosomal form and development time from complex clines and geographic races in the grasshopper *Caledia captiva* (Orthoptera: Acridadae). *Biological Journal of the Linnean Society* **59**, 243–259.

Gross BL, Henk AD, Richards CM, Fazio G, and Volk GM (2014). Genetic diversity in *Malus xdomestica* (Rosaceae) through time in response to domestication. *American Journal of Botany* **101**, 1770–1779.

Gross BL, Kane NC, Lexer C, Ludwig F, Rosenthal DM, Donovan LA, and Rieseberg LH (2004). Reconstructing the origin of *Helianthus deserticola*: Survival and selection on the desert floor. *American Naturalist* **164**, 145–156.

Grossen C, Keller L, and Biebach I, The International Goat Genome Consortium and Croll D (2014). Introgression from domestic goat generated variation at the major histocompatibility complex of Alpine ibex. *PLoS Genetics* **10**, e1004438.

Grossowicz M, Sivan N, and Heller J (2003). *Melanopsis* from the Pleistocene of the Hula Valley (Gastropoda: Cerithioidea). *Israel Journal of Earth Sciences* **52**, 221–234.

Grummer JA, Bryson RW Jr, and Reeder TW (2014). Species delimitation using Bayes factors: Simulations and application to the *Sceloporus scalaris* species group (Squamata: Phrynosomatidae). *Systematic Biology* **63**, 119–133.

Guevara EE and Steiper ME (2014). Molecular phylogenetic analysis of the Papionina using concatenation and species tree methods. *Journal of Human Evolution* **66**, 18–28.

Hagen RH and Scriber JM (1989). Sex-linked diapause, color, and allozyme loci in *Papilio glaucus*: Linkage analysis and significance in a hybrid zone. *Journal of Heredity* **80**, 179–185.

Hahn MW, White BJ, Muir CD, and Besansky NJ (2012). No evidence for biased co-transmission of speciation islands in *Anopheles gambiae*. *Philosophical Transactions of the Royal Society of London B* **367**, 374–384.

Hailer F, Kutschera VE, Hallström BM, Klassert D, Fain SR, Leonard JA, *et al.* (2012). Nuclear genomic sequences reveal that polar bears are an old and distinct bear lineage. *Science* **336**, 344–347.

Halas D and Simons AM (2014). Cryptic speciation reversal in the *Etheostoma zonale* (Teleostei: Percidae) species group, with an examination of the effect of recombination and introgression on species tree inference. *Molecular Phylogenetics and Evolution* **70**, 13–28.

Halbert ND and Derr JN (2007). A comprehensive evaluation of cattle introgression into US federal bison herds. *Journal of Heredity* **98**, 1–12.

Halbert ND, Ward TJ, Schnabel RD, Taylor JF, and Derr JN (2005). Conservation genomics: disequilibrium mapping of domestic cattle chromosomal segments in North American bison populations. *Molecular Ecology* **14**, 2343–2362.

Haldane JBS (1948). The theory of a cline. *Journal of Genetics* **48**, 277–284.

Hale TL (1991). Genetic basis of virulence in *Shigella* species. *Microbiological Reviews* **55**, 206–224.

Hall RJ, Hastings A, and Ayres DR (2006). Explaining the explosion: Modelling hybrid invasions. *Proceedings of the Royal Society of London B* **273**, 1385–1389.

Hallström BM, Schneider A, Zoller S, and Janke A (2011). A genomic approach to examine the complex evolution of Laurasiatherian mammals. *PLoS ONE* **6**, e28199.

Hamilton JA, Lexer C, and Aitken SN (2013). Differential introgression reveals candidate genes for selection across a spruce (*Picea sitchensis* × *P. glauca*) hybrid zone. *New Phytologist* **197**, 927–938.

Hamlin JAP and Arnold ML (2014). Determining population structure and hybridization for two Louisiana Iris species. *Ecology and Evolution* **4**, 743–755.

Hammer MF, Woerner AE, Mendez FL, Watkins JC, and Wall JD (2011). Genetic evidence for archaic admixture in Africa. *Proceedings of the National Academy of Sciences USA* **108**, 15123–15128.

Hamrick JL and Allard RW (1972). Microgeographical variation in allozyme frequencies in *Avena barbata*. *Proceedings of the National Academy of Sciences USA* **69**, 2100–2104.

Hao W, Richardson AO, Zheng Y, and Palmer JD (2010). Gorgeous mosaic of mitochondrial genes created by horizontal transfer and gene conversion. *Proceedings of the National Academy of Sciences USA* **107**, 21576–21581.

Hapke A, Zinner D, and Zischler H (2001). Mitochondrial DNA variation in Eritrean hamadryas baboons (*Papio hamadryas hamadryas*): life history influences population genetic structure. *Behavioral Ecology and Sociobiology* **50**, 483–492.

Harbert RS, Brown AHD, and Doyle JJ (2014). Climate niche modeling in the perennial *Glycine* (Leguminosae) allopolyploid complex. *American Journal of Botany* **101**, 710–721.

Hardy C, Callou C, Vigne J-D, Casane D, Dennebouy N, Mounolou J-C, and Monnerot M (1995). Rabbit mitochondrial DNA diversity from prehistoric to modern times. *Journal of Molecular Evolution* **40**, 227–237.

Harris SA, Robinson JP, and Juniper BE (2002). Genetic clues to the origin of the apple. *Trends in Genetics* **18**, 426–430.

Harrison RG (1986). Pattern and process in a narrow hybrid zone. *Heredity* **56**, 337–349.

Harrison RG (1990). Hybrid zones: Windows on evolutionary process. *Oxford Surveys in Evolutionary Biology* **7**, 69–128.

Harrison RG (2012). The language of speciation. *Evolution* **66**, 3643–3657.

Harrison JS and Burton RS (2006). Tracing hybrid incompatibilities to single amino acid substitutions. *Molecular Biology and Evolution* **23**, 559–564.

Harrison JS and Edmands S (2006). Chromosomal basis of viability differences in *Tigriopus californicus* interpopulation hybrids. *Journal of Evolutionary Biology* **19**, 2040–2051.

Harry AV, Morgan JAT, Ovenden JR, Tobin AJ, Welch DJ, and Simpfendorfer CA (2012). Comparison of the reproductive ecology of two sympatric blacktip sharks (*Carcharhinus limbatus* and *Carcharhinus tilstoni*) off north-eastern Australia with species identification inferred from vertebral counts. *Journal of Fish Biology* **81**, 1225–1233.

Haus T, Ferguson B, Rogers J, Doxiadis G, Certa U, Rose NJ, et al. (2014). Genome typing of nonhuman primate models: implications for biomedical research. *Trends in Genetics* **30**, 482–487.

Hawks J (2013). Significance of Neandertal and Denisovan genomes in human evolution. *Annual Review of Anthropology* **42**, 433–449.

Hawks JD and Wolpoff MH (2001). The Accretion model of Neandertal evolution. *Evolution* **55**, 1474–1485.

Hayakawa T, Aki I, Varki A, Satta Y, and Takahata N (2006). Fixation of the human-specific CMP-*N*-Acetylneuraminic Acid Hydroxylase pseudogene and implications of haplotype diversity for human evolution. *Genetics* **172**, 1139–1146.

Hearty PJ (2010). Chronostratigraphy and morphological changes in *Cerion* land snail shells over the past 130 ka on Long Island, Bahamas. *Quaternary Geochronology* **5**, 50–64.

Hearty PJ and Schellenberg SA (2008). Integrated Late Quaternary chronostratigraphy for San Salvador Island, Bahamas: Patterns and trends of morphological change in the land snail *Cerion*. *Palaeoecology* **267**, 41–58.

Hebert PDN and Finston TL (2001). Macrogeographic patterns of breeding system diversity in the *Daphnia pulex* group from the United States and Mexico. *Heredity* **87**, 153–161.

Heckman KL, Mariani CL, Rasoloarison R, and Yoder AD (2007). Multiple nuclear loci reveal patterns of incomplete lineage sorting and complex species history within western mouse lemurs (*Microcebus*). *Molecular Phylogenetics and Evolution* **43**, 353–367.

Hedrick PW (2009). Conservation genetics and North American bison (*Bison bison*). *Journal of Heredity* **100**, 411–420.

Heeney JL, Dalgleish AG, and Weiss RA (2006). Origins of HIV and the evolution of resistance to AIDS. *Science* **313**, 462–466.

Hegarty MJ, Batstone T, Barker GL, Edwards KJ, Abbott RJ, and Hiscock SJ (2011). Nonadditive changes to cytosine methylation as a consequence of hybridization and genome duplication in *Senecio* (Asteraceae). *Molecular Ecology* **20**, 105–113.

Heiser CB Jr (1947). Hybridization between the sunflower species *Helianthus annuus* and *H. petiolaris*. *Evolution* **1**, 249–262.

Heiser CB Jr (1951a). Hybridization in the annual sunflowers: *Helianthus annuus* X *H. argophyllus*. *American Naturalist* **85**, 65–72.

Heiser CB Jr (1951b). Hybridization in the annual sunflowers: *Helianthus annuus* X *H. debilis* var *cucumerifolius*. *Evolution* **5**, 42–51.

Heiser CB Jr (1979). Origins of some New World plants. *Annual Review of Ecology and Systematics* **10**, 309–326.

Heiser CB Jr, Martin WC, and Smith DM (1962). Species crosses in *Helianthus*: I. diploid species. *Brittonia* **14**, 137–147.

Heiser CB Jr, Smith DM, Clevenger SB, and Martin WC, Jr (1969). The North American sunflowers (*Helianthus*). *Memoirs of the Torrey Botanical Club* **22**, 1–213.

Heled J and Drummond AJ (2010). Bayesian inference of species trees from multilocus data. *Molecular Biology and Evolution* **27**, 570–580.

Heller J (2007). A historic biogeography of the aquatic fauna of the Levant. *Biological Journal of the Linnean Society* **92**, 625–639.

Heller J, Mordan P, Ben-Ami F, and Sivan N (2005). Conchometrics, systematics and distribution of *Melanopsis* (Mollusca: Gastropoda) in the Levant. *Zoological Journal of the Linnean Society* **144**, 229–260.

Heller J and Sivan N (2002). *Melanopsis* from the Pleistocene site of 'Ubeidiya, Jordan Valley: Direct evidence of early hybridization (Gastropoda: Cerithioidea). *Biological Journal of the Linnean Society* **75**, 39–57.

Hellriegel B and Reyer H-U (2000). Factors influencing the composition of mixed populations of a hemiclonal hybrid and its sexual host. *Journal of Evolutionary Biology* **13**, 906–918.

Hendrix RW, Lawrence JG, Hatfull GF, and Casjens S (2000). The origins and ongoing evolution of viruses. *Trends in Microbiology* **8**, 504–508.

Hennig W (1966). *Phylogenetic Systematics*. University of Illinois Press, Urbana, Illinois.

Henning F and Meyer A (2014). The evolutionary genomics of cichlid fishes: Explosive speciation and adaptation in the postgenomic era. *Annual Review of Genomics and Human Genetics* **15**, 417–441.

Herder F, Nolte AW, Pfaender J, Schwarzer J, Hadiaty RK, and Schliewen UK (2006). Adaptive radiation and hybridization in Wallace's Dreamponds: Evidence from sailfin silversides in the Malili Lakes of Sulawesi. *Proceedings of the Royal Society of London B* **273**, 2209–2217.

Heredia SM and Ellstrand NC (2014). Novel seed protection in the recently evolved invasive, California wild radish, a hybrid *Raphanus* sp. (Brassicaceae). *American Journal of Botany* **101**, 2043–2051.

Hermann K, Klahre U, Moser M, Sheehan H, Mandel T, and Kuhlemeier C (2013). Tight genetic linkage of prezygotic barrier loci creates a multifunctional speciation island in *Petunia*. *Current Biology* **23**, 873–877.

Hermansen JS, Haas F, Trier CN, Bailey RI, Nederbragt AJ, et al. (2014). Hybrid speciation through sorting of parental incompatibilities in Italian sparrows. *Molecular Ecology* **23**, 5831–5842.

Hermansen JS, Sæther SA, Elgvin TO, Borge T, Hjelle E, and Sætre G-P (2011). Hybrid speciation in sparrows I: Phenotypic intermediacy, genetic admixture and barriers to gene flow. *Molecular Ecology* **20**, 3812–3822.

Herrig DK, Modrick AJ, Brud E, and Llopart A (2014). Introgression in the *Drosophila subobscura*—*D. madeirensis* sister species: Evidence of gene flow in nuclear genes despite mitochondrial differentiation. *Evolution* **68**, 705–719.

Hess WR, Rocap G, Ting CS, Larimer F, Stilwagen S, Lamerdin J, and Chisholm SW (2001). The photosynthesis apparatus of *Prochlorococcus*: Insights through comparative genomics. *Photosynthesis Research* **70**, 53–71.

Hewitt GM (1975). A new hypothesis for the origin of the parthogenetic grasshopper *Moraba virgo*. *Heredity* **34**, 117–136.

Hey J (2003). Speciation and inversions: Chimps and humans. *BioEssays* **25**, 825–828.

Hey J (2010). The divergence of chimpanzee species and subspecies as revealed in multipopulation isolation-with-migration analyses. *Molecular Biology and Evolution* **27**, 921–933.

Hey J and Nielsen R (2004). Multilocus methods for estimating population sizes, migration rates and divergence time, with applications to the divergence of *Drosophila pseudoobscura* and *D. persimilis*. *Genetics* **167**, 747–760.

Hey J, Waples RS, Arnold ML, Butlin RK, and Harrison RG (2003). Understanding and confronting species uncertainty in biology and conservation. *Trends in Ecology and Evolution* **18**, 597–603.

Hey J, Won Y-J, Sivasundar A, Nielsen R, and Markert JA (2004). Using nuclear haplotypes with microsatellites to study gene flow between recently separated cichlid species. *Molecular Ecology* **13**, 909–919.

Higham T, Douka K, Wood R, Bronk Ramsey C, Brock F, Basell L, *et al.* (2014). The timing and spatiotemporal patterning of Neanderthal disappearance. *Nature* **512**, 306–309.

Hillis DM, Moritz C, Porter CA, and Baker RJ (1991). Evidence for biased gene conversion in concerted evolution of ribosomal DNA. *Science* **251**, 308–310.

Hinchliffe SJ, Isherwood KE, Stabler RA, Prentice MB, Rakin A, Nichols RA, *et al.* (2003). Application of DNA microarrays to study the evolutionary genomics of *Yersinia pestis* and *Yersinia pseudotuberculosis*. *Genome Research* **13**, 2018–2029.

Hines HM, Counterman BA, Papa R, de Moura PA, Cardoso MZ, Linares M, *et al.* (2011). Wing patterning gene redefines the mimetic history of *Heliconius* butterflies. *Proceedings of the National Academy of Sciences USA* **108**, 19666–19671.

Hinnebusch BJ, Rudolph AE, Cherepanov P, Dixon JE, Schwan TG, and Forsberg A (2002). Role of Yersinia murine toxin in survival of *Yersinia pestis* in the midgut of the flea vector. *Science* **296**, 733–735.

Hinsinger DD, Basak J, Gaudeul M, Cruaud C, Bertolino P, Frascaria-Lacoste N, and Bousquet J (2013). The phylogeny and biogeographic history of ashes (*Fraxinus*, Oleaceae) highlight the roles of migration and vicariance in the diversification of temperate trees. *PLoS ONE* **8**, e80431.

Hinsinger DD, Gaudeul M, Couloux A, Bousquet J, and Frascaria-Lacoste N (2014). The phylogeography of Eurasian Fraxinus species reveals ancient transcontinental reticulation. *Molecular Phylogenetics and Evolution* **77**, 223–237.

Hird H, Chisholm J, and Brown J (2005). The detection of commercial duck species in food using a single probe-multiple species-specific primer real-time PCR assay. *European Food Research and Technology* **221**, 559–563.

Hird S and Sullivan J (2009). Assessment of gene flow across a hybrid zone in red-tailed chipmunks (*Tamias ruficaudus*). *Molecular Ecology* **18**, 3097–3109.

Ho L, Cortés-Ortiz L, Dias PAD, Canales-Espinosa D, Kitchen DM, and Bergman TJ (2014). Effect of ancestry on behavioral variation in two species of howler monkeys (*Alouatta pigra* and *A. palliata*) and their hybrids. *American Journal of Primatology* **76**, 855–867.

Hodges SA, Burke JM, and Arnold ML (1996). Natural formation of *Iris* hybrids: experimental evidence on the establishment of hybrid zones. *Evolution* **50**, 2504–2509.

Hoekstra HE, Hirschmann RJ, Bundey RA, Insel PA, and Crossland JP (2006). A single amino acid mutation contributes to adaptive beach mouse color pattern. *Science* **313**, 101–104.

Hoffmann FG, Owen JG, and Baker RJ (2003). mtDNA perspective of chromosomal diversification and hybridization in Peters' tent-making bat (*Uroderma bilobatum*: Phyllostomidae). *Molecular Ecology* **12**, 2981–2993.

Hohenlohe PA, Day MD, Amish SJ, Miller MR, Kamps-Hughes N, Boyer MC, *et al.* (2013). Genomic patterns of introgression in rainbow and westslope cutthroat trout illuminated by overlapping paired-end RAD sequencing. *Molecular Ecology* **22**, 3002–3013.

Holland LZ, Albalat R, Azumi K, Benito-Gutiérrez E, Blow MJ, Bronner-Fraser M, *et al.* (2008). The amphioxus genome illuminates vertebrate origins and cephalochordate biology. *Genome Research* **18**, 1100–1111.

Holliday TW (2003). Species concepts, reticulation, and human evolution. *Current Anthropology* **44**, 653–660.

Holliday TW (2010). Review of *Reticulate Evolution and Humans Origins and Ecology*. *American Journal of Physical Anthropology* **141**, 668–669.

Holsinger KE and Weir BS (2009). Genetics in geographically structured populations: defining, estimating and interpreting FST. *Nature Reviews Genetics* **10**, 639–650.

Holt KE, Baker S, Weill F-X, Holmes EC, Kitchen A, Yu J, *et al.* (2012). *Shigella sonnei* genome sequencing and phylogenetic analysis indicate recent global dissemination from Europe. *Nature Genetics* **44**, 1056–1059.

Holt KE, Nga TVT, Thanh DP, Vinh H, Kim DW, Tra MPV, *et al.et al.* (2013). Tracking the establishment of local endemic populations of an emergent enteric pathogen. *Proceedings of the National Academy of Sciences USA* **110**, 17522–17527.

Honeycutt RL and Wilkinson P (1989). Electrophoretic variation in the parthenogenetic grasshopper *Warramaba virgo* and its sexual relatives. *Evolution* **43**, 1027–1044.

Hotopp JCD (2011). Horizontal gene transfer between bacteria and animals. *Trends in Genetics* **27**, 157–163.

Hotopp JCD, Clark ME, Oliveira DCSG, Foster JM, Fischer P, Muñoz Torres MC, *et al.* (2007).Widespread lateral gene transfer from intracellular bacteria to multicellular eukaryotes. *Science* **317**, 1753–1756.

Hotz H, Beerli P, and Spolsky C (1992). Mitochondrial DNA reveals formation of nonhybrid frogs by natural matings between hemiclonal hybrids. *Molecular Biology and Evolution* **9**, 610–620.

Howard DJ (1982). *Speciation and coexistence in a group of closely related ground crickets*. Ph.D. Dissertation, Yale University, New Haven, Connecticut.

Howard DJ (1986). A zone of overlap and hybridization between two ground cricket species. *Evolution* **40**, 34–43.

Howard DJ (1993). Reinforcement: origin, dynamics, and fate of an evolutionary hypothesis. In RG Harrison, ed., *Hybrid Zones and the Evolutionary Process*, pp. 46–69. Oxford University Press, Oxford.

Hoyer BH, van de Velde NW, Goodman M, and Roberts RB (1972). Examination of hominid evolution by DNA sequence homology. *Journal of Human Evolution* **1**, 645–649.

Hu W-S and Temin HM (1990). Genetic consequences of packaging two RNA genomes in one retroviral particle: Pseudodiploidy and high rate of genetic recombination. *Proceedings of the National Academy of Sciences USA* **87**, 1556–1560.

Huang LO, Labbe A, and Infante-Rivard C (2013). Transmission ratio distortion: review of concept and implications for genetic association studies. *Human Genetics* **132**, 245–263.

Hubbs C L (1955). Hybridization between fish species in nature. *Systematic Zoology* **4**, 1–20.

Hubbs CL and Hubbs LC (1932). Apparent parthenogenesis in nature, in a form of fish of hybrid origin. *Science* **76**, 628–630.

Hubby JL and Lewontin RC (1966). A molecular approach to the study of genic heterozygosity in natural populations. I. The number of alleles at different loci in *Drosophila pseudoobscura*. *Genetics* **54**, 577–594.

Huber KT, van Iersel L, Moulton V, and Wu T (2015). How much information is needed to infer reticulate evolutionary histories?. *Systematic Biology* **64**, 102–111.

Hubricht L and Anderson E (1941). Vicinism in *Tradescantia*. *American Journal of Botany* **28**, 957.

Hudson AG, Vonlanthen P, and Seehausen O (2011). Rapid parallel adaptive radiations from a single hybridogenic ancestral population. *Proceedings of the Royal Society of London B* **278**, 58–66.

Huerta-Sánchez E, Jin X, Asan, Bianba Z, Peter BM, Vinckenbosch N, *et al.* (2014). Altitude adaptation in Tibetans caused by introgression of Denisovan-like DNA. *Nature* **512**, 194–197.

Hufford MB, Lubinksy P, Pyhäjärvi T, Devengenzo MT, Ellstrand NC, and Ross-Ibarra J (2013). The genomic signature of crop-wild introgression in maize. *PLoS Genetics* **9**, e1003477.

Hufford MB, Martínez-Meyer E, Gaut BS, Eguiarte LE, and Tenaillon MI (2012) Inferences from the historical distribution of wild and domesticated maize provide ecological and evolutionary insight. *PLoS ONE* **7**, e47659.

Hughes AL and Friedman R (2004). Patterns of sequence divergence in 5′ intergenic spacers and linked coding regions in 10 species of pathogenic bacteria reveal distinct recombinational histories. *Genetics* **168**, 1795–1803.

Hughes AL, Friedman R, and Murray M (2002). Genome-wide pattern of synonymous nucleotide substitution in two complete genomes of Mycobacterium tuberculosis. *Emerging Infectious Diseases* **8**, 1342–1346.

Hunt WG and Selander RK (1973). Biochemical genetics of hybridisation in European house mice. *Heredity* **31**, 11–33.

Huson DH and Bryant D (2006). Application of phylogenetic networks in evolutionary studies. *Molecular Biology and Evolution* **23**, 254–267.

Hutchinson JF (2001). The biology and evolution of HIV. *Annual Review of Anthropology* **30**, 85–108.

Huxley J (1938). Clines: an auxiliary taxonomic principle. *Nature* **142**, 219–220.

Hwang AS, Northrup SL, Alexander JK, Vo KT, and Edmands S (2011). Long-term experimental hybrid swarms between moderately incompatible *Tigriopus californicus* populations: Hybrid inferiority in early generations yields to hybrid superiority in later generations. *Conservation Genetics* **12**, 895–909.

Hwang AS, Northrup SL, Peterson DL, Kim Y, and Edmands S (2012). Long-term experimental hybrid swarms between nearly incompatible *Tigriopus californicus* populations: persistent fitness problems and assimilation by the superior population. *Conservation Genetics* **13**, 567–579.

Ingman M, Kaessmann H, Pääbo S, and Gyllensten U (2000). Mitochondrial genome variation and the origin of modern humans. *Nature* **408**, 708–713.

Innan H and Watanabe H (2006). The effect of gene flow on the coalescent time in the human-chimpanzee ancestral population. *Molecular Biology and Evolution* **23**, 1040–1047.

Jackson AP, Eastwood H, Bell SM, Adu J, Toomes C, Carr IM, *et al.* (2002). Identification of *microcephalin*, a protein implicated in determining the size of the human brain. *American Journal of Human Genetics* **71**, 136–142.

Jackson HD, Steane DA, Potts BM, and Vaillancourt RE (1999). Chloroplast DNA evidence for reticulate evolution in *Eucalyptus* (Myrtaceae). *Molecular Ecology* **8**, 739–751.

Jackson JA and Tinsley RC (2003). Parasite infectivity to hybridising host species: a link between hybrid resistance and allopolyploid speciation? *International Journal of Parasitology* **33**, 137–144.

Jaillon O, Aury J-M, Noel B, Policriti A, Clepet C, Casagrande A, *et al.* (2007). The grapevine genome sequence suggests ancestral hexaploidization in major angiosperm phyla. *Nature* **449**, 463–467.

James JK and Abbott RJ (2005). Recent, allopatric, homoploid hybrid speciation: The origin of *Senecio squalidus* (Asteraceae) in the British Isles from a hybrid zone on Mount Etna, Sicily. *Evolution* 59, 2533–2547.

Jankowski T and Straile D (2004). Allochronic differentiation among *Daphnia* species, hybrids and backcrosses: The importance of sexual reproduction for population dynamics and genetic architecture. *Journal of Evolutionary Biology* **17**, 312–321.

Janoušek V, Wang L, Luzynski K, Dufková P, Vyskočilová MM, Nachman MW, *et al.* (2012). Genome-wide architecture of reproductive isolation in a naturally occurring hybrid zone between *Mus musculus musculus* and *M. m. domesticus*. *Molecular Ecology* **21**, 3032–3047.

Jeandroz G, Faivre-Rampant F, Pugin A, Bousquet J, and Bervillé A (1995). Organization of nuclear ribosomal DNA and species-specific polymorphism in closely related *Fraxinus excelsior* and *F. oxyphylla*. *Theoretical and Applied Genetics* **91**, 885–892.

Jegouic S, Joffret M-L, Blanchard C, Riquet FB, Perret C, Pelletier I, *et al.* (2009). Recombination between polioviruses and co-circulating coxsackie A viruses: Role in the emergence of pathogenic vaccine-derived polioviruses. *PLoS Pathogens* **5**, e1000412.

Jenkins TM, Babcock CS, Geiser DM, and Anderson WW (1996). Cytoplasmic incompatibility and mating preference in Colombian *Drosophila pseudoobscura*. *Genetics* **142**, 189–194.

Jiang C-X, Wright RJ, El-Zik K, and Paterson AH (1998). Polyploid formation created unique avenues for response to selection in *Gossypium* (cotton). *Proceedings of the National Academy of Sciences USA* **95**, 4419–4424.

Jiggins CD, Linares M, Naisbit RE, Salazar C, Yang ZH, and Mallet J (2001). Sex-linked hybrid sterility in a butterfly. *Evolution* **55**, 1631–1638.

Jiggins CD, Salazar C, Linares M, and Mavarez J (2008). Hybrid trait speciation and *Heliconius* butterflies. *Philosophical Transactions of the Royal Society of London B* **363**, 3047–3054.

Joachim BL and Schlupp I (2012). Mating preferences of Amazon mollies (*Poecilia formosa*) in multi-host populations. *Behaviour* **149**, 233–249.

Joffret M-L, Jégouic S, Bessaud M, Balanant J, Tran C, Caro V, *et al.* (2012). Common and diverse features of cocirculating type 2 and 3 recombinant vaccine-derived polioviruses isolated from patients with poliomyelitis and healthy children. *Journal of Infectious Diseases* **205**, 1363–1373.

Johnsgard PA (1960). Hybridization in the Anatidae and its taxonomic implications. *Condor* **62**, 25–33.

Johnsgard PA (1967). Sympatry changes and hybridization incidence in Mallards and Black Ducks. *American Midland Naturalist* **77**, 51–63.

Johnson C (2009). *The Dark Horse: A Walt Longmire Mystery*. Viking, New York.

Johnson WE and O'Brien SJ (1997). Phylogenetic reconstruction of the Felidae using 16S rRNA and NADH-5 mitochondrial genes. *Journal of Molecular Evolution* **44** (**Supplement 1**), S98–S116.

Johnson WE, Onorato DP, Roelke ME, Land ED, Cunningham M, Belden RC, *et al.* (2010). Genetic restoration of the Florida panther. *Science* **329**, 1641–1645.

Johnson WE, Slattery JP, Eizirik E, Kim J-H, Raymond MM, Bonacic C, *et al.* (1999). Disparate phylogeographic patterns of molecular genetic variation in four closely related South American small cat species. *Molecular Ecology* **8**, S79–S94.

Johnston RF (1969). Taxonomy of house sparrows and their allies in the Mediterranean basin. *Condor* **71**, 129–139.

Johnston JA, Wesselingh RA, Bouck AC, Donovan LA, and Arnold ML (2001). Intimately linked or hardly speaking? The relationship between genotype and environmental gradients in a Louisiana Iris hybrid population. *Molecular Ecology* **10**, 673–681.

Jolly CJ (2001). A proper study for mankind: Analogies from the papionin monkeys and their implications for human evolution. *Yearbook of Physical Anthropology* **44**, 177–204.

Jolly CJ (2009). Mixed signals: Reticulation in human and primate evolution. *Evolutionary Anthropology* **18**, 275–281.

Jolly CJ, Burrell AS, Phillips-Conroy JE, Bergey C, and Rogers J (2011). Kinda baboons (*Papio kindae*) and grayfoot chacma baboons (*P. ursinus griseipes*) hybridize in the Kafue River valley, Zambia. *American Journal of Primatology* **73**, 291–303.

Jolly CJ, Wooley-Barker T, Beyene S, Disotell TR, and Phillips-Conroy JE (1997). Intergeneric hybrid baboons. *International Journal of Primatology* **18**, 597–627.

Joly S, McLenachan PA, and Lockhart PJ (2009). A statistical approach for distinguishing hybridization and incomplete lineage sorting. *American Naturalist* **174**, E54–E70.

Jones EP, Van Der Kooij J, Solheim R, and Searle JB (2010). Norwegian house mice (*Mus musculus musculus/domesticus*): distributions, routes of colonization and patterns of hybridization. *Molecular Ecology* **19**, 5252–5264.

Jónsson H, Schubert M, Seguin-Orlando A, Ginolhac A, Petersen L, Fumagalli M, *et al.* (2014). Speciation with gene flow in equids despite extensive chromosomal plasticity. *Proceedings of the National Academy of Sciences USA* **111**, 18655–18660.

Joron M, Jiggins CD, Papanicolaou A, and McMillan WO (2006). *Heliconius* wing patterns: An evo-devo model for understanding phenotypic diversity. *Heredity* **97**, 157–167.

Jourda C, Cardi C, Mbéguié-A-Mbéguié D, Bocs S, Garsmeur O, D'Hont A, and Yahiaoui N (2014). Expansion of banana (*Musa acuminata*) gene families involved in ethylene biosynthesis and signaling after lineage-specific whole-genome duplications. *New Phytologist* **202**, 986–1000.

Joyce DA, Lunt DH, Bills R, Turner GF, Katongo C, Duftner N, *et al.* (2005). An extant cichlid fish radiation emerged in an extinct Pleistocene lake. *Nature* **435**, 90–95.

Joyce DA, Lunt DH, Genner MJ, Turner GF, Bills R, and Seehausen O (2011). Repeated colonization and hybridization in Lake Malawi cichlids. *Current Biology* **21**, R108–R109.

Kadereit JW, Uribe-Convers S, Westberg E, and Comes HP (2006). Reciprocal hybridization at different times between *Senecio flavus* and *Senecio glaucus* gave rise to two polyploid species in north Africa and south-west Asia. *New Phytologist* **169**, 431–441.

Kaessmann H and Pääbo S (2002). The genetical history of humans and the great apes. *Journal of Internal Medicine* **251**, 1–18.

Kaessmann H, Wiebe V, and Pääbo S (1999). Extensive nuclear DNA sequence diversity among chimpanzees. *Science* **286**, 1159–1162.

Kaessmann H, Wiebe V, Weiss G, and Pääbo S (2001). Great ape DNA sequences reveal a reduced diversity and an expansion in humans. *Nature Genetics* **27**, 155–156.

Kalmar A and Currie DJ (2010). The completeness of the continental fossil record and its impact on patterns of diversification. *Paleobiology* **36**, 51–60.

Kandun IN, Wibisono H, Sedyaningsih ER, Yusharmen, Hadisoedarsuno W, Purba W, *et al.* (2006). Three Indonesian clusters of H5N1 virus infection in 2005. *New England Journal of Medicine* **355**, 2186–2194.

Kane NC, King MG, Barker MS, Raduski A, Karrenberg S, Yatabe Y, *et al.* (2009). Comparative genomic and population genetic analyses indicate highly porous genomes and high levels of gene flow between divergent *Helianthus* species. *Evolution* **63**, 2061–2075.

Kaneshiro KY (1990). Natural hybridization in *Drosophila*, with special reference to species from Hawaii. *Canadian Journal of Zoology* **68**, 1800–1805.

Karanth KP (2008). Primate numts and reticulate evolution of capped and golden leaf monkeys (Primates: Colobinae). *Journal of Biosciences* **33**, 761–770.

Karanth KP (2010). Molecular systematics and conservation of the langurs and leaf monkeys of South Asia. *Journal of Genetics* **89**, 393–399.

Karanth KP, Singh L, Collura RV, and Stewart C-B (2008). Molecular phylogeny and biogeography of langurs and leaf monkeys of South Asia (Primates: Colobinae). *Molecular Phylogenetics and Evolution* **46**, 683–694.

Karberg KA, Olsen GJ, and Davis JJ (2011). Similarity of genes horizontally acquired by *Escherichia coli* and *Salmonella enterica* is evidence of a supraspecies pangenome. *Proceedings of the National Academy of Sciences USA* **108**, 20154–20159.

Karlin EF, Boles SB, Ricca M, Temsch EM, Greilhuber J, and Shaw AJ (2009). Three-genome mosses: complex double allopolyploid origins for triploid gametophytes in *Sphagnum*. *Molecular Ecology* **18**, 1439–1454.

Kaur P, Banga S, Kumar N, Gupta S, Akhatar J, and Banga SS (2014). Polyphyletic origin of *Brassica juncea* with B. *rapa* and B. *nigra* (Brassicaceae) participating as cytoplasm donor parents in independent hybridization events. *American Journal of Botany* **101**, 1157–1166.

Kautt AF, Elmer KR, and Meyer A (2012). Genomic signatures of divergent selection and speciation patterns in a "natural experiment," the young parallel radiations of Nicaraguan crater lake cichlid fishes. *Molecular Ecology* **21**, 4770–4786.

Kearney MR (2003). Why is sex so unpopular in the Australian desert? *Trends in Ecology and Evolution* **18**, 605–607.

Kearney M (2005). Hybridization, glaciation and geographical parthenogenesis. *Trends in Ecology and Evolution* **20**, 495–502.

Kearney M and Blacket MJ (2008). The evolution of sexual and parthenogenetic *Warramaba*: A window onto Plio-Pleistocene diversification processes in an arid biome. *Molecular Ecology* **17**, 5257–5275.

Kearney M and Blacket MJ, Strasburg JL, and Moritz C (2006). Waves of parthenogenesis in the desert: evidence for the parallel loss of sex in a grasshopper and a gecko from Australia. *Molecular Ecology* **15**, 1743–1748.

Kearney M and Shine R (2004a). Developmental success, stability, and plasticity in closely related parthenogenetic and sexual lizards (*Heteronotia*, Gekkonidae). *Evolution* **58**, 1560–1572.

Kearney M and Shine R (2004b). Morphological and physiological correlates of hybrid parthenogenesis. *American Naturalist* **164**, 803–813.

Kearney M, Wahl R, and Autumn K (2005). Increased capacity for sustained locomotion at low temperature in parthenogenetic geckos of hybrid origin. *Physiological and Biochemical Zoology* **78**, 316–324.

Keeling PJ and Palmer JD (2008). Horizontal gene transfer in eukaryotic evolution. *Nature Reviews Genetics* **9**, 605–618.

Kelaita MA and Cortés-Ortiz L (2013). Morphological variation of genetically confirmed *Alouatta pigra* x A. *palliata* hybrids from a natural hybrid zone in Tabasco, Mexico. *American Journal of Physical Anthropology* **150**, 223–234.

Keller C, Roos C, Groeneveld LF, Fischer J, and Zinner D (2010). Introgressive hybridization in southern African baboons shapes patterns of mtDNA variation. *American Journal of Physical Anthropology* **142**, 125–136.

Keller SR and Taylor DR (2010). Genomic admixture increases fitness during a biological invasion. *Journal of Evolutionary Biology* **23**, 1720–1731.

Keller I, Wagner CE, Greuter L, Mwaiko S, Selz OM, Sivasundar A, *et al.* (2013). Population genomic signatures of divergent adaptation, gene flow and hybrid speciation in the rapid radiation of Lake Victoria cichlid fishes. *Molecular Ecology* **22**, 2848–2863.

Keller B, Wolinska J, Tellenbach C, and Spaak P (2007). Reproductive isolation keeps hybridizing *Daphnia* species distinct. *Limnology and Oceanography* **52**, 984–991.

Kelly JK and Noor MAF (1996). Speciation by reinforcement: A model derived from studies of *Drosophila*. *Genetics* **143**, 1485–1497.

Kelly BP, Whiteley A, and Tallmon D (2010). The arctic melting pot. *Nature* **468**, 891.

Kew OM, Sutter RW, de Gourville EM, Dowdle WR, and Pallansch MA (2005). Vaccine-derived polioviruses and the endgame strategy for global polio eradication. *Annual Review of Microbiology* **59**, 587–635.

Key KHL (1968). The concept of stasipatric speciation. *Systematic Zoology* **17**, 14–22.

Key KHL (1976). A generic and suprageneric classification of the Morabinae (Orthoptera: Eumastacidae), with description of the type species and a bibliography of the subfamily. *Australian Journal of Zoology Supplementary Series* **24**, 1–185.

Khrameeva EE, Bozek K, He L, Yan Z, Jiang X, Wei Y, *et al.* (2014). Neanderthal ancestry drives evolution of lipid catabolism in contemporary Europeans. *Nature Communications* **5**, 3584.

Kiang YT and Hamrick JL (1978). Reproductive isolation in the *Mimulus guttatus-M. nasutus* complex. *American Midland Naturalist* **100**, 269–276.

Kiang YT and Libby WJ (1972). Maintenance of a lethal in a natural population of *Mimulus guttatus*. *American Naturalist* **106**, 351–367.

Kijas JMH and Andersson L (2001). A phylogenetic study of the origin of the domestic pig estimated from the near-complete mtDNA genome. *Journal of Molecular Evolution* **52**, 302–308.

Kilpatrick ST and Rand DM (1995). Conditional hitchhiking of mitochondrial DNA: Frequency changes of *Drosophila melanogaster* mitochondrial DNA variants depend on nuclear genetic background. *Genetics* **141**, 1113–1124.

Kim M, Cui M-L, Cubas P, Gillies A, Lee K, Chapman MA, *et al.* (2008). Regulatory genes control a key morphological and ecological trait transferred between species. *Science* **322**, 1116–1119.

Kim G, LeBlanc ML, Wafula EK, dePamphilis CW, and Westwood JH (2014). Genomic-scale exchange of mRNA between a parasitic plant and its hosts. *Science* **345**, 808–811.

Kim Y-K, Ruiz-García M, Alvarez D, Phillips DR, and Anderson WW (2012). Sexual isolation between North American and Bogota strains of *Drosophila pseudoobscura*. *Behavior Genetics* **42**, 472–482.

King M-C and Wilson AC (1975). Evolution at two levels in humans and chimpanzees. *Science* **188**, 107–116.

Kirkpatrick NC, Mahmoudian A, O'Rourke D, and Noormohammadi AH (2006). Differentiation of infectious Laryngotracheitis virus isolates by restriction fragment length polymorphic analysis of polymerase chain reaction products amplified from multiple genes. *Avian Diseases* **50**, 28–34.

Kirkpatrick M and Ravigné V (2002). Speciation by natural and sexual selection: Models and experiments. *American Naturalist* **159**, S22–S35.

Knie N, Polsakiewicz M, and Knoop V (2015). Horizontal gene transfer of Chlamydial-like tRNA genes into early vascular plant mitochondria. *Molecular Biology and Evolution* **32**, 629–634.

Knowles LL (2009). Estimating species trees: Methods of phylogenetic analysis when there is incongruence across genes. *Systematic Biology* **58**, 463–467.

Knowles LL and Maddison WP (2002). Statistical phylogeography. *Molecular Ecology* **11**, 2623–2635.

Knox RB (1984). Pollen-pistil interactions. In H.F. Linskens and J. Heslop-Harrison, eds., *Encyclopedia of Plant Physiology*, pp. 508–608, Springer-Verlag, Berlin.

Kobel HR and Du Pasquier L (1986). Genetics of polyploid *Xenopus*. *Trends in Genetics* **2**, 310–315.

Koblmüller S, Duftner N, Sefc KM, Aibara M, Stipacek M, Blanc M, Egger B, and Sturmbauer C (2007). Reticulate phylogeny of gastropod-shell-breeding cichlids from Lake Tanganyika—the result of repeated introgressive hybridization. *BMC Evolutionary Biology* **7**, 7.

Koblmüller S, Egger B Sturmbauer C, and Sefc KM (2010). Rapid radiation, ancient incomplete lineage sorting and ancient hybridization in the endemic Lake Tanganyika cichlid tribe Tropheini. *Molecular Phylogenetics and Evolution* **55**, 318–334.

Kocher TD (2004). Adaptive evolution and explosive speciation: The cichlid fish model. *Nature Reviews Genetics* **5**, 288–298.

Kocher TD and Sage RD (1986). Further genetic Analyses of a hybrid zone between leopard frogs (*Rana pipiens* complex) in central Texas. *Evolution* **40**, 21–33.

Koh J, Chen S, Zhu N, Yu F, Soltis PS, and Soltis DE (2012). Comparative proteomics of the recently and recurrently formed natural allopolyploid *Tragopogon mirus* (Asteraceae) and its parents. *New Phytologist* **196**, 292–305.

Kohlmann B, Nix H, and Shaw DD (1988). Environmental predictions and distributional limits of chromosomal taxa in the Australian grasshopper *Caledia captiva* (F.). *Oecologia* **75**, 483–493.

Konopka RJ and Benzer S (1971). Clock mutants of *Drosophila melanogaster*. *Proceedings of the National Academy of Sciences USA* **68**, 2112–2116.

Konstantinidis KT, Serres MH, Romine MF, Rodrigues JLM, Auchtung J, and Lee-Ann McCue L-A (2009). Comparative systems biology across an evolutionary gradient within the *Shewanella* genus. *Proceedings of the National Academy of Sciences USA* **106**, 15909–15914.

Konstantinidis K and Tiedje JM (2005). Genomic insights that advance the species definition for prokaryotes. *Proceedings of the National Academy of Sciences USA* **102**, 2567–2572.

Koonin EV, Wolf YI, Nagasaki K, and Dolja VV (2008). The Big Bang of picorna-like virus evolution antedates the radiation of eukaryotic supergroups. *Nature Reviews Microbiology* **6**, 925–939.

Koopman KF (1950). Natural selection for reproductive isolation between *Drosophila pseudoobscura* and *Drosophila persimilis*. *Evolution* **4**, 135–148.

Kopaliani N, Shakarashvili M, Gurielidze Z, Qurkhuli T, and Tarkhnishvili D (2014). Gene flow between wolf and shepherd dog populations in Georgia (Caucasus). *Journal of Heredity* **105**, 345–353.

Kotloff KL, Winickoff JP, Ivanoff B, Clemens JD, Swerdlow DL, Sansonetti PJ, et al. (1999). Global burden of *Shigella* infections: implications for vaccine development and implementation of control strategies. *Bulletin of the World Health Organization* **77**, 651–666.

Kovach RP, Muhlfeld CC, Boyer MC, Lowe WH, Allendorf FW, and Luikart G (2015). Dispersal and selection mediate hybridization between a native and invasive species. *Proceedings of the Royal Society of London B* **282**, 20142454.

Kraus RHS, Kerstens HHD, van Hooft P, Megens H-J, Elmberg J, Tsvey A, et al. (2012). Widespread horizontal genomic exchange does not erode species barriers among sympatric ducks. *BMC Evolutionary Biology* **12**, 45.

Kraus FB, Szentgyörgyi H, Rożej E, Rhode M, Moroń D, Woyciechowski M, and Moritz RFA (2011). Greenhouse bumblebees (*Bombus terrestris*) spread their genes into the wild. *Conservation Genetics* **12**, 187–192.

Kronforst MR, Young LG, Blume LM, and Gilbert LE (2006). Multilocus analyses of admixture and introgression among hybridizing *Heliconius* butterflies. *Evolution* **60**, 1254–1268.

Kronforst MR, Young LG, and Gilbert LE (2007). Reinforcement of mate preference among hybridizing *Heliconius* butterflies. *Journal of Evolutionary Biology* **20**, 278–285.

Kruse KC and Dunlap DG (1976). Serum albumins and hybridization in two species of the *Rana pipiens* complex in the north central United States. *Copeia* **1976**, 394–396.

Kubatko LS (2009). Identifying hybridization events in the presence of coalescence via model selection. *Systematic Biology* **58**, 478–488.

Kuhn A, Ong YM, Cheng C-Y, Wong TY, Quake SR, and Burkholder WF (2014). Linkage disequilibrium and signatures of positive selection around LINE-1 retrotransposons in the human genome. *Proceedings of the National Academy of Sciences USA* **111**, 8131–8136.

Kulikova IV, Drovetski SV, Gibson DD, Harrigan RJ, Rohwer S, Sorenson MD, et al. (2005). Phylogeography of the mallard (*Anas platyrhynchos*): Hybridization, dispersal, and lineage sorting contribute to complex geographic structure. *Auk* **122**, 949–965.

Kulikova LA, McAlister MB, Ogden KL, Larkin MJ, and O'Hanlon JF (2002). Analysis of bacteria contaminating ultrapure water in industrial systems. *Applied and Environmental Microbiology* **68**, 1548–1555.

Kulikova IV, Zhuravlev YN, and McCracken KG (2004). Asymmetric hybridization and sex-biased gene flow between eastern spot-billed ducks (*Anas zonorhyncha*) and mallards (*A. platyrhynchus*) in the Russian Far East. *Auk* **121**, 930–949.

Kunte K, Shea C, Aardema ML, Scriber JM, Juenger TE, Gilbert LE, and Kronforst MR (2011). Sex chromosome mosaicism and hybrid speciation among tiger swallowtail butterflies. *PLoS Genetics* **7**, e1002274.

Kurland CG (2005). What tangled web: barriers to rampant horizontal gene transfer. *BioEssays* **27**, 741–747.

Kutschera VE, Bidon T, Hailer F, Rodi JL, Fain SR, and Janke A (2014). Bears in a forest of gene trees: Phylogenetic inference is complicated by incomplete lineage sorting and gene flow. *Molecular Biology and Evolution* **31**, 2004–2017.

Kyriacou CP and Hall JC (1980). Circadian rhythm mutations in *Drosophila melanogaster* affect short-term fluctuations in the male's courtship song. *Proceedings of the National Academy of Sciences USA* **77**, 6729–6733.

Lachance J, Vernot B, Elbers CC, Ferwerda B, Froment A, Bodo J-M, et al. (2012). Evolutionary history and adaptation from high-coverage whole-genome sequences of diverse African hunter-gatherers. *Cell* **150**, 457–469.

Lack D (1947). *Darwin's finches*. Cambridge University Press, Cambridge.

Ladner JT and Palumbi SR (2012). Extensive sympatry, cryptic diversity and introgression throughout the geographic distribution of two coral species complexes. *Molecular Ecology* **21**, 2224–2238.

Lalueza-Fox C and Gilbert MTP (2011). Paleogenomics of archaic hominins. *Current Biology* **21**, R1002–R1009.

Lalueza-Fox C, Krause J, Caramelli D, Catalano G, Milani L, Lourdes M, et al. (2006). Mitochondrial DNA of an Iberian Neandertal suggests a population affinity with other European Neandertals. *Current Biology* **16**, R629–R630.

Lam TT-Y, Wang J, Shen Y, Zhou B, Duan L, Cheung C-L, et al. (2013). The genesis and source of the H7N9 influenza viruses causing human infections in China. *Nature* **502**, 241–244.

Lamichhaney S, Berglund J, Sällman Almén M, Maqbool K, Grabherr M, Martinez-Barrio A, *et al.* (2015). Evolution of Darwin's finches and their beaks revealed by genome sequencing. *Nature* **518**, 371–375.

Lampert KP, Lamatsch DK, Epplen JT, and Schartl M (2005). Evidence for a monophyletic origin of triploid clones of the Amazon molly, *Poecilia formosa*. *Evolution* **59**, 881–889.

Lampert KP and Schartl M (2008). The origin and evolution of a unisexual hybrid: *Poecilia formosa*. *Proceedings of the Royal Society of London B* **363**, 2901–2909.

Lancaster ML, Bradshaw CJA, Goldsworthy SD, and Sunnucks P (2007). Lower reproductive success in hybrid fur seal males indicates fitness costs to hybridization. *Molecular Ecology* **16**, 3187–3197.

Lancaster ML, Goldsworthy SD, and Sunnucks P (2010). Two behavioural traits promote fine-scale species segregation and moderate hybridisation in a recovering sympatric fur seal population. *BMC Evolutionary Biology* **10**, 143.

Lande R (1982). A quantitative genetic theory of life history evolution. *Ecology* **63**, 607–615.

Lanzaro GC, Touré YT, Carnahan J, Zheng L, Dolo G, Traoré S, *et al.* (1998). Complexities in the genetic structure of *Anopheles gambiae* populations in west Africa as revealed by microsatellite DNA analysis. *Proceedings of the National Academy of Sciences USA* **95**, 14260–14265.

Larsen PA, Marchán-Rivadeneira MR, and Baker RJ (2010). Natural hybridization generates mammalian lineage with species characteristics. *Proceedings of the National Academy of Sciences USA* **107**, 11447–11452.

Larson G, Cucchi T, Fijita M, Matisoo-Smith E, Robins J, Anderson A, *et al.* (2007). Phylogeny and ancient DNA of *Sus* provides insights into neolithic expansion in Island Southeast Asia and Oceania. *Proceedings of the National Academy of Sciences USA* **104**, 4834–4839.

Larson G, Dobney K, Albarella U, Fang M, Matisoo-Smith E, Robins J, *et al.* (2005). Worldwide phylogeography of wild boar reveals multiple centers of pig domestication. *Science* **307**, 1618–1621.

Larson G and Fuller DQ (2014). The evolution of animal domestication. *Annual Review of Ecology, Evolution, and Systematics* **45**, 115–136.

Larson SG, Jungers WL, Morwood MJ, Sutikna T, Jatmiko, Saptomo EW, *et al.* (2007). *Homo floresiensis* and the evolution of the hominin shoulder. *Journal of Human Evolution* **53**, 718–731.

Latorre-Margalef N, Tolf, C, Grosbois V, Avril A, Bengtsson D, Wille M, *et al.* (2014). Long-term variation in influenza A virus prevalence and subtype diversity in migratory mallards in northern Europe. *Proceedings of the Royal Society of London B* **281**, 20140098.

Lavretsky P, McCracken KG, and Peters JL (2014). Phylogenetics of a recent radiation in the mallards and allies (Aves: *Anas*): Inferences from a genomic transect and the multispecies coalescent. *Molecular Phylogenetics and Evolution* **70**, 402–411.

Lawton-Rauh A, Robichaux RH, and Purugganan MD (2003). Diversity and divergence patterns in regulatory genes suggest differential gene flow in recently derived species of the Hawaiian silversword alliance adaptive radiation (Asteraceae). *Molecular Ecology* **16**, 3995–4013.

Lawton-Rauh A, Robichaux RH, and Purugganan MD (2007). Patterns of nucleotide variation in homoeologous regulatory genes in the allotetraploid Hawaiian silversword alliance (Asteraceae). *Molecular Ecology* **12**, 1301–1313.

Lazaridis I, Patterson N, Mittnik A, Renaud G, Mallick S, Kirsanow K, *et al.* (2014). Ancient human genomes suggest three ancestral populations for present-day Europeans. *Nature* **513**, 409–413.

Leaché AD, Harris RB, Rannala B, and Yang Z (2014). The influence of gene flow on species tree estimation: A simulation study. *Systematic Biology* **63**, 17–30.

Lecis R, Pierpaoli M, and Birò ZS (2006). Bayesian analyses of admixture in wild and domestic cats (*Felis silvestris*) using linked microsatellite loci. *Molecular Ecology* **15**, 119–131.

Le Comber SC and Smith C (2004). Polyploidy in fishes: Patterns and processes. *Biological Journal of the Linnean Society* **82**, 431–442.

Lee S-W, Devlin JM, Markham JF, Noormohammadi AH, Browning GF, Ficorilli NP, *et al.* (2011). Comparative analysis of the complete genome sequences of two Australian origin live attenuated vaccines of infectious laryngotracheitis virus. *Vaccine* **29**, 9583–9587.

Lee S-W, Markham PF, Coppo MJC, Legione AR, Markham JF, Noormohammadi AH, *et al.* (2012). Attenuated vaccines can recombine to form virulent field viruses. *Science* **337**, 188.

Lee Y, Marsden CD, Norris LC, Collier TC, Main BJ, Fofana A, *et al.* (2013). Spatiotemporal dynamics of gene flow and hybrid fitness between the M and S forms of the malaria mosquito, *Anopheles gambiae*. *Proceedings of the National Academy of Sciences USA* **110**, 19854–19859.

Lehman N, Eisenhawer A, Hansen K, Mech LD, Peterson RO, Gogan PJP, and Wayne RK (1991). Introgression of coyote mitochondrial DNA into sympatric North American gray wolf populations. *Evolution* **45**, 104–119.

Lemeunier F and Ashburner M (1984). Relationships within the *melanogaster* species subgroup of the genus *Drosophila* (*Sophophora*). IV. The chromosomes of two new species. *Chromosoma* **89**, 343–351.

Lerdau M and Wickham JD (2011). Non-natives: Four risk factors. *Nature* **475**, 36–37.

Leuenberger J, Gander A, Schmidt BR, and Perrin N (2014). Are invasive marsh frogs (*Pelophylax ridibundus*) replacing the native *P. lessonae/P. esculentus* hybridogenetic complex in western Europe? Genetic evidence from a field study. *Conservation Genetics* **15**, 869–878.

Levin BA, Freyhof J, Lajbner Z, Perea S, Abdoli A, Gaffaroğlu M, *et al.* (2012). Phylogenetic relationships of the algae scraping cyprinid genus *Capoeta* (Teleostei: Cyprinidae). *Molecular Phylogenetics and Evolution* **62**, 542–549.

Levins R and Lewontin R (1985). *The Dialectical Biologist*. Harvard University Press, Cambridge, Massachusetts.

Lewis CS (2013). *The Allegory of Love*. Cambridge University Press, New York.

Lewontin RC (1974). *The Genetic Basis of Evolutionary Change*. Columbia University Press, New York.

Lewontin RC and Birch LC (1966). Hybridization as a source of variation for adaptation to new environments. *Evolution* **20**, 315–336.

Lexer C, Welch ME, Durphy JL, and Rieseberg LH (2003). Natural selection for salt tolerance quantitative trait loci (QTLs) in wild sunflower hybrids: implications for the origin of *Helianthus paradoxus*, a diploid hybrid species. *Molecular Ecology* **12**, 1225–1235.

Lexer C and Widmer A (2008). The genic view of plant speciation: Recent progress and emerging questions. *Philosophical Transactions of the Royal Society of London B* **363**, 3023–3036.

Li Y, Stocks M, Hemmilä S, Källman T, Zhu H, Zhou Y, *et al.* (2010). Demographic histories of four spruce (*Picea*) species of the Qinghai-Tibetan Plateau and neighboring areas inferred from multiple nuclear loci. *Molecular Biology and Evolution* **27**, 1001–1014.

Li F-W, Villarreal JC, Kelly S, Rothfels CJ, Melkonian M, Frangedakis E, *et al.* (2014). Horizontal transfer of an adaptive chimeric photoreceptor from bryophytes to ferns. *Proceedings of the National Academy of Sciences USA* **108**, 6672–6677.

Li X, Xing J, Li B, Yu F, Lan X, and Liu J (2013). Phylogenetic analysis reveals the coexistence of interfamily and interspecies horizontal gene transfer in *Streptococcus thermophilus* strains isolated from the same yoghurt. *Molecular Phylogenetics and Evolution* **69**, 286–292.

Li H-F, Zhu W-Q, Song W-T, Shu J-T, Han W, and Chen K-W (2010). Origin and genetic diversity of Chinese domestic ducks. *Molecular Phylogenetics and Evolution* **57**, 634–640.

Libkind D, Hittinger CT, Valerio E, Gonçalves C, Dover J, Johnston M, *et al.* (2011). Microbe domestication and the identification of the wild genetic stock of lager-brewing yeast. *Proceedings of the National Academy of Sciences USA* **108**, 14539–14544.

Liedigk R, Roos C, Brameier M, and Zinner D (2014). Mitogenomics of the Old World monkey tribe Papionini. *BMC Evolutionary Biology* **14**, 176.

Liedigk R, Yang M, Jablonski NG, Momberg F, Geissmann T, Lwin N, *et al.* (2012). Evolutionary history of the odd-nosed monkeys and the phylogenetic position of the newly described Myanmar snub-nosed monkey *Rhinopithecus strykeri*. *PLoS One* **7**, e37418.

Lima-Mendez G, Van Helden J, Toussaint A, and Leplae R (2008). Reticulate representation of evolutionary and functional relationships between phage genomes. *Molecular Biology and Evolution* **25**, 762–777.

Lindell D, Sullivan MB, Johnson ZI, Tolonen AC, Rohwer F, and Chisholm SW (2004). Transfer of photosynthesis genes to and from *Prochlorococcus* viruses. *Proceedings of the National Academy of Sciences USA* **101**, 11013–11018.

Lindtke D, Buerkle CA, Barbará T, Heinze B, Castiglione S, Bartha D, and Lexer C (2012). Recombinant hybrids retain heterozygosity at many loci: New insights into the genomics of reproductive isolation in *Populus*. *Molecular Ecology* **21**, 5042–5058.

Lindtke D, Gompert Z, Lexer C, and Buerkle CA (2014). Unexpected ancestry of *Populus* seedlings from a hybrid zone implies a large role for postzygotic selection in the maintenance of species. *Molecular Ecology* **23**, 4316–4330.

Linnaeus C (1760). *Disquisitio de Sexu Plantarum*. Academy of Sciences, St. Petersburg.

Lipman MJ, Chester M, Soltis PS, and Soltis DE (2013). Natural hybrids between *Tragopogon mirus* and *T. miscellus* (Asteraceae): A new perspective on karyotypic changes following hybridization at the polyploid level. *American Journal of Botany* **100**, 2016–2022.

Lister AM and Sher AV (2001). The origin and evolution of the woolly mammoth. *Science* **294**, 1094–1097.

Lister AM, Sher AV, van Essen H, and Wei G (2005). The pattern and process of mammoth evolution in Eurasia. *Quarternary International* **126–128**, 49–64.

Lister AM and Stuart AJ (2010). The West Runton mammoth (*Mammuthus trogontherii*) and its evolutionary significance. *Quarternary International* **228**, 180–209.

Littlejohn MJ and Oldham RS (1968). *Rana pipiens* complex: Mating call structure and taxonomy. *Science* **162**, 1003–1005.

Liu H, Nolla HA, and Campbell L (1997). *Prochlorococcus* growth rate and contribution to primary production in the equatorial and subtropical North Pacific Ocean. *Aquatic Microbial Ecology* **12**, 39–47.

Liu KJ, Steinberg E, Yozzo A, Song Y, Kohn MH, and Nakhleh L (2015). Interspecific introgressive origin of genomic diversity in the house mouse. *Proceedings of the National Academy of Sciences USA* **112**, 196–201.

Liu J, Yu L, Arnold ML, Wu C-H, Wu S-F, Lu X, *et al.* (2011). Reticulate evolution: Frequent introgressive hybridization among Chinese hares (genus *Lepus*) revealed by analyses of multiple mitochondrial and nuclear DNA loci. *BMC Evolutionary Biology* **11**, 223.

Llopart A, Herrig D, Brud E, and Stecklein Z (2014). Sequential adaptive introgression of the mitochondrial genome in *Drosophila yakuba* and *Drosophila santomea*. *Molecular Ecology* **23**, 1124–1136.

Llopart A, Lachaise D, and Coyne JA (2005a). An anomalous hybrid zone in *Drosophila*. *Evolution* **59**, 2602–2607.

Llopart A, Lachaise D, and Coyne JA (2005b). Multilocus analysis of introgression between two sympatric sister species of *Drosophila: Drosophila yakuba* and *D. santomea*. *Genetics* **171**, 197–210.

Lo EYY and Donoghue MJ (2012). Expanded phylogenetic and dating analyses of the apples and their relatives (Pyreae, Rosaceae). *Molecular Phylogenetics and Evolution* **63**, 230–243.

Lockwood JD, Aleksić JM, Zou J, Wang J, Liu J, and Renner SS (2013). A new phylogeny for the genus *Picea* from plastid, mitochondrial, and nuclear sequences. *Molecular Phylogenetics and Evolution* **69**, 717–727.

Loh Y-HE, Bezault E, Muenzel FM, Roberts RB, Swofford R, Barluenga M, *et al.* (2013). Origins of shared genetic variation in African cichlids. *Molecular Biology and Evolution* **30**, 906–917.

Lohse K and Frantz LAF (2014). Neandertal admixture in Eurasia confirmed by maximum-likelihood analysis of three genomes. *Genetics* **196**, 1241–1251.

Lomax BH, Hilton J, Bateman RM, Upchurch GR, Lake JA, Leitch IJ, *et al.* (2014). Reconstructing relative genome size of vascular plants through geological time. *New Phytologist* **201**, 636–644.

Long JC (1991). The genetic structure of admixed populations. *Genetics* **127**, 417–428.

Looker KJ, Garnett GP, and Schmid GP (2008). An estimate of the global prevalence and incidence of herpes simplex virus type 2 infection. *Bulletin of the World Health Organization* **86**, 737–816.

Lopes MS, Mendonça D, dos Santos MR, Eiras-Dias JE, and da Câmara Machado A (2009). New insights on the genetic basis of Portuguese grapevine and on grapevine domestication. *Genome* **52**, 790–800.

Lorch PD and Servidio MR (2005). Postmating-prezygotic isolation is not an important source of selection for reinforcement within and between species in *Drosophila pseudoobscura* and *D. persimilis*. *Evolution* **59**, 1039–1045.

Lordkipanidze D, Jashashvili T, Vekua A, Ponce de Leon MS, Zollikofer CPE, Rightmire GP, *et al.* (2007). Postcranial evidence from early *Homo* from Dmanisi, Georgia. *Nature* **449**, 305–310.

Lotsy JP (1916). *Evolution by Means of Hybridization*. Martinus Nijhoff, The Hague.

Lotsy JP (1931). On the species of the taxonomist in its relation to evolution. *Genetica* **13**, 1–16.

Lowe PR (1936). The finches of the Galápagos in relation to Darwin's conception of species. *Ibis* **6**, 310–321.

Lowe AJ and Abbott RJ (2000). Routes of origin of two recently evolved hybrid taxa: *Senecio vulgaris* var. *hibernicus* and York Radiate Groundsel (Asteraceae). *American Journal of Botany* **87**, 1159–1167.

Lowery RK, Uribe G, Jimenez EB, Weiss MA, Herrera KJ, Regueiro M, and Herrera RJ (2013). Neanderthal and Denisova genetic affinities with contemporary humans: Introgression versus common ancestral polymorphisms. *Gene* **530**, 83–94.

Lowry DB and Willis JH (2010). A widespread chromosomal inversion polymorphism contributes to a major life-history transition, local adaptation, and reproductive isolation. *PLoS Biology* **8**, e1000500.

Lu J, Li W-H, and Wu C-I (2003). Comment on "Chromosomal speciation and molecular divergence—Accelerated evolution in rearranged chromosomes." *Science* **302**, 988b.

Lucek K, Lemoine M, and Seehausen O (2014). Contemporary ecotypic divergence during a recent range expansion was facilitated by adaptive introgression. *Journal of Evolutionary Biology* **27**, 2233–2248.

Ludwig F, Rosenthal DM, Johnston JA, Kane N, Gross BL, Lexer C, *et al.* (2004). Selection on leaf ecophysiological traits in a desert hybrid *Helianthus* species and early-generation hybrids. *Evolution* **58**, 2682–2692.

Lundsgaard-Hansen B, Matthews B, and Seehausen O (2014). Ecological speciation and phenotypic plasticity affect ecosystems. *Ecology* **95**, 2723–2735.

Luo J, Gao Y, Ma W, Bi X-y, Wang S-y, Wang J, *et al.* (2014). Tempo and mode of recurrent polyploidization in the *Carassius auratus* species complex (Cypriniformes, Cyprinidae). *Heredity* **112**, 415–427.

Luttikhuizen PC, Drent J, Peijnenburg KTCA, van der Veer HW, and Johannesson K (2012). Genetic architecture in a marine hybrid zone: Comparing outlier detection and genomic clines analysis in the bivalve *Macoma balthic*a. *Molecular Ecology* **21**, 3048–3061.

Lynch Alfaro JW, Boubli JP, Paim FP, Ribas CC, da Silva MNF, Messias MR, *et al.* (2015). Biogeography of squirrel monkeys (genus *Saimiri*): South-central Amazon origin and rapid pan-Amazonian diversification of a lowland primate. *Molecular Phylogenetics and Evolution* **82**, 436–454.

Lynch M (1984). The genetic structure of a cyclical parthenogen. *Evolution* **38**, 186–203.

Ma X-F, Szmidt AE, and Wang X-R (2006). Genetic structure and evolutionary history of a diploid hybrid pine

Pinus densata inferred from the nucleotide variation at seven gene loci. *Molecular Biology and Evolution* **23**, 807–816.

Mable BK (2013). Polyploids and hybrids in changing environments: winners or losers in the struggle for adaptation? *Heredity* **110**, 95–96.

Mable BK, Alexandrou MA, and Taylor MI (2011). Genome duplication in amphibians and fish: an extended synthesis. *Journal of Zoology* **284**, 151–182.

Machado CA, Haselkorn TS, and Noor MAF (2007). Evaluation of the genomic extent of effects of fixed inversion differences on intraspecific variation and interspecific gene flow in *Drosophila pseudoobscura* and *D. persimilis*. *Genetics* **175**, 1289–1306.

Machado CA and Hey J (2003). The causes of phylogenetic conflict in a classic *Drosophila* species group. *Proceedings of the Royal Society of London B* **270**, 1193–1202.

Machado CA, Kliman RM, Markert JA, and Hey J (2002). Inferring the history of speciation from multilocus DNA sequence data: The case of *Drosophila pseudoobscura* and close relatives. *Molecular Biology and Evolution* **19**, 472–488.

MacRae AF and Anderson WW (1988). Evidence for non-neutrality of mitochondrial DNA haplotypes in *Drosophila pseudoobscura*. *Genetics* **120**, 485–494.

Maddison WP (1997). Gene trees in species trees. *Systematic Biology* **46**, 523–536.

Mailund T, Halager AE, Westergaard M, Dutheil JY, Munch K, Anderson LN, *et al.* (2012). A new isolation with migration model along complete genomes infers very different divergence processes among closely related great ape species. *PLoS Genetics* **8**, e1003125.

Mallet J (1986). Hybrid zones of *Heliconius* butterflies in Panama and the stability and movement of warning colour clines. *Heredity* **56**, 191–202.

Mallet J (2005). Hybridization as an invasion of the genome. *Trends in Ecology and Evolution* **20**, 229–237.

Mallet J (2007). Hybrid speciation. *Nature* **446**, 279–283.

Malukiewicz J, Boere V, Fuzessy LF, Grativol AD, French JA, de Oliveira e Silva I, *et al.* (2014). Hybridization effects and genetic diversity of the common and black-tufted marmoset (*Callithrix jacchus* and *Callithrix penicillata*) mitochondrial control region. *American Journal of Physical Anthropology* **155**, 522–536.

Mandáková T, Marhold K, and Lysak MA (2014). The widespread crucifer species *Cardamine flexuosa* is an allotetraploid with a conserved subgenomic structure. *New Phytologist* **201**, 982–992.

Mank JE, Carlson JE, and Brittingham MC (2004). A century of hybridization: Decreasing genetic distance between American black ducks and mallards. *Conservation Genetics* **5**, 395–403.

Mann NH, Cook A, Millard A, Bailey S, and Clokie M (2003). Bacterial photosynthesis genes in a virus. *Nature* **424**, 741.

Mao J-F and Wang X-R (2011). Distinct niche divergence characterizes the homoploid hybrid speciation of *Pinus densata* on the Tibetan Plateau. *American Naturalist* **177**, 424–439.

Marchant AD (1988). Apparent introgression of mitochondrial DNA across a narrow hybrid zone in the *Caledia captiva* species-complex. *Heredity* **60**, 39–46.

Marchant AD, Arnold ML, and Wilkinson P (1988). Gene flow across a chromosomal tension zone I. Relicts of ancient hybridization. *Heredity* **61**, 321–328.

Marchant AD and Shaw DD (1993). Contrasting patterns of geographic variation shown by mtDNA and karyotype organization in two subspecies of *Caledia captiva* (Orthoptera). *Molecular Biology and Evolution* **10**, 855–872.

Marco A and Marín I (2009). *CGIN1*: A retroviral contribution to mammalian genomes. *Molecular Biology and Evolution* **26**, 2167–2170.

Marquès-Bonet T, Cáceres M, Bertranpetit J, Preuss TM, Thomas JW, and Navarro A (2004). Chromosomal rearrangements and the genomic distribution of gene expression divergence in humans and chimpanzees. *Trends in Genetics* **20**, 524–529.

Marshall JC, Arévalo E, Benavides E, Sites JL, and Sites JW Jr (2006). Delimiting species: Comparing methods for Mendelian characters using lizards of the *Sceloporus grammicus* (Squamata: Phrynosomatidae) complex. *Evolution* **60**, 1050–1065.

Martin NH, Bouck AC, and Arnold ML (2005). Loci affecting long-term hybrid survivability in Louisiana Irises: Implications for reproductive isolation and introgression. *Evolution* **59**, 2116–2124.

Martin NH, Bouck AC, and Arnold ML (2006). Detecting adaptive trait introgression between *Iris fulva* and *I. brevicaulis* in highly selective field conditions. *Genetics* **172**, 2481–2489.

Martin NH, Bouck AC, and Arnold ML (2007). The genetic architecture of reproductive isolation in Louisiana Irises: Flowering phenology. *Genetics* **175**, 1803–1812.

Martin SH, Dasmahapatra KK, Nadeau NJ, Salazar C, Walters JR, Simpson F, *et al.* (2013). Genome-wide evidence for speciation with gene flow in *Heliconius* butterflies. *Genome Research* **23**, 1817–1828.

Martin SH, Davey JW, and Jiggins CD (2015). Evaluating the use of ABBA–BABA statistics to locate introgressed loci. *Molecular Biology and Evolution* **32**, 244–257.

Martin KJ and Holland PWH (2014). Enigmatic orthology relationships between *Hox* clusters of the African butterflyfish and other Teleosts following ancient whole-genome duplication. *Molecular Biology and Evolution* **31**, 2592–2611.

Martin NH, Sapir Y, and Arnold ML (2008). The genetic architecture of reproductive isolation in Louisiana Irises: Pollination syndromes and pollinator preferences. *Evolution* **62**, 740–752.

Martin CH and Wainwright PC (2013). Multiple fitness peaks on the adaptive landscape drive adaptive radiation in the wild. *Science* **339**, 208–211.

Martin NH and Willis JH (2007). Ecological divergence associated with mating system causes nearly complete reproductive isolation between sympatric *Mimulus* species. *Evolution* **61**, 68–82.

Martin NH and Willis JH (2010). Geographical variation in postzygotic isolation and its genetic basis within and between two *Mimulus* species. *Philosophical Transactions of the Royal Society of London B* **365**, 2469–2478.

Maschinski J, Sirkin E, and Fant J (2010). Using genetic and morphological analysis to distinguish endangered taxa from their hybrids with the cultivated exotic pest plant *Lantana strigocamara* (syn: *Lantana camara*). *Conservation Genetics* **11**, 1607–1621.

Masterson J (1994). Stomatal size in fossil plants: evidence for polyploidy in majority of angiosperms. *Science* **264**, 421–424.

Matsudaira K, Reichard UH, Malaivijitnond S, and Ishida T (2013). Molecular evidence for the introgression between *Hylobates lar* and *H. pileatus* in the wild. *Primates* **54**, 33–37.

Matsumura S, Inoshima Y, and Ishiguro N (2014). Reconstructing the colonization history of lost wolf lineages by the analysis of the mitochondrial genome. *Molecular Phylogenetics and Evolution* **80**, 105–112.

Matsuoka Y, Vigouroux Y, Goodman MM, Sanchez J, Buckler E, and Doebley J (2002). A single domestication for maize shown by multilocus microsatellite genotyping. *Proceedings of the National Academy of Sciences USA* **99**, 6080–6084.

Matute DR and Ayroles JF (2014). Hybridization occurs between *Drosophila simulans* and *D. sechellia* in the Seychelles archipelago. *Journal of Evolutionary Biology* **27**, 1057–1068.

Matute DR, Gavin-Smyth J, and Liu G (2014). Variable post-zygotic isolation in *Drosophila melanogaster*/*D. simulans* hybrids. *Journal of Evolutionary Biology* **27**, 1691–1705.

Mavárez J and Linares M (2008). Homoploid hybrid speciation in animals. *Molecular Ecology* **17**, 4181–4185.

Mavárez J, Salazar CA, Bermingham E, Salcedo C, Jiggins CD, and Linares M (2006). Speciation by hybridization in *Heliconius* butterflies. *Nature* **441**, 868–871.

May RM, Endler JA, and McMurtrie RE (1975). Gene frequency clines in the presence of selection opposed by gene flow. *American Naturalist* **109**, 659–676.

Mayer WE, Schuster LN, Bartelmes G, Dieterich C, and Sommer RJ (2011). Horizontal gene transfer of microbial cellulases into nematode genomes is associated with functional assimilation and gene turnover. *BMC Evolutionary Biology* **11**, 13.

Maynar Smith J (1966). Sympatric speciation. *The American Naturalist* **100**, 637–650.

Mayr E (1942). *Systematics and the Origin of Species*. Columbia University Press, New York.

Mayr E (1963). *Animal Species and Evolution*. Belknap Press, Cambridge, Massachusetts.

Mayr E (1992). A local flora and the biological species concept. *American Journal of Botany* **79**, 222–238.

Mayr E (2004). 80 years of watching the evolutionary scenery. *Science* **305**, 46–47.

McBreen K and Lockhart PJ (2006). Reconstructing reticulate evolutionary histories of plants. *Trends in Plant Sciences* **11**, 398–404.

McCue ME, Bannasch DL, Petersen JL, Gurr J, Bailey E, Binns MM, *et al.* (2012). A high density SNP array for the domestic horse and extant Perissodactyla: Utility for association mapping, genetic diversity, and phylogeny studies. *PLoS Genetics* **8**, e1002451.

McDermott SR and Noor MAF (2011). Genetics of hybrid male sterility among strains and species in the *Drosophila pseudoobscura* species group. *Evolution* **65**, 1969–1978.

McDermott SR and Noor MAF (2012). Mapping of within-species segregation distortion in *Drosophila persimilis* and hybrid sterility between *D. persimilis* and *D. pseudoobscura*. *Journal of Evolutionary Biology* **25**, 2023–2032.

McDonald TR, Dietrich FS, and Lutzoni F (2012). Multiple horizontal gene transfers of ammonium transporters/ammonia permeases from prokaryotes to eukaryotes: Toward a new functional and evolutionary classification. *Molecular Biology and Evolution* **29**, 51–60.

McDonald DB, Parchman TL, Bower MR, Hubert WA, and Rahel FJ (2008). An introduced and a native vertebrate hybridize to form a genetic bridge to a second native species. *Proceedings of the National Academy of Sciences USA* **105**, 10837–10842.

McGovern PE, Hartung U, Badler V, Glusker DL, and Exner LJ (1997). The beginnings of winemaking and viniculture in the ancient Near East and Egypt. *Expedition* **39**, 3–21.

McGraw JB and Caswell H (1996). Estimation of individual fitness from life-history data. *American Naturalist* **147**, 47–64.

McGraw EA, Li J, Selander RK, and Whittam TS (1999). Molecular evolution and mosaic structure of α, β, and γ intimins of pathogenic *Escherichia coli*. *Molecular Biology and Evolution* **16**, 12–22.

McKinnon GE, Vaillancourt RE, Steane DA, and Potts BM (2004). The rare silver gum, *Eucalyptus cordata*, is leaving its trace in the organellar gene pool of *Eucalyptus globulus*. *Molecular Ecology* **13**, 3751–3762.

McManus KF, Kelley JL, Song S, Veeramah KR, Woerner AE, Stevison LS, *et al.* (2015). Inference of gorilla demographic and selective history from whole-genome sequence data. *Molecular Biology and Evolution* **32**, 600–612.

Mecham JS (1968). Evidence of reproductive isolation between two populations of the frog, *Rana pipiens*, in Arizona. *Southwestern Naturalist* **13**, 35–44.

Medina RA and García-Sastre A (2011). Influenza A viruses: new research developments. *Nature Reviews Microbiology* **9**, 590–603.

Meerow AW, Gideon M, Kuhn DN, Mopper S, and Nakamura K (2011). The genetic mosaic of *Iris* series *Hexagonae* in Florida: Inferences on the Holocene history of the Louisiana irises and the anthropogenic effects on their distribution. *International Journal of Plant Sciences* **172**, 1026–1052.

Meerow AW, Gideon M, Kuhn DN, Motamayor JC, and Nakamura K (2007). Genetic structure and gene flow among south Florida populations of *Iris hexagona* Walt. (Iridaceae) assessed with 19 microsatellite DNA loci. *International Journal of Plant Sciences* **168**, 1291–1309.

Meirmans PG, Lamothe M, Gros-Louis M-C, Khasa D, Périnet P, Bousquet J, and Isabel N (2010). Complex patterns of hybridization between exotic and native North American poplar species. *American Journal of Botany* **97**, 1688–1697.

Meléndez-Ackerman E (1997). Patterns of color and nectar variation across an *Ipomopsis* (Polemoniaceae) hybrid zone. *American Journal of Botany* **84**, 41–47.

Meléndez-Ackerman E and Campbell DR (1998). Adaptive significance of flower color and inter-trait correlations in an *Ipomopsis* hybrid zone. *Evolution* **52**, 1293–1303.

Meléndez-Ackerman E, Campbell DR, and Waser NM (1997). Hummingbird behavior and mechanisms of selection on flower color in *Ipomopsis*. *Ecology* **78**, 2532–2541.

Mellars P, Gravina B, and Ramsey CB (2007). Confirmation of Neanderthal/modern human interstratification at the Chatelperronian type-site. *Proceedings of the National Academy of Sciences USA* **104**, 3657–3662.

Melo-Ferreira J, Alves PC, Freitas H, Ferrand N, and Boursot P (2009). The genomic legacy from the extinct *Lepus timidus* to the three hare species of Iberia: Contrast between mtDNA, sex chromosomes and autosomes. *Molecular Ecology* **18**, 2643–2658.

Melo-Ferreira J, Alves PC, Rocha J, Ferrand N, and Boursot P (2011). Interspecific X-chromosome and mitochondrial DNA introgression in the Iberian hare: Selection or allele surfing? *Evolution* **65**, 1956–1968.

Melo-Ferreira J, Boursot P, Carneiro M, Esteves PJ, Farelo L, and Alves PC (2012). Recurrent introgression of mitochondrial DNA among hares (*Lepus* spp.) revealed by species-tree inference and coalescent simulations. *Systematic Biology* **61**, 367–381.

Melo-Ferreira J, Boursot P, Randi E, Kryukov A, Suchentrunk F, Ferrand N, and Alves PC (2007). The rise and fall of the mountain hare (*Lepus timidus*) during Pleistocene glaciations: Expansion and retreat with hybridization in the Iberian Peninsula. *Molecular Ecology* **16**, 605–618.

Melo-Ferreira J, Boursot P, Suchentrunk F, Ferrand N, and Alves PC (2005). Invasion from the cold past: Extensive introgression of mountain hare (*Lepus timidus*) mitochondrial DNA into three other hare species in northern Iberia. *Molecular Ecology* **14**, 2459–2464.

Melo-Ferreira J, Farelo L, Freitas H, Suchentrunk F, Boursot P, and Alves PC (2014a). Home-loving boreal hare mitochondria survived several invasions in Iberia: The relative roles of recurrent hybridisation and allele surfing. *Heredity* **112**, 265–273.

Melo-Ferreira J, Seixas FA, Cheng E, Mills LS, and Alves PC (2014b). The hidden history of the snowshoe hare, *Lepus americanus*: Extensive mitochondrial DNA introgression inferred from multilocus genetic variation. *Molecular Ecology* **23**, 4617–4630.

Mendes C, Felix R, Sousa A-M, Lamego J, Charlwood D, do Rosário VE, *et al.* (2010). Molecular evolution of the three short PGRPs of the malaria vectors *Anopheles gambiae* and *Anopheles arabiensis* in East Africa. *BMC Evolutionary Biology* **10**, 9.

Mendez FL, Watkins JC, and Hammer MF (2012). Global genetic variation at *OAS1* provides evidence of archaic admixture in Melanesian populations. *Molecular Biology and Evolution* **29**, 1513–1520.

Mendez FL, Watkins JC, and Hammer MF (2013). Neandertal origin of genetic variation at the cluster of *OAS* immunity genes. *Molecular Biology and Evolution* **30**, 798–801.

Meng C and Kubatko LS (2009). Detecting hybrid speciation in the presence of incomplete lineage sorting using gene tree incongruence: A model. *Theoretical Population Biology* **75**, 35–45.

Meraner A, Venturi A, Ficetola GF, Rossi S, Candiotto A, and Gandolfi A (2013). Massive invasion of exotic *Barbus barbus* and introgressive hybridization with endemic *Barbus plebejus* in Northern Italy: where, how and why? *Molecular Ecology* **22**, 5295–5312.

Mercês MP, Lynch Alfaro JW, Ferreira WAS, Harada ML, and Júnior JSS (2015). Morphology and mitochondrial phylogenetics reveal that the Amazon River separates two eastern squirrel monkey species: *Saimiri sciureus* and *S. collinsi*. *Molecular Phylogenetics and Evolution* **82**, 426–435.

Mercure A, Ralls K, Koepfli KP, and Wayne RK (1993). Genetic subdivisions among small canids: Mitochondrial DNA differentiation of swift, kit, and arctic foxes. *Evolution* **47**, 1313–1328.

Mergeay J, Aguilera X, Declerck S, Petrusek A, Huyse T, and De Meester L (2008). The genetic legacy of polyploid Bolivian *Daphnia*: The tropical Andes as a source for the North and South American *D. pulicaria* complex. *Molecular Ecology* **17**, 1789–1800.

Mérot C, Mavárez J, Evin A, Dasmahapatra KK, Mallet J, Lamas G, and Joron M (2013). Genetic differentiation without mimicry shift in a pair of hybridizing *Heliconius* species (Lepidoptera: Nymphalidae). *Biological Journal of the Linnean Society* **109**, 830–847.

Merrill RM, Van Schooten B, Scott JA, and Jiggins CD (2011). Pervasive genetic associations between traits causing reproductive isolation in *Heliconius* butterflies. *Proceedings of the Royal Society of London B* **278**, 511–518.

Metcalf JL, Siegle MR, and Martin AP (2008). Hybridization dynamics between Colorado's native cutthroat trout and introduced rainbow trout. *Journal of Heredity* **99**, 149–156.

Metzgar JS, Alverson ER, Chen S, Vaganov AV, and Ickert-Bond SM (2013). Diversification and reticulation in the circumboreal fern genus *Cryptogramma*. *Molecular Phylogenetics and Evolution* **67**, 589–599.

Metzker ML (2010). Sequencing technologies—the next generation. *Nature Reviews Genetics* **11**, 31–46.

Meyer M, Fu Q, Aximu-Petri A, Glocke I, Nickel B, Arsuaga J-L, et al. (2014). A mitochondrial genome sequence of a hominin from Sima de los Huesos. *Nature* **505**, 403–406.

Meyer RS, Karol KG, Little DP, Nee MH, and Litt A (2012). Phylogeographic relationships among Asian eggplants and new perspectives on eggplant domestication. *Molecular Phylogenetics and Evolution* **63**, 685–701.

Meyer M, Kircher M, Gansauge M-T, Li H, Racimo F, Mallick S, et al. (2012). A high-coverage genome sequence from an archaic Denisovan individual. *Science* **338**, 222–226.

Meyer A, Kocher TD, Basasibwaki P, and Wilson AC (1990). Monophyletic origin of Lake Victoria cichlid fishes suggested by mitochondrial DNA sequences. *Nature* **347**, 550–553.

Miao Y-W, Peng M-S, Wu G-S, Ouyang Y-N, Yang Z-Y, Yu N, et al. (2013). Chicken domestication: An updated perspective based on mitochondrial genomes. *Heredity* **110**, 277–282.

Miglia KJ, McArthur ED, Redman RS, Rodriguez RJ, Zak JC, and Freeman DC (2007). Genotype, soil type, and locale effects on reciprocal transplant vigor, endophyte growth, and microbial functional diversity of a narrow sagebrush hybrid zone in Salt Creek Canyon, Utah. *American Journal of Botany* **94**, 425–436.

Mikulíček P, Kautman M, Demovič B, and Janko K (2014). When a clonal genome finds its way back to a sexual species: Evidence from ongoing but rare introgression in the hybridogenetic water frog complex. *Journal of Evolutionary Biology* **27**, 628–642.

Milián-García Y, Ramos-Targarona R, Pérez-Fleitas E, Sosa-Rodríguez G, Guerra-Manchena L, Alonso-Tabet M, et al. (2015). Genetic evidence of hybridization between the critically endangered Cuban crocodile and the American crocodile: Implications for population history and in situ/ex situ conservation. *Heredity* **114**, 272–280.

Millard A, Clokie MRJ, Shub DA, and Mann NH (2004). Genetic organization of the *psbAD* region in phages infecting marine *Synechococcus* strains. *Proceedings of the National Academy of Sciences USA* **101**, 11007–11012.

Miller CR, Adams JR, and Waits LP (2003). Pedigree-based assignment tests for reversing coyote (*Canis latrans*) introgression into the wild red wolf (*Canis rufus*) population. *Molecular Ecology* **12**, 3287–3301.

Miller ES, Heidelberg JF, Eisen JA, Nelson WC, Durkin AS, Ciecko A, et al. (2003a). Complete genome sequence of the broad-host-range vibriophage KVP40: comparative genomics of a T4-related bacteriophage. *Journal of Bacteriology* **185**, 5220–5233.

Miller ES, Kutter E, Mosig G, Arisaka F, Kunisawa T, and Rüger W (2003b). Bacteriophage T4 genome. *Microbiology and Molecular Biology Reviews* **67**, 86–156.

Miller W, Schuster SC, Welch AJ, Ratan A, Bedoya-Reina OC, Zhao F, et al. (2012). Polar and brown bear genomes reveal ancient admixture and demographic footprints of past climate change. *Proceedings of the National Academy of Sciences USA* **109**, E2382–E2390.

Mir C, Zerjal T, Combes V, Dumas F, Madur D, Bedoya C, et al. (2013). Out of America: Tracing the genetic footprints of the global diffusion of maize. *Theoretical and Applied Genetics* **126**, 2671–2682.

Mira A, Ochman H, and Moran NA (2001). Deletional bias and the evolution of bacterial genomes. *Trends in Genetics* **17**, 589–596.

Mishler BD and Donoghue MJ (1982). Species concepts: a case for pluralism. *Systematic Zoology* **31**, 491–503.

Modliszewski JL and Willis JH (2012). Allotetraploid *Mimulus sookensis* are highly interfertile despite independent origins. *Molecular Ecology* **21**, 5280–5298.

Molina MC and Naranjo CA (1987). Cytogenetic studies in the genus *Zea*. *Theoretical and Applied Genetics* **73**, 542–550.

Mona S, Randi E, and Tommaseo-Ponzetta M (2007). Evolutionary history of the genus *Sus* inferred from cytochrome b sequences. *Molecular Phylogenetics and Evolution* **45**, 757–762.

Montague MJ, Li G, Gandolfi B, Khan R, Aken BL, Searle SMJ, *et al.* (2014). Comparative analysis of the domestic cat genome reveals genetic signatures underlying feline biology and domestication. *Proceedings of the National Academy of Sciences USA* **111**, 17230–17235.

Montelongo T and Gómez-Zurita J (2015). Nonrandom patterns of genetic admixture expose the complex historical hybrid origin of unisexual leaf beetle species in the genus *Calligrapha*. *American Naturalist* **185**, 113–134.

Monzón J, Kays R, and Dykhuizen DE (2014). Assessment of coyote-wolf-dog admixture using ancestry-informative diagnostic SNPs. *Molecular Ecology* **23**, 182–197.

Moore JA (1939). Temperature tolerance and rates of development in the eggs of amphibian. *Ecology* **20**, 459–478.

Moore JA (1944). Geographic variation in *Rana pipiens* Schreber of eastern North America. *Bulletin of the American Museum of Natural History* **82**, 345–370.

Moore JA (1946a). Hybridization between *Rana palustris* and different geographical forms of *Rana pipiens*. *Proceedings of the National Academy of Sciences USA* **32**, 209–212.

Moore JA (1946b). Incipient intraspecific isolating mechanisms in *Rana pipiens*. *Genetics* **31**, 304–326.

Moore JA (1947). Hybridization between *Rana pipiens* from Vermont and eastern Mexico. *Proceedings of the National Academy of Sciences USA* **33**, 72–75.

Moore WS (1977). An evaluation of narrow hybrid zones in vertebrates. *The Quarterly Review of Biology* **52**, 263–277.

Moore WS and Buchanan DB (1985). Stability of the Northern Flicker hybrid zone in historical times: Implications for adaptive speciation theory. *Evolution* **39**, 135–151.

Moore WS and Koenig WD (1986). Comparative reproductive success of yellow-shafted, red-shafted, and hybrid flickers across a hybrid zone. *The Auk* **103**, 42–51.

Moore WS and Price JT (1993). Nature of selection in the northern flicker hybrid zone and its implications for speciation theory, pp. 196–225. In R.G. Harrison, ed., *Hybrid Zones and the Evolutionary Process*, Oxford University Press, Oxford.

Morales L and Dujon B (2012). Evolutionary role of interspecies hybridization and genetic exchanges in yeasts. *Microbiology and Molecular Biology Reviews* **76**, 721.

Moran C (1979). The structure of the hybrid zone in *Caledia captiva*. *Heredity* **42**, 13–32.

Moran C and Shaw DD (1977). Population cytogenetics of the genus *Caledia* (Orthoptera: Acridinae). *Chromosoma* **63**, 181–204.

Moran C, Wilkinson P, and Shaw DD (1980). Allozyme variation across a narrow hybrid zone in the grasshopper, *Caledia captiva*. *Heredity* **44**, 69–81.

Moreira D (2000). Multiple independent horizontal transfers of informational genes from bacteria to plasmids and phages: Implications for the origin of bacterial replication machinery. *Molecular Microbiology* **35**, 1–5.

Morgan JAT, Harry AV, Welch DJ, Street R, White J, Geraghty PT, *et al.* (2012). Detection of interspecies hybridisation in Chondrichthyes: Hybrids and hybrid offspring between Australian (*Carcharhinus tilstoni*) and common (*C. limbatus*) blacktip shark found in an Australian fishery. *Conservation Genetics* **13**, 455–463.

Morgan K, O'Loughlin SM, Chen B, Linton Y-V, Thongwat D, Somboon P, *et al.* (2011). Comparative phylogeography reveals a shared impact of pleistocene environmental change in shaping genetic diversity within nine *Anopheles* mosquito species across the Indo-Burma biodiversity hotspot. *Molecular Ecology* **20**, 4533–4549.

Morin PA, Moore JJ, Chakraborty R, Jin L, Goodall J, and Woodruff DS (1994). Kin selection, social structure, gene flow, and the evolution of chimpanzees. *Science* **265**, 1193–1201.

Moritz C (1983). Parthenogenesis in the endemic Australian lizard *Heteronotia binoei* (Gekkonidae). *Science* **220**, 735–737.

Moritz C (1991). The origin and evolution of parthenogenesis in *Heteronotia binoei* (Gekkonidae): Evidence for recent and localized origins of widespread clones. *Genetics* **129**, 211–219.

Moritz C, Donnellan S, Adams M, and Baverstock PR (1989). The origin and evolution of parthenogenesis in *Heteronotia binoei* (Gekkonidae): Extensive genotypic diversity among parthenogens. *Evolution* **43**, 994–1003.

Morrell PL, Gonzales AM, Meyer KKT, and Clegg MT (2014). Resequencing data indicate a modest effect of domestication on diversity in barley: A cultigen with multiple origins. *Journal of Heredity* **105**, 253–264.

Morwood MJ, Soejono RP, Roberts RG, Sutikna T, Turney CSM, Westaway KE, *et al.* (2004). Archaeology and age of a new hominin from Flores in eastern Indonesia. *Nature* **431**, 1087–1091.

Moya A, Holmes EC, and González-Candelas F (2004). The population genetics and evolutionary epidemiology of RNA viruses. *Nature Reviews Microbiology* **2**, 1–10.

Muir CC, Galdikas BMF, and Beckenbach AT (2000). mtDNA sequence diversity of orangutans from the islands of Borneo and Sumatra. *Journal of Molecular Evolution* **51**, 471–480.

Muñoz-Fuentes V, Vilà C, Green AJ, Negro JJ, and Sorenson MD (2007). Hybridization between white-headed ducks and introduced ruddy ducks in Spain. *Molecular Ecology* **16**, 629–638.

Muschick M, Nosil P, Roesti M, Dittmann MT, Harmon L, and Salzburger W (2014). Testing the stages model in the adaptive radiation of cichlid fishes in East African Lake Tanganyika. *Proceedings of the Royal Society of London B* **281**, 20140605.

Nadeau NJ, Martin SH, Kozak KM, Salazar C, Dasmahapatra KK, Davey JW, *et al.* (2013). Genome-wide patterns of divergence and gene flow across a butterfly radiation. *Molecular Ecology* **22**, 814–826.

Nag KSC, Pramod P, and Karanth KP (2011a). Natural range extension, sampling artifact, or human mediated translocations? Range limits of Northern type *Semnopithecus entellus* (Dufresne, 1797) (Primates: Cercopithecidae: Colobinae) in peninsular India. *Journal of Threatened Taxa* **3**, 2028–2032.

Nag KSC, Pramod P, and Karanth KP (2011b). Taxonomic implications of a field study of morphotypes of Hanuman langurs (*Semnopithecus entellus*) in peninsular India. *International Journal of Primatology* **32**, 830–848.

Nagai H and Roy CR (2003). Show me the substrates: Modulation of host cell function by type IV secretion systems. *Cellular Microbiology* **5**, 373–383.

Naisbit RE, Jiggins CD, Linares M, Salazar C, and Mallet J (2002). Hybrid sterility, Haldane's rule and speciation in *Heliconius cydno* and *H. melpomene*. *Genetics* **161**, 1517–1526.

Nason JD, Ellstrand NC, and Arnold ML (1992). Patterns of hybridization and introgression in populations of oaks, manzanitas and irises. *American Journal of Botany* **79**, 101–111.

National Agricultural Statistics Service (2007). Trout production. 17 pp. United States Department of Agriculture.

Navarro A and Barton NH (2003a). Accumulating postzygotic isolation genes in parapatry: A new twist on chromosomal speciation. *Evolution* **57**, 447–459.

Navarro A and Barton NH (2003b). Chromosomal speciation and molecular divergence—accelerated evolution in rearranged chromosomes. *Science* **300**, 321–324.

Neafsey DE, Lawniczak MKN, Park DJ, Redmond SN, Coulibaly MB, Traoré SF, *et al.* (2010). SNP genotyping defines complex gene-flow boundaries among African malaria vector mosquitoes. *Science* **330**, 514–517.

Neaves WB and Baumann P (2011). Unisexual reproduction among vertebrates. *Trends in Genetics* **27**, 81–88.

Neaves LE, Zenger KR, Cooper DW, and Eldridge MDB (2010). Molecular detection of hybridization between sympatric kangaroo species in south-eastern Australia. *Heredity* **104**, 502–512.

Nelson-Sathi S, Sousa FL, Roettger M, Lozada-Chávez N, Thiergart T, Janssen A, *et al.* (2015). Origins of major archaeal clades correspond to gene acquisitions from bacteria. *Nature* **517**, 77–80.

Nesbø CL, Dlutek M, and Doolittle WF (2006). Recombination in *Thermotoga*: Implications for species concepts and biogeography. *Genetics* **172**, 759–769.

Nevado B, Fazalova V, Backeljau T, Hanssens M, and Verheyen E (2011). Repeated unidirectional introgression of nuclear and mitochondrial DNA between four congeneric Tanganyikan cichlids. *Molecular Biology and Evolution* **28**, 2253–2267.

Ng DW-K, Miller M, Yu HH, Huang T-Y, Kim E-D, Xie Q, *et al.* (2014). A role for CHH methylation in the parent-of-origin effect on altered circadian rhythms and biomass heterosis in *Arabidopsis* intraspecific hybrids. *Plant Cell* **26**, 2430–2440.

Nice CC, Gompert Z, Fordyce JA, Forister ML, Lucas, LK, and Buerkle CA (2013). Hybrid speciation and independent evolution in lineages of alpine butterflies. *Evolution* **67**, 1055–1068.

Nichols P, Genner MJ, van Oosterhout C, Smith A, Parsons P, Sungani H, *et al.* (2015). Secondary contact seeds phenotypic novelty in cichlid fishes. *Proceedings of the Royal Society of London B* **282**, 20142272.

Nielsen R and Beaumont MA (2009). Statistical inferences in phylogeography. *Molecular Ecology* **18**, 1034–1047.

Nielsen R and Wakeley J (2001). Distinguishing migration from isolation: A Markov Chain Monte Carlo approach. *Genetics* **158**, 885–896.

Nietlisbach P, Wandeler P, Parker PG, Grant PR, Grant BR, Keller LF, and Hoeck PEA (2013). Hybrid ancestry of an island subspecies of Galápagos mockingbird explains discordant gene trees. *Molecular Phylogenetics and Evolution* **69**, 581–592.

Nikiforova SV, Cavalieri D, Velasco R, and Goremykin V (2013). Phylogenetic analysis of 47 chloroplast genomes clarifies the contribution of wild species to the domesticated apple maternal line. *Molecular Biology and Evolution* **30**, 1751–1760.

Nikolaidis N, Doran N, and Cosgrove DJ (2014). Plant expansins in bacteria and fungi: Evolution by horizontal gene transfer and independent domain fusion. *Molecular Biology and Evolution* **31**, 376–386.

Nishibori M, Shimogiri T, Hayashi T, and Yasue H (2005). Molecular evidence for hybridization of species in the genus *Gallus* except for *Gallus varius*. *Animal Genetics* **36**, 367–375.

Niskanen AK, Hagstrom E, Lohi H, Ruokonen M, Esparza-Salas R, Aspi J, and Savolainen P (2013). MHC variability supports dog domestication from a large number of wolves: high diversity in Asia. *Heredity* **110**, 80–85.

Nixon KC and Wheeler QD (1990). An amplification of the phylogenetic species concept. *Cladistics* **6**, 211–223.

Noda-García L, Camacho-Zarco AR, Medina-Ruíz S, Gaytán P, Carrillo-Tripp M, Fülöp V, and Barona-Gomez F (2013). Evolution of substrate specificity in a recipient's enzyme following horizontal gene transfer. *Molecular Biology and Evolution* **30**, 2024–2034.

Noor MAF (1995). Speciation driven by natural selection in *Drosophila*. *Nature* **375**, 674–675.

Noor MAF, Grams KL, Bertucci LA, Almendarez Y, Reiland J, and Smith KR (2001). The genetics of reproductive isolation and the potential for gene exchange between *Drosophila pseudoobscura* and *D. persimilis* via backcross hybrid males. *Evolution* **55**, 512–521.

Norén K, Dalén L, Kvaløy K, and Angerbjörn A (2005). Detection of farm fox and hybrid genotypes among wild arctic foxes in Scandinavia. *Conservation Genetics* **6**, 885–894.

Norris LC, Main BJ, Lee Y, Collier TC, Fofana A, Cornel AJ, and Lanzaro GC (2015). Adaptive introgression in an African malaria mosquito coincident with the increased usage of insecticide-treated bed nets. *Proceedings of the National Academy of Sciences USA* **112**, 815–820.

Nosil P and Feder JL (2012). Genomic divergence during speciation: Causes and consequences. *Philosophical Transactions of the Royal Society of London B* **367**, 332–342.

Noutsos C, Borevitz JO, and Hodges SA (2014). Gene flow between nascent species: Geographic, genotypic and phenotypic differentiation within and between *Aquilegia formosa* and *A. pubescens*. *Molecular Ecology* **23**, 5589–5598.

Nowak R (1994). Mining treasures from "Junk DNA." *Science* **263**, 608–610.

Nunes MDS, Wengel PO-T, Kreissl M, and Schlötterer C (2010). Multiple hybridization events between *Drosophila simulans* and *Drosophila mauritiana* are supported by mtDNA introgression. *Molecular Ecology* **19**, 4695–4707.

Nussberger B, Wandeler P, Weber D, and Keller LF (2014). Monitoring introgression in European wildcats in the Swiss Jura. *Conservation Genetics* **15**, 1219–1230.

Nwakanma DC, Neafsey DE, Jawara M, Adiamoh M, Lund E, Rodrigues A, *et al.* (2013). Breakdown in the process of incipient speciation in *Anopheles gambiae*. *Genetics* **193**, 1221–1231.

O'Brien SJ and Mayr E (1991). Bureaucratic mischief: Recognizing endangered species and subspecies. *Science* **251**, 1187–1188.

Ochman H, Lawrence JG, and Groisman EA (2000). Lateral gene transfer and the nature of bacterial innovation. *Nature* **405**, 299–304.

O'Connor TJ, Adepoju Y, Boyd D, and Isberg RR (2011). Minimization of the *Legionella pneumophila* genome reveals chromosomal regions involved in host range expansion. *Proceedings of the National Academy of Sciences USA* **108**, 14733–14740.

Ohta T and Dover GA (1983). Population genetics of multigene families that are dispersed into two or more chromosomes. *Proceedings of the National Academy of Sciences USA* **80**, 4079–4083.

O'hUigin C, Satta Y, Takahata N, and Klein J (2002). Contribution of homoplasy and of ancestral polymorphism to the evolution of genes in anthropoid primates. *Molecular Biology and Evolution* **19**, 1501–1513.

O'Keefe KJ, Silander OK, McCreery H, Weinreich DM, Wright KM, Chao L, *et al.* (2010). Geographic differences in sexual reassortment in RNA phage. *Evolution* **64**, 3010–3023.

Oliveira R, Godinho R, Randi E, Ferrand N, and Alves PC (2007). Molecular analysis of hybridisation between wild and domestic cats (*Felis silvestris*) in Portugal: Implications for conservation. *Conservation Genetics* **9**, 1–11.

Olmstead RG (1995). Species concepts and plesiomorphic species. *Systematic Botany* **20**, 623–630.

O'Loughlin SM, Magesa S, Mbogo C, Mosha F, Midega J, Lomas S, and Burt A (2014). Genomic analyses of three malaria vectors reveals extensive shared polymorphism but contrasting population histories. *Molecular Biology and Evolution* **31**, 889–902.

Olson JR, Cooper SJ, Swanson DL, Braun MJ, and Williams JB (2010). The relationship of metabolic performance and distribution in black-capped and Carolina chickadees. *Physiological and Biochemical Zoology* **83**, 263–275.

Omer S, Kovacs A, Mazor Y, and Gophna U (2010). Integration of a foreign gene into a native complex does not impair fitness in an experimental model of lateral gene transfer. *Molecular Biology and Evolution* **27**, 2441–2445.

Oneal E, Lowry DB, Wright KM, Zhu Z, and Willis JH (2014). Divergent population structure and climate associations of a chromosomal inversion polymorphism across the *Mimulus guttatus* species complex. *Molecular Ecologist* **23**, 2844–2860.

Onyabe DY and Conn JE (2001). Genetic differentiation of the malaria vector *Anopheles gambiae* across Nigeria suggests that selection limits gene flow. *Heredity* **87**, 647–658.

Orlando L, Darlu P, Toussaint M, Bonjean D, Otte M, and Hänni C (2006). Revisiting Neandertal diversity with a 100,000 year old mtDNA sequence. *Current Biology* **16**, R400–R402.

Orozco-Terwengel P, Andreone F, Louis E Jr, and Vences M (2013). Mitochondrial introgressive hybridization following a demographic expansion in the tomato frogs of Madagascar, genus *Dyscophus*. *Molecular Ecology* **22**, 6074–6090.

Orr HA (1990). "Why polyploidy is rarer in animals than in plants" revisited. *American Naturalist* **136**, 759–770.

Orsini L, Schwenk K, De Meester L, Colbourne JK, Pfrender ME, and Weider LJ (2013). The evolutionary time machine: using dormant propagules to forecast how populations can adapt to changing environments. *Trends in Ecology and Evolution* **28**, 274–282.

Ortíz-Barrientos D, Counterman BA, Noor MAF (2004). The genetics of speciation by reinforcement. *PLoS Biology* **2**, e416.

Ortíz-Barrientos D and Noor MAF (2005). Evidence for a one-allele assortative mating locus. *Science* **310**, 1467.

Osada N and Wu C-I (2005). Inferring the mode of speciation from genomic data: A study of the great apes. *Genetics* **169**, 259–264.

Østbye K, Bernatchez L, Næsje TF, Himberg K-JM, and Hindar K (2005). Evolutionary history of the European whitefish *Coregonus lavaretus* (L.) species complex as inferred from mtDNA phylogeography and gill-raker numbers. *Molecular Ecology* **14**, 4371–4387.

Ottoni C, Flink LG, Evin A, Georg C, De Cupere B, Van Neer W, *et al.* (2013). Pig domestication and human-mediated dispersal in western Eurasia revealed through ancient DNA and geometric morphometrics. *Molecular Biology and Evolution* **30**, 824–832.

Ouborg NJ, Pertoldi C, Loeschcke V, Bijlsma R(K), and Hedrick PW (2010). Conservation genetics in transition to conservation genomics. *Trends in Genetics* **26**, 177–187.

Ovenden JR, Morgan JAT, Kashiwagi T, Broderick D, and Salini J (2010). Towards better management of Australia's shark fishery: Genetic analyses reveal unexpected ratios of cryptic blacktip species *Carcharhinus tilstoni* and *C. limbatus*. *Marine and Freshwater Research* **61**, 253–262.

Owens GL and Rieseberg LH (2014). Hybrid incompatibility is acquired faster in annual than in perennial species of sunflower and tarweed. *Evolution* **68**, 893–900.

Pääbo S (2003). The mosaic that is our genome. *Nature* **421**, 409–412.

Pääbo S, Poinar H, Serre D, Jaenicke-Després V, Hebler J, Rohland N, *et al.* (2004). Genetic analyses from ancient DNA. *Annual Review of Genetics* **38**, 645–679.

Pagès M, Bazin E, Galan M, Chaval Y, Claude J, Herbreteau V, *et al.* (2013). Cytonuclear discordance among Southeast Asian black rats (*Rattus rattus* complex). *Molecular Ecology* **22**, 1019–1034.

Palenik B, Brahamsha B, Larimer FW, Land M, Hauser L, Chain P, *et al.* (2003). The genome of a motile marine *Synechococcus*. *Nature* **424**, 1037–1042.

Pallotta M, Schnurbusch T, Hayes J, Hay A, Baumann U, Paull J, *et al.* (2014). Molecular basis of adaptation to high soil boron in wheat landraces and elite cultivars. *Nature* **514**, 88–91.

Palmer EJ (1948). Hybrid oaks of North America. *Journal of the Arnold Arboretum* **29**, 1–48.

Pang J-F, Kluetsch C, Zou X-J, Zhang A-b, Luo L-Y, Angleby H, *et al.* (2009). mtDNA data indicate a single origin for dogs south of Yangtze River, less than 16,300 years ago, from numerous wolves. *Molecular Biology and Evolution* **26**, 2849–2864.

Papadopulos AST, Baker WJ, Crayn D, Butlin RK, Kynast RG, Hutton I, and Savolainen V (2011). Speciation with gene flow on Lord Howe Island. *Proceedings of the National Academy of Sciences USA* **108**, 13188–13193.

Papadopulos AST, Kaye M, Devaux C, Hipperson H, Lighten J, Dunning LT, *et al.* (2014). Evaluation of genetic isolation within an island flora reveals unusually widespread local adaptation and supports sympatric speciation. *Philosophical Transactions of the Royal Society of London B* **369**, 20130342.

Papadopulos AST, Price Z, Devaux C, Hipperson H, Smadja CM, Hutton I, Baker WJ, Butlin RK, and Savolainen V (2013). A comparative analysis of the mechanisms underlying speciation on Lord Howe Island. *Journal of Evolutionary Biology* **26**, 733–745.

Papke RT, Zhaxybayeva O, Feil EJ, Sommerfeld K, Muise D, and Doolittle WF (2007). Searching for species in haloarchaea. *Proceedings of the National Academy of Sciences USA* **104**, 14092–14097.

Paraskevis D, Lemey P, Salemi M, Suchard M, Van de Peer Y, and Vandamme A-M (2003). Analysis of the evolutionary relationships of HIV-1 and SIVcpz sequences using Bayesian inference: implications for the origin of HIV-1. *Molecular Biology and Evolution* **20**, 1986–1996.

Pardo-Diaz C, Salazar C, Baxter SW, Merot C, Figueiredo-Ready W, Joron M, *et al.* (2012). Adaptive introgression across species boundaries in *Heliconius* butterflies. *PLoS Genetics* **8**, e1002752.

Park C and Zhang J (2012). High expression hampers horizontal gene transfer. *Genome Biology and Evolution* **4**, 523–532.

Parkhill J, Wren BW, Thomson NR, Titball RW, Holden MTG, Prentice MB, *et al.* (2001). Genome sequence of *Yersinia pestis*, the causative agent of plague. *Nature* **413**, 523–527.

Parmakelis A, Slotman MA, Marshall JC, Awono-Ambene PH, Antonio-Nkondjio C, Simard F, *et al.* (2008). The molecular evolution of four anti-malarial immune genes in the *Anopheles gambiae* species complex. *BMC Evolutionary Biology* **8**, 79.

Parris MJ (1999). Hybridization in leopard frogs (*Rana pipiens* complex): Larval fitness components in single genotype populations and mixtures. *Evolution* **53**, 1872–1883.

Parris MJ (2000). Experimental analysis of hybridization in leopard frogs (Anurae: Ranidae): Larval performance in desiccating environments. *Copeia* **2000**, 11–19.

Parris MJ (2001a). High larval performance of leopard frog hybrids: Effects of environment-dependent selection. *Ecology* **82**, 3001–3009.

Parris MJ (2001b). Hybridization in leopard frogs (*Rana pipiens* complex): Terrestrial performance of newly metamorphosed hybrid and parental genotypes in field enclosures. *Canadian Journal of Zoology* **79**, 1552–1558.

Parris MJ (2001c). Hybridization in leopard frogs (*Rana pipiens* complex): Variation in interspecific hybrid larval fitness components along a natural contact zone. *Evolutionary Ecology Research* **3**, 91–105.

Parris MJ (2004). Hybrid response to pathogen infection in interspecific crosses between two amphibian species (Anura: Ranidae). *Evolutionary Ecology Research* **6**, 457–471.

Parris MJ, Laird CW, and Semlitsch RD (2001). Reptiles differential predation on experimental populations of parental and hybrid leopard frog (*Rana blairi* and *Rana sphenocephala*) larvae. *Journal of Herpetology* **35**, 479–485.

Parris MJ, Semlitsch RD, and Sage RD (1999). Experimental analysis of the evolutionary potential of hybridization in leopard frogs (Anura: Ranidae). *Journal of Evolutionary Biology* **12**, 662–671.

Partensky F, Hess WR, and Vaulot D (1999). *Prochlorococcus*, a marine photosynthetic prokaryote of global significance. *Microbiology and Molecular Biology Reviews* **63**, 106–127.

Patel S, Schell T, Eifert C, Feldmeyer B, and Pfenninger M (2015). Characterizing a hybrid zone between a cryptic species pair of freshwater snails. *Molecular Ecology* **24**, 643–655.

Paterson HEH (1985). The recognition concept of species. In ES Vrba, ed., *Species and Speciation*, pp. 21–29, Transvaal Museum Monograph No. 4, Transvaal Museum, Pretoria.

Paterson AH, Wendel JF, Gundlach H, Guo H, Jenkins J, Jin D, *et al.* (2012). Repeated polyploidization of *Gossypium* genomes and the evolution of spinnable cotton fibres. *Nature* **492**, 423–427.

Patterson N, Richter DJ, Gnerre S, Lander ES, and Reich D (2006). Genetic evidence for complex speciation of humans and chimpanzees. *Nature* **441**, 1103–1108.

Pauchet Y and Heckel DG (2013). The genome of the mustard leaf beetle encodes two active xylanases originally acquired from bacteria through horizontal gene transfer. *Proceedings of the Royal Society of London B* **280**, 20131021.

Payseur BA, Krenz JG, and Nachman MW (2004). Differential patterns of introgression across the X chromosome in a hybrid zone between two species of house mice. *Evolution* **58**, 2064–2078.

Pedulla ML, Ford ME, Houtz JM, Karthikeyan T, Wadsworth C, Lewis JA, *et al.* (2003). Origins of highly mosaic mycobacteriophage genomes. *Cell* **113**, 171–182.

Pegoraro M, Gesto JS, Kyriacou CP, and Tauber E (2014). Role for circadian clock Genes in seasonal timing: Testing the Bünning hypothesis. *PLoS Genetics* **10**, e1004603.

Pelser PB, Abbott RJ, Comes HP, Milton JJ, Möller M, Looseley ME, *et al.* (2012). The genetic ghost of an invasion past: colonization and extinction revealed by historical hybridization in *Senecio*. *Molecular Ecology* **21**, 369–387.

Pelser PB, Kennedy AH, Tepe EJ, Shidler JB, Nordenstam B, Kadereit JW, and Watson LE (2010). Patterns and causes of incongruence between plastid and nuclear Senecioneae (Asteraceae) phylogenies. *American Journal of Botany* **97**, 856–873.

Percy DM, Argus GW, Cronk QC, Fazekas AJ, Kesanakurti PR, Burgess KS, *et al.* (2014). Understanding the spectacular failure of DNA barcoding in willows (Salix): Does this result from a trans-specific selective sweep? *Molecular Ecology* **23**, 4737–4756.

Pereira RJ, Barreto FS, and Burton RS (2014). Ecological novelty by hybridization: Experimental evidence for increased thermal tolerance by transgressive segregation in *Tigriopus californicus*. *Evolution* **68**, 204–215.

Peris D, Lopes CA, Belloch C, Querol A, and Barrio E (2012). Comparative genomics among *Saccharomyces cerevisiae* x *Saccharomyces kudriavzevii* natural hybrid strains isolated from wine and beer reveals different origins. *BMC Genomics* **13**, 407.

Peris D, Sylvester K, Libkind D, Gonçalves P, Sampaio JP, Alexander WG, and Hittinger CT (2014). Population structure and reticulate evolution of *Saccharomyces eubayanus* and its lager-brewing hybrids. *Molecular Ecology* **23**, 2031–2045.

Perry RD and Fetherston JD (1997). *Yersinia pestis*—etiologic agent of plague. *Clinical Microbiology Reviews* **10**, 35–66.

Pesole G, Sbisá E, Preparata G, and Saccone C (1992). The evolution of the mitochondrial D-loop region and the origin of modern man. *Molecular Biology and Evolution* **9**, 587–598.

Peters JL, McCracken KG, Zhuravlev YN, Lu Y, Wilson RE, Johnson KP, and Omland KE (2005). Phylogenetics of wigeons and allies (Anatidae: *Anas*): The importance of sampling multiple loci and multiple individuals. *Molecular Phylogenetics and Evolution* **35**, 209–224.

Peters JL and Omland KE (2007). Population structure and mitochondrial polyphyly in North American gadwalls (*Anas strepera*). *Auk* **124**, 444–462.

Peters JL, Sonsthagen SA, Lavretsky P, Rezsutek M, Johnson WP, and McCracken KG (2014a). Interspecific hybridization contributes to high genetic diversity and apparent effective population size in an endemic population of mottled ducks (*Anas fulvigula maculosa*). *Conservation Genetics* **15**, 509–520.

Peters JL, Winker K, Millam KC, Lavretsky P, Kulikova I, Wilson RE, *et al.* (2014b). Mito-nuclear discord in six congeneric lineages of Holarctic ducks (genus *Anas*). *Molecular Ecology* **23**, 2961–2974.

Peters JL, Zhuravlev Y, Fefelov I, Logie A, and Omland KE (2007). Nuclear loci and coalescent methods support ancient hybridization as cause of mitochondrial paraphyly between Gadwall and Falcated duck (*Anas* spp.). *Evolution* **61**, 1992–2006.

Phadnis N (2011). Genetic architecture of male sterility and segregation distortion in *Drosophila pseudoobscura* Bogota-USA hybrids. *Genetics* **189**, 1001–1009.

Phares CR, Wangroongsarb P, Chantra S, Paveenkitiporn W, Tondella M-L, Benson RF, *et al.* (2007). Epidemiology of severe pneumonia caused by *Legionella longbeachae*, *Mycoplasma pneumoniae*, and *Chlamydia pneumoniae*: 1-year, population-based surveillance for severe pneumonia in Thailand. *Clinical Infectious Diseases* **45**, e147–155.

Phifer-Rixey M, Bomhoff M, and Nachman MW (2014). Genome-wide patterns of differentiation among house mouse subspecies. *Genetics* **198**, 283–297.

Pickrell JK and Reich D (2014). Toward a new history and geography of human genes informed by ancient DNA. *Trends in Genetics* **30**, 377–389.

Pierpaoli M, Biro ZS, Herrmann M, Hupe K, Fernandes M, Ragni B, *et al.* (2003). Genetic distinction of wildcat (*Felis silvestris*) populations in Europe, and hybridization with domestic cats in Hungary. *Molecular Ecology* **12**, 2585–2598.

Pietri C, Alves PC, and Melo-Ferreira JM (2011). Hares in Corsica: High prevalence of Lepus corsicanus and hybridization with introduced *L. europaeus* and *L. granatensis*. *European Journal of Wildlife Research* **57**, 313–321.

Pigliucci M (2003). Species as family resemblance concepts: The (dis-)solution of the species problem? *BioEssays* **25**, 596–602.

Pinhasi R, Higham TFG, Golovanova LV, and Doronichev VB (2011). Revised age of late Neanderthal occupation and the end of the Middle Paleolithic in the northern Caucasus. *Proceedings of the National Academy of Sciences USA* **108**, 8611–8616.

Pinho C and Hey J (2010). Divergence with gene flow: Models and data. *Annual Review of Ecology, Evolution and Systematics* **41**, 215–230.

Pirie MD, Humphreys AM, Barker NP, and Linder HP (2009). Reticulation, data combination, and inferring evolutionary history: An example from Danthonioideae (Poaceae). *Systematic Biology* **58**, 612–628.

Placyk JS Jr, Fitzpatrick BM, Casper GS, Small RL, Reynolds RG, Noble DWA, *et al.* (2012). Hybridization between two gartersnake species (*Thamnophis*) of conservation concern: A threat or an important natural interaction? *Conservation Genetics* **13**, 649–663.

Plagnol V and Wall JD (2006). Possible ancestral structure in human populations. *PLoS Genetics* **2**, 0972–0979.

Plantier J-C, Leoz M, Dickerson JE, De Oliveira F, Cordonnier F, Lemée V, *et al.* (2009). A new human immunodeficiency virus derived from gorillas. *Nature Medicine* **15**, 871–872.

Plénet S, Joly P, Hervant F, Fromont E, and Grolet O (2005). Are hybridogenetic complexes structured by habitat in water frogs? *Journal of Evolutionary Biology* **18**, 1575–1586.

Plénet S, Pagano A, Joly P, and Fouillet P (2000). Variation of plastic responses to oxygen availability within the hybridogenetic *Rana esculenta* complex. *Journal of Evolutionary Biology* **13**, 20–28.

Plötner J, Uzzell T, Beerli P, Spolsky C, Ohst T, Litvinchuk SN, *et al.* (2008). Widespread unidirectional transfer of mitochondrial DNA: A case in western Palaearctic water frogs. *Journal of Evolutionary Biology* **21**, 668–681.

Poelstra JW, Ellegren H, and Wolf JBW (2013). An extensive candidate gene approach to speciation: Diversity, divergence and linkage disequilibrium in candidate pigmentation genes across the European crow hybrid zone. *Heredity* **111**, 467–473.

Poelstra JW, Vijay N, Bossu CM, Lantz H, Ryll B, Müller I, *et al.* (2014). The genomic landscape underlying phenotypic integrity in the face of gene flow in crows. *Science* **344**, 1410–1414.

Polz MF, Alm EJ, and Hanage WP (2013). Horizontal gene transfer and the evolution of bacterial and archaeal population structure. *Trends in Genetics* **29**, 170–175.

Potter D, Eriksson T, Evans RC, Oh S, Smedmark JEE, Morgan DR, *et al.* (2007). Phylogeny and classification of Rosaceae. *Plant Systematics and Evolution* **266**, 5–43.

Potts BM and Reid JB (1988). Hybridization as a dispersal mechanism. *Evolution* **42**, 1245–1255.

Potts BM and Reid JB (1990). The evolutionary significance of hybridization in *Eucalyptus*. *Evolution* **44**, 2151–2152.

Powell JR (1983). Interspecific cytoplasmic gene flow in the absence of nuclear gene flow: Evidence from *Drosophila*. *Proceedings of the National Academy of Sciences USA* **80**, 492–495.

Powell JR (1991). Monophyly/paraphyly/polyphyly and gene/species trees: An example from *Drosophila*. *Molecular Biology and Evolution* **8**, 892–896.

Prada C and Hellberg ME (2014). Strong natural selection on juveniles maintains a narrow adult hybrid zone in a broadcast spawner. *American Naturalist* **184**, 702–713.

Prado-Martinez J, Sudmant PH, Kidd JM, Li H, Kelley JL, Lorente-Galdos B, *et al.* (2013). Great ape genetic diversity and population history. *Nature* **499**, 471–475.

Prager EM and Wilson AC (1975). Slow evolutionary loss of the potential for interspecific hybridization in birds: A manifestation of slow regulatory evolution.

Proceedings of the National Academy of Sciences USA **72**, 200–204.

Prentis PJ, White EM, Radford IJ, Lowe AJ, and Clarke AR (2007). Can hybridization cause local extinction: A case for demographic swamping of the Australian native *Senecio pinnatifolius* by the invasive *Senecio madagascariensis*? *New Phytologist* **176**, 902–912.

Presgraves DC and Yi SV (2009). Doubts about complex speciation between humans and chimpanzees. *Trends in Ecology and Evolution* **24**, 533–540.

Price DK and Muir C (2008). Conservation implications of hybridization in Hawaiian picture-winged *Drosophila*. *Molecular Phylogenetics and Evolution* **47**, 1217–1226.

Prince VE and Pickett FB (2002). Splitting pairs: The diverging fates of duplicated genes. *Nature Reviews Genetics* **3**, 827–837.

Pritchard VL and Edmands S (2013). The genomic trajectory of hybrid swarms: Outcomes of repeated crosses between populations of *Tigriopus californicus*. *Evolution* **67**, 774–791.

Pritchard VL, Knutson VL, Lee M, Zieba J, and Edmands S (2013). Fitness and morphological outcomes of many generations of hybridization in the copepod *Tigriopus californicus*. *Journal of Evolutionary Biology* **26**, 416–433.

Pritchard JK, Stephens M, and Donnelly P (2000). Inference of population structure using multilocus genotype data. *Genetics* **155**, 945–959.

Promislow DEL, Jung CF, and Arnold ML (2001). Age-specific fitness components in hybrids of *Drosophila pseudoobscura* and *D. persimilis*. *Journal of Heredity* **92**, 30–37.

Prüfer K, Munch K, Hellmann I, Akagi K, Miller JR, Walenz B, et al. (2012). The bonobo genome compared with the chimpanzee and human genomes. *Nature* **486**, 527–531.

Prüfer K, Racimo F, Patterson N, Sankararaman S, Sawyer S, Heinze A, et al. (2014). The complete genome sequence of a Neanderthal from the Altai Mountains. *Nature* **505**, 43–49.

Pruvost NBM, Mikulíček P, Choleva L, and Reyer H-U (2015). Contrasting reproductive strategies of triploid hybrid males in vertebrate mating systems. *Journal of Evolutionary Biology* **28**, 189–204.

Pu J, Wang S, Yin Y, Zhang G, Carter RA, Wang J, et al. (2015). Evolution of the H9N2 influenza genotype that facilitated the genesis of the novel H7N9 virus. *Proceedings of the National Academy of Sciences USA* **112**, 548–553.

Pujolar JM, Jacobsen MW, Als TD, Frydenberg J, Magnussen E, Jónsson B, et al. (2014). Assessing patterns of hybridization between North Atlantic eels using diagnostic single-nucleotide polymorphisms. *Heredity* **112**, 627–637.

Pupo GM, Lan R, and Reeves PR (2000). Multiple independent origins of *Shigella* clones of *Escherichia coli* and convergent evolution of many of their characteristics. *Proceedings of the National Academy of Sciences USA* **97**, 10567–10572.

Queney G, Ferrand N, Weiss S, Mougel F, and Monnerot M (2001). Stationary distributions of microsatellite loci between divergent population groups of the European Rabbit (*Oryctolagus cuniculus*). *Molecular Biology and Evolution* **18**, 2169–2178.

Rabosky DL and Matute DR (2013). Macroevolutionary speciation rates are decoupled from the evolution of intrinsic reproductive isolation in *Drosophila* and birds. *Proceedings of the National Academy of Sciences USA* **110**, 15354–15359.

Ramírez O, Burgos-Paz W, Casas E, Ballester M, Bianco E, Olalde I, et al. (2015). Genome data from a sixteenth century pig illuminate modern breed relationships. *Heredity* **114**, 175–184.

Ramírez O, Ojeda A, Tomàs A, Gallardo D, Huang LS, Folch JM, et al. (2009). Integrating Y-chromosome, mitochondrial, and autosomal data to analyze the origin of pig breeds. *Molecular Biology and Evolution* **26**, 2061–2072.

Ramos-Onsins SE, Burgos-Paz W, Manunza A, and Amills M (2014). Mining the pig genome to investigate the domestication process. *Heredity* **113**, 471–484.

Ramsey J, Bradshaw HD Jr, and Schemske DW (2003). Components of reproductive isolation between the monkeyflowers *Mimulus lewisii* and *M. cardinalis* (Phrymaceae). *Evolution* **57**, 1520–1534.

Ran J-H, Wei X-X, and Wang X-Q (2006). Molecular phylogeny and biogeography of *Picea* (Pinaceae): Implications for phylogeographical studies using cytoplasmic haplotypes. *Molecular Phylogenetics and Evolution* **41**, 405–419.

Randi E (2008). Detecting hybridization between wild species and their domesticated relatives. *Molecular Ecology* **17**, 285–293.

Randi E and Lucchini V (2002). Detecting rare introgression of domestic dog genes into wild wolf (*Canis lupus*) populations by Bayesian admixture analyses of microsatellite variation. *Conservation Genetics* **3**, 31–45.

Randi E, Lucchini V, Fjeldsø Christensen M, Mucci N, Funk SM, Dolf G, and Loeschcke V (2000). Mitochondrial DNA variability in Italian and East European wolves: Detecting the consequences of small population size and hybridization. *Conservation Biology* **14**, 464–473.

Randi E, Pierpaoli M, Beaumont M, Ragni B, and Sforzi A (2001). Genetic identification of wild and domestic cats (*Felis silvestris*) and their hybrids using Bayesian clustering methods. *Molecular Biology and Evolution* **18**, 1679–1693.

Randolph LF (1966). *Iris nelsonii*, a new species of Louisiana iris of hybrid origin. *Baileya* **14**, 143–169.

Randolph LF, Nelson IS, and Plaisted RL (1967). Negative evidence of introgression affecting the stability of Louisiana *Iris* species. *Cornell University Agricultural Experiment Station Memoir* **398**, 1–56.

Rankin DJ, Rocha EPC, and Brown SP (2011). What traits are carried on mobile genetic elements, and why? *Heredity* **106**, 1–10.

Rasmussen M, Guo X, Wang Y, Lohmueller KE, Rasmussen S, Albrechtsen A, *et al.* (2011). An Aboriginal Australian genome reveals separate human dispersals into Asia. *Science* **334**, 94–98.

Rasmussen JB, Robinson MD, Hontela A, and Heath DD (2012). Metabolic traits of westslope cutthroat trout, introduced rainbow trout and their hybrids in an ecotonal hybrid zone along an elevation gradient. *Biological Journal of the Linnean Society* **105**, 56–72.

Raven PH (1975). The bases of angiosperm phylogeny: Cytology. *Annals of the Missouri Botanical Garden* **62**, 724–764.

Raz Y and Tannenbaum E (2010). The influence of horizontal gene transfer on the mean fitness of unicellular populations in static environments. *Genetics* **185**, 327–337.

Razafindratsimandresy R, Joffret M-L, Rabemanantsoa S, Andriamamonjy S, Heraud J-M, and Delpeyroux F (2013). Recombinant vaccine-derived polioviruses in healthy children, Madagascar. *Emerging Infectious Diseases* **19**, 1008–1010.

Reich D, Green RE, Kircher M, Krause J, Patterson N, Durand EY, *et al.* (2010). Genetic history of an archaic hominin group from Denisova Cave in Siberia. *Nature* **468**, 1053–1060.

Reid N, Demboski JR, and Sullivan J (2012). Phylogeny estimation of the radiation of western North American chipmunks (*Tamias*) in the face of introgression using reproductive protein genes. *Systematic Biology* **61**, 44–62.

Reidenbach KR, Neafsey DE, Costantini C, Sagnon N'F, Simard F, Ragland GJ, *et al.* (2012). Patterns of genomic differentiation between ecologically differentiated M and S forms of *Anopheles gambiae* in West and Central Africa. *Genome Biology and Evolution* **4**, 1202–1212.

Renaut S, Nolte AW, Rogers SM, Derome N, and Bernatchez L (2011). SNP signatures of selection on standing genetic variation and their association with adaptive phenotypes along gradients of ecological speciation in lake whitefish species pairs (*Coregonus* spp.). *Molecular Ecology* **20**, 545–559.

Renaut S, Rowe HC, Ungerer MC, and Rieseberg LH (2014). Genomics of homoploid hybrid speciation: Diversity and transcriptional activity of LTR retrotransposons in hybrid sunflowers. *Philosophical Transactions of the Royal Society of London B* **369**, 20130345.

Reudink MW, Mech SG, and Curry RL (2006). Extrapair paternity and mate choice in a chickadee hybrid zone. *Behavioral Ecology* **17**, 56–62.

Reuter S, Connor TR, Barquist L, Walker D, Feltwell T, Harris SR, *et al.* (2014). Parallel independent evolution of pathogenicity within the genus *Yersinia*. *Proceedings of the National Academy of Sciences USA* **111**, 6768–6773.

Rezaei HR, Naderi S, Chintauan-Marquier IC, Taberlet P, Virk AT, Naghash HR, *et al.* (2010). Evolution and taxonomy of the wild species of the genus *Ovis* (Mammalia, Artiodactyla, Bovidae). *Molecular Phylogenetics and Evolution* **54**, 315–326.

Rheindt FE, Fujita MK, Wilton PR, and Edwards SV (2014). Introgression and phenotypic assimilation in *Zimmerius* flycatchers (Tyrannidae): Population genetic and phylogenetic inferences from genome-wide SNPs. *Systematic Biology* **63**, 134–152.

Rhymer JM and Simberloff D (1996). Extinction by hybridization and introgression. *Annual Review of Ecology and Systematics* **27**, 83–109.

Rice DW, Alverson AJ, Richardson AO, Young GJ, Sanchez-Puerta MV, Munzinger J, *et al.* (2013). Horizontal transfer of entire genomes via mitochondrial fusion in the Angiosperm *Amborella*. *Science* **342**, 1468–1473.

Richards TA, Soanes DM, Foster PG, Leonard G, Thornton CR, and Talbot NJ (2009). Phylogenomic analysis demonstrates a pattern of rare and ancient horizontal gene transfer between plants and fungi. *Plant Cell* **21**, 1897–1911.

Richter M and Rosselló-Móra R (2009). Shifting the genomic gold standard for the prokaryotic species definition. *Proceedings of the National Academy of Sciences USA* **106**, 19126–19131.

Rieseberg LH (1991). Homoploid reticulate evolution in *Helianthus* (Asteraceae): evidence from ribosomal genes. *American Journal of Botany* **78**, 1218–1237.

Rieseberg LH (1997). Hybrid origins of plant species. *Annual Review of Ecology and Systematics* **28**, 359–389.

Rieseberg LH, Carter R, and Zona S (1990). Molecular tests of the hypothesized hybrid origin of two diploid *Helianthus* species (Asteraceae). *Evolution* **44**, 1498–1511.

Rieseberg LH and Livingstone K (2003). Chromosomal speciation in primates. *Science* **300**, 267–268.

Rieseberg LH, Raymond O, Rosenthal DM, Lai Z, Lingstone K, Nakazato T, *et al.* (2003). Major ecological transitions in wild sunflowers facilitated by hybridization. *Science* **301**, 1211–1216.

Rieseberg LH, Sinervo B, Linder CR, Ungerer MC, and Arias DM (1996). Role of gene interactions in hybrid speciation: Evidence from ancient and experimental hybrids. *Science* **272**, 741–745.

Rieseberg LH, Soltis DE, and Palmer JD (1988). A molecular reexamination of introgression between *Helianthus annuus* and *H. bolanderi* (Compositae). *Evolution* **42**, 227–238.

Riginos C and Cunningham CW (2007). Hybridization in postglacial marine habitats. *Molecular Ecology* **16**, 3971–3972.

Riley HP (1938). A character analysis of colonies of *Iris fulva, Iris hexagona* var. *giganticaerulea* and natural hybrids. *American Journal of Botany* **25**, 727–738.

Riley HP (1939). Pollen fertility in Iris and its bearing on the hybrid origin of some of Small's "species." *Journal of Heredity* **30**, 481–483.

Riley HP (1942). Development of the embryo sac of *Iris fulva* and *I. hexagona* var *giganticaerulea*. *Transactions of the American Microscopical Society* **61**, 328–335.

Riley HP (1943a). Cell size in developing ovaries of *Iris hexagona* var *giganticaerulea*. *American Journal of Botany* **30**, 356–361.

Riley HP (1943b). Development and relative growth in ovaries of *Iris fulva* and *I. hexagona* var *giganticaerulea*. *American Journal of Botany* **29**, 323–331.

Robbins TR, Walker LE, Gorospe KD, Karl SA, Schrey AW, McCoy ED, and Mushinsky HR (2014). Rise and fall of a hybrid zone: Implications for the roles of aggression, mate choice, and secondary succession. *Journal of Heredity* **105**, 226–236.

Roberton SI, Bell DJ, Smith GJD, Nicholls JM, Chan KH, Nguyen DT, *et al.* (2006). Avian influenza H5N1 in viverrids: Implications for wildlife health and conservation. *Proceedings of the Royal Society of London B* **273**, 1729–1732.

Roberts DG, Gray CA, West RJ, and Ayre DJ (2010). Marine genetic swamping: hybrids replace an obligately estuarine fish. *Molecular Ecology* **19**, 508–520.

Roberts JA, Vo HD, Fujita MK, Moritz C, and Kearney M (2012). Physiological implications of genomic state in parthenogenetic lizards of reciprocal hybrid origin. *Journal of Evolutionary Biology* **25**, 252–263.

Robertson HM (1983). Mating behavior and the evolution of *Drosophila mauritiana*. *Evolution* **37**, 1283–1293.

Robichaux RH, Carr GD, Liebman M, and Pearcy RW (1990). Adaptive radiation of the Hawaiian silversword alliance (Compositae-Madiinae): Ecological, morphological, and physiological diversity. *Annals of the Missouri Botanical Garden* **77**, 64–72.

Robinson JD, Bunnefeld L, Hearn J, Stone GN, and Hickerson MJ (2014). ABC inference of multi-population divergence with admixture from unphased population genomic data. *Molecular Ecology* **23**, 4458–4471.

Rocap G, Larimer FW, Lamerdin J, Malfatti S, Chain P, Ahlgren NA, *et al.* (2003). Genome divergence in two *Prochlorococcus* ecotypes reflects oceanic niche differentiation. *Nature* **424**, 1042–1047.

Rogers MB, Downing T, Smith BA, Imamura H, Sanders M, Svobodova M, *et al.* (2014). Genomic confirmation of hybridisation and recent inbreeding in a vector-isolated *Leishmania* population. *PLoS Genetics* **10**, e1004092.

Rogers J and Gibbs RA (2014). Comparative primate genomics: Emerging patterns of genome content and dynamics. *Nature Reviews Genetics* **15**, 347–359.

Rohwer F, Segall A, Steward G, Seguritan V, Breitbart M, Wolven F, and Azam F (2000). The complete genomic sequence of the marine phage Roseophage SIO1 shares homology with nonmarine phages. *Limnology and Oceanography* **45**, 408–418.

Rokyta DR and Wichman HA (2009). Genic incompatibilities in two hybrid bacteriophages. *Molecular Biology and Evolution* **26**, 2831–2839.

Roos C, Zinner D, Kubatko LS, Schwarz C, Yang M, Meyer D, *et al.* (2011). Nuclear versus mitochondrial DNA: Evidence for hybridization in colobine monkeys. *BMC Evolutionary Biology* **11**, 77.

Rose EG, Brand CL, and Wilkinson GS (2014). Rapid evolution of asymmetric reproductive incompatibilities in stalk-eyed flies. *Evolution* **68**, 384–396.

Rosenthal DM, Rieseberg LH, and Donovan LA (2005). Re-creating ancient hybrid species' complex phenotypes from early-generation synthetic hybrids: Three examples using wild sunflowers. *American Naturalist* **166**, 26–41.

Rosselló-Móra R and Amann R (2001). The species concept for prokaryotes. *FEMS Microbiology Reviews* **25**, 39–67.

Rosser N, Dasmahapatra KK, and Mallet J (2014). Stable *Heliconius* butterfly hybrid zones are correlated with a local rainfall peak at the edge of the Amazon basin. *Evolution* **68**, 3470–3484.

Ross-Ibarra J, Tenaillon M, and Gaut BS (2009). Historical divergence and gene flow in the genus *Zea*. *Genetics* **181**, 1399–1413.

Roullier C, Kambouo R, Paofa J, McKey D, and Lebot V (2013). On the origin of sweet potato (Ipomoea batatas (L.) Lam.) genetic diversity in New Guinea, a secondary centre of diversity. *Heredity* **110**, 594–604.

Roux C and Pannell JR (2015). Inferring the mode of origin of polyploid species from next-generation sequence data. *Molecular Ecology* **24**, 1047–1059.

Rowbotham TJ (1980). Preliminary report on the pathogenicity of *Legionella pneumophila* for freshwater and soil amoebae. *Journal of Clinical Pathology* **33**, 1179–1183.

Roy MS, Geffen E, Smith D, Ostrander EA, and Wayne RK (1994). Patterns of differentiation and hybridization in North American wolflike canids, revealed by analysis of microsatellite loci. *Molecular Biology and Evolution* **11**, 553–570.

Ruegg K, Anderson EC, Boone J, Pouls J, and Smith TB (2014). A role for migration-linked genes and genomic islands in divergence of a songbird. *Molecular Ecology* **23**, 4757–4769.

Ruhsam M, Hollingsworth PM, and Ennos RA (2013). Patterns of mating, generation of diversity, and fitness of offspring in a *Geum* hybrid swarm. *Evolution* **67**, 2728–2740.

Ruibal R (1955). A study of altitudinal races in *Rana pipiens*. *Evolution* **9**, 322–338.

Ruiz-García M, Pinedo-Castro M, and Shostell JM (2014). How many genera and species of woolly monkeys (Atelidae, Platyrrhine, Primates) are there? The first molecular analysis of *Lagothrix flavicauda*, an endemic Peruvian primate species. *Molecular Phylogenetics and Evolution* **79**, 179–198.

Ryan ME, Johnson JR, and Fitzpatrick BM (2009). Invasive hybrid tiger salamander genotypes impact native amphibians. *Proceedings of the National Academy of Sciences USA* **111**, 15786–15791.

Sabehi G, Shaulov L, Silver DH, Yanai I, Harel A, and Lindell D (2012). A novel lineage of myoviruses infecting cyanobacteria is widespread in the oceans. *Proceedings of the National Academy of Sciences USA* **106**, 11166–11171.

Sacks BN, Moore M, Statham MJ, and Wittmer HU (2011). A restricted hybrid zone between native and introduced red fox (*Vulpes vulpes*) populations suggests reproductive barriers and competitive exclusion. *Molecular Ecology* **20**, 326–341.

Sage RD and Selander RK (1979). Hybridization between species of the *Rana pipiens* complex in central Texas. *Evolution* **33**, 1069–1088.

Sakowski EG, Munsell EV, Hyatt M, Kress W, Williamson SJ, Nasko DJ, et al. (2014). Ribonucleotide reductases reveal novel viral diversity and predict biological and ecological features of unknown marine viruses. *Proceedings of the National Academy of Sciences USA* **111**, 15786–15791.

Salazar C, Baxter SW, Pardo-Diaz C, Wu G, Surridge A, Linares M, et al. (2010). Genetic evidence for hybrid trait speciation in *Heliconius* butterflies. *PLoS Genetics* **6**, e1000930.

Salazar C, Jiggins CD, Arias CF, Tobler A, Bermingham E, and Linares M (2005). Hybrid incompatibility is consistent with a hybrid origin of *Heliconius heurippa* Hewitson from its close relatives, *Heliconius cydno* Doubleday and *Heliconius melpomene* Linnaeus. *Journal of Evolutionary Biology* **18**, 247–256.

Salem A-H, Ray DA, Xing J, Callinan PA, Myers JS, Hedges DJ, et al. (2003). Alu elements and hominid phylogenetics. *Proceedings of the National Academy of Sciences USA* **100**, 12787–12791.

Salzburger W, Baric S, and Sturmbauer C (2002). Speciation via introgressive hybridization in East African cichlids? *Molecular Ecology* **11**, 619–625.

Salzburger W, Niederstätter H, Brandstätter A, Berger B, Parson W, Snoeks J, and Sturmbauer C (2006). Colour-assortative mating among populations of *Tropheus moorii*, a cichlid fish from Lake Tanganyika, East Africa. *Proceedings of the Royal Society of London B* **273**, 257–266.

Salzburger W, Van Bocxlaer B, and Cohen AS (2014). Ecology and evolution of the African Great Lakes and their faunas. *Annual Review of Ecology, Evolution, and Systematics* **45**, 519–545.

Sambatti JBM, Strasburg JL, Ortiz-Barrientos D, Baack EJ, and Rieseberg LH (2012). Reconciling extremely strong barriers with high levels of gene exchange in annual sunflowers. *Evolution* **66**, 1459–1473.

Sampson JF and Byrne M (2012). Genetic diversity and multiple origins of polyploid *Atriplex nummularia* Lindl. (Chenopodiaceae). *Biological Journal of the Linnean Society* **105**, 218–230.

Sandkam BA, Joy JB, Watson CT, Gonzalez-Bendiksen P, Gabor CR, and Breden F (2013). Hybridization leads to sensory repertoire expansion in a gynogenetic fish, the Amazon molly (*Poecilia formosa*): A test of the hybrid-sensory expansion hypothesis. *Evolution* **67**, 120–130.

Sangster G (2014). The application of species criteria in avian taxonomy and its implications for the debate over species concepts. *Biological Reviews* **89**, 199–214.

Sankararaman S, Mallick S, Dannemann M, Prüfer K, Kelso J, Pääbo S, et al. (2014). The genomic landscape of Neanderthal ancestry in present-day humans. *Nature* **507**, 354–357.

Sankararaman S, Patterson N, Li H, Pääbo S, and Reich D (2012). The Date of interbreeding between Neandertals and modern humans. *PLoS Genetics* **8**, e1002947.

Santos ME, Braasch I, Boileau N, Meyer BS, Sauteur L, Böhne A, et al. (2014). The evolution of cichlid fish eggspots is linked with a cis-regulatory change. *Nature Communications* **5**, 5149.

Santucci F, Nascetti G, and Bullini L (1996). Hybrid zones between two genetically differentiated forms of the pond frog *Rana lessonae* in southern Italy. *Journal of Evolutionary Biology* **9**, 429–450.

Sarich VM and Wilson AC (1967). Immunological time scale for hominid evolution. *Science* **158**, 1200–1203.

Savolainen V, Anstett M-C, Lexer C, Hutton I, Clarkson JJ, Norup MV, et al. (2006). Sympatric speciation in palms on an oceanic island. *Nature* **441**, 210–213.

Savolainen-Kopra C and Blomqvist S (2010). Mechanisms of genetic variation in polioviruses. *Reviews in Medical Virology* **20**, 358–371.

Scally A, Dutheil JY, Hillier LW, Jordan GE, Goodhead I, Herrero J, et al. (2012). Insights into hominid evolution from the gorilla genome sequence. *Nature* **483**, 169–175.

Scandura M, Iacolina L, Cossu A, and Apollonio M (2011). Effects of human perturbation on the genetic make-up of an island population: the case of the Sardinian wild boar. *Heredity* **106**, 1012–1020.

Scandura M, Iacolina L, Crestanello B, Pecchioli E, Di Benedetto MF, Russo V, et al. (2008). Ancient vs. recent processes as factors shaping the genetic variation of the European wild boar: Are the effects of the last glaciation still detectable? *Molecular Ecology* **17**, 1745–1762.

Schaeffer SW and Anderson WW (2005). Mechanisms of genetic exchange within the chromosomal inversions of *Drosophila pseudoobscura*. *Genetics* **171**, 1729–1739.

Schaeffer SW, Goetting-Minesky MP, Kovacevic M, Peoples JR, Graybill JL, Miller JM, et al. (2003). Evolutionary genomics of inversions in *Drosophila pseudoobscura*: Evidence for epistasis. *Proceedings of the National Academy of Sciences USA* **100**, 8319–8324.

Schaeffer SW and Miller EL (1993). Estimates of linkage disequilibrium and the recombination parameter determined from segregating nucleotide sites in the alcohol dehydrogenase region of *Drosophila pseudoobscura*. *Genetics* **135**, 541–552.

Schemske DW (2000). Understanding the origin of species. *Evolution* 54, 1069–1073.

Schemske DW and Bradshaw HD Jr (1999). Pollinator preference and the evolution of floral traits in monkey-flowers (*Mimulus*). *Proceedings of the National Academy of Sciences USA* **96**, 11910–11915.

Schlebusch CM, Skoglund P, Sjödin P, Gattepaille LM, Hernandez D, Jay F, et al. (2012). Genomic variation in seven Khoe-San groups reveals adaptation and complex African history. *Science* **338**, 374–379.

Schluter D (2000). *The Ecology of Adaptive Radiation*. Oxford University Press, Oxford.

Schmeller DS, Seitz A, Crivelli A, and Veith M (2005). Crossing species' range borders: interspecies gene exchange mediated by hybridogenesis. *Proceedings of the Royal Society of London Series B* **272**, 1625–1631.

Schubert M, Jonsson H, Chang D, Der Sarkissian C, Ermini L, Ginolhac A, et al. (2014). Prehistoric genomes reveal the genetic foundation and cost of horse domestication. *Proceedings of the National Academy of Sciences USA* **111**, E5661–E5669.

Schwartz JH and Tattersall I (2010). Fossil evidence for the origin of *Homo sapiens*. *Yearbook of Physical Anthropology* **53**, 94–121.

Schwarzer J, Misof B, and Schliewen UK (2012a). Speciation within genomic networks: a case study based on *Steatocranus* cichlids of the lower Congo rapids. *Journal of Evolutionary Biology* **25**, 138–148.

Schwarzer J, Swartz ER, Vreven E, Snoeks J, Cotterill FPD, Misof B, and Schliewen UK (2012b). Repeated trans-watershed hybridization among haplochromine cichlids (Cichlidae) was triggered by Neogene landscape evolution. *Proceedings of the Royal Society of London B* **279**, 4389–4398.

Scopece G, Croce A, Lexer C, and Cozzolino S (2013). Components of reproductive isolation between *Orchis mascula* and *Orchis pauciflora*. *Evolution* **67**, 2083–2093.

Seehausen O (2004). Hybridization and adaptive radiation. *Trends in Ecology and Evolution* **19**, 198–207.

Seehausen O (2006). African cichlid fish: A model system in adaptive radiation research. *Proceedings of the Royal Society of London B* **273**, 1987–1998.

Seehausen O, Butlin RK, Keller I, Wagner CE, Boughman JW, Hohenlohe PA, et al. (2014). Genomics and the origin of species. *Nature Reviews Genetics* **15**, 176–192.

Seehausen O, Koetsier E, Schneider MV, Chapman LJ, Chapman CA, Knight ME, et al. (2003). Nuclear markers reveal unexpected genetic variation and a Congolese–Nilotic origin of the Lake Victoria cichlid species flock. *Proceedings of the Royal Society of London B* **270**, 129–137.

Seehausen O, Takimoto G, Roy D, and Jokela J (2006). Speciation reversal and biodiversity dynamics with hybridization in changing environments. *Molecular Ecology* **17**, 30–44.

Seehausen O, Terai Y, Magalhaes IS, Carleton KL, Mrosso HDJ, Miyagi R, et al. (2008). Speciation through sensory drive in cichlid fish. *Nature* **455**, 620–626.

Seehausen O, van Alphen JJM, and Witte F (1997). Cichlid fish diversity threatened by eutrophication that curbs sexual selection. *Science* **277**, 1808–1811.

Seehausen O and Wagner CE (2014). Speciation if freshwater fishes. *Annual Review of Ecology, Evolution, and Systematics* **45**, 621–651.

Segal G, Purcell M, and Shuman HA (1998). Host cell killing and bacterial conjugation require overlapping sets of genes within a 22-kb region of the *Legionella pneumophila* genome. *Proceedings of the National Academy of Sciences USA* **95**, 1669–1674.

Segal G, Russo JJ and Shuman HA (1999). Relationships between a new type IV secretion system and the *icm/dot* virulence system of *Legionella pneumophila*. *Molecular Microbiology* **34**, 799–809.

Seguin-Orlando A, Korneliussen TS, Sikora M, Malaspinas A-S, Manica A, Moltke I, et al. (2014). Genomic structure in Europeans dating back at least 36,200 years. *Science* **346**, 1113–1118.

Selz OM, Lucek K, Young KA, and Seehausen O (2014a). Relaxed trait covariance in interspecific cichlid hybrids predicts morphological diversity in adaptive radiations. *Journal of Evolutionary Biology* **27**, 11–24.

Selz OM, Thommen R, Maan ME, and Seehausen O (2014b). Behavioural isolation may facilitate homoploid hybrid speciation in cichlid fish. *Journal of Evolutionary Biology* **27**, 275–289.

Semlitsch RD (1993a). Adaptive genetic variation in growth and development of tadpoles of the hybridogenetic *Rana esculenta* complex. *Evolution* **47**, 1805–1818.

Semlitsch RD (1993b). Asymmetric competition in mixed populations of tadpoles of the hybridogenetic *Rana esculenta* complex. *Evolution* **47**, 510–519.

Semlitsch RD, Hotz H, and Guex G-D (1997). Competition among tadpoles of coexisting hemiclones of hybridogenetic *Rana esculenta*: Support for the frozen niche variation model. *Evolution* **51**, 1249–1261.

Semlitsch RD, Pickle J, Parris MJ, and Sage RD (1999). Jumping performance and short-term repeatability of newly metamorphosed hybrid and parental leopard frogs (*Rana sphenocephala* and *Rana blairi*). *Canadian Journal of Zoology* **77**, 748–754.

Semlitsch RD and Reyer H-U (1992). Performance of tadpoles from the hybridogenetic *Rana esculenta* complex: Interactions with pond drying and interspecific competition. *Evolution* **46**, 665–676.

Semlitsch RD, Schmiedehausen S, Hotz H, and Beerli P (1996). Genetic compatibility between sexual and clonal genomes in local populations of the hybridogenetic *Rana esculenta*. *Evolutionary Ecology* **10**, 531–543.

Sémon M and Wolfe KH (2008). Preferential subfunctionalization of slow-evolving genes after allopolyploidization in *Xenopus laevis*. *Proceedings of the National Academy of Sciences USA* **105**, 8333–8338.

Serre D, Langaney A, Chech M, Teschler-Nicola M, Paunovic M, Mennecier P, *et al.* (2004). No evidence of Neandertal mtDNA contribution to early modern humans. *PLoS Biology* **2**, 0313–0317.

Servedio MR and Noor MAF (2003). The role of reinforcement in speciation: Theory and data. *Annual Review of Ecology, Evolution and Systematics* **34**, 339–364.

Shaw DD (1976). Population cytogenetics of the genus *Caledia* (Orthoptera: Acridinae) I. Inter- and intraspecific karyotype diversity. *Chromosoma* **54**, 221–243.

Shaw DD and Coates DJ (1983). Chromosomal variation and the concept of the coadapted genome—A direct cytological assessment. In PE Brandham and MD Bennett, eds., *Kew Chromosome Conference II*, pp. 207–216. George Allen and Unwin, London.

Shaw DD, Coates DJ, and Arnold ML (1988). Complex patterns of chromosomal variation along a latitudinal cline in the grasshopper *Caledia captiva*. *Genome* **30**, 108–117.

Shaw DD, Coates DJ, Arnold ML, and Wilkinson P (1985). Temporal variation in the chromosomal structure of a hybrid zone and its relationship to karyotypic repatterning. *Heredity* **55**, 293–306.

Shaw DD, Marchant AD, Arnold ML, Contreras N, and Kohlmann B (1990). The control of gene flow across a narrow hybrid zone: A selective role for chromosomal rearrangement? *Canadian Journal of Zoology* **68**, 1761–1769.

Shaw DD, Marchant AD, Contreras N, Arnold ML, Groeters F, and Kohlmann BC (1993). Genomic and environmental determinants of a narrow hybrid zone: cause or coincidence? In RG Harrison, ed., *Hybrid Zones and the Evolutionary Process*, pp. 165–195. Oxford University Press, Oxford.

Shaw KL and Mendelson TC (2013). The targets of selection during reinforcement. *Journal of Evolutionary Biology* **26**, 286–287.

Shaw DD, Moran C, and Wilkinson P (1980). Chromosomal reorganization, geographic differentiation and the mechanism of speciation in the genus *Caledia*. In RL Blackman, GM Hewitt, and M Ashburner, eds., *Insect Cytogenetics*, pp. 171–194, Blackwell Scientific Publications, Oxford.

Shaw DD, Webb GC, and Wilkinson P (1976). Population cytogenetics of the genus *Caledia* (Orthoptera: Acridinae) II. Variation in the pattern of C-banding. *Chromosoma* **56**, 169–190.

Shaw DD and Wilkinson P (1978). "Homologies" between non-homologous chromosomes in the grasshopper *Caledia captiva*. *Chromosoma* **68**, 241–259.

Shaw DD, Wilkinson P, and Coates DJ (1982). The chromosomal component of reproductive isolation in the grasshopper *Caledia captiva* II. The relative viabilities of recombinant and non-recombinant chromosomes during embryogenesis. *Chromosoma* **86**, 533–549.

Shaw DD, Wilkinson P, and Coates DJ (1983). Increased chromosomal mutation rate after hybridization between two subspecies of grasshoppers. *Science* **220**, 1165–1167.

Shaw DD, Wilkinson P, and Moran C (1979). A comparison of chromosomal and allozymal variation across a narrow hybrid zone in the grasshopper *Caledia captiva*. *Chromosoma* **75**, 333–351.

Shen H, He H, Li J, Chen W, Wang X, *et al.* (2012). Genome-wide analysis of DNA methylation and gene expression changes in two *Arabidopsis* ecotypes and their reciprocal hybrids. *Plant Cell* **24**, 875–892.

Sherman-Broyles S, Bombarely A, Powell AF, Doyle JL, Egan AN, Coate JE, and Doyle JJ (2014). The wild side of a major crop: Soybean's perennial cousins from Down Under. *American Journal of Botany* **101**, 1651–1665.

Shoemaker DD, Ross KG, and Arnold ML (1996). Genetic structure and evolution of a fire ant hybrid zone. *Evolution* **50**, 1958–1976.

Shotake T (1981). Population genetical study of natural hybridization between *Papio anubis* and *P. hamadryas*. *Primates* **22**, 285–308.

Shotake T, Nozawa K, and Tanabe Y (1977). Blood protein variations in baboons I. Gene exchange and genetic distance between *Papio anubis, Papio hamadryas* and their hybrid. *Japanese Journal of Genetics* **52**, 223–237.

Shriner D, Rodrigo AG, Nickle DC, and Mullins JI (2004). Pervasive genomic recombination of HIV-1 in vivo. *Genetics* **167**, 1573–1583.

Sigel EM, Windham MD, and Pryer KM (2014). Evidence for reciprocal origins in *Polypodium hesperium* (Polypodiaceae): A fern model system for investigating how multiple origins shape allopolyploid genomes. *American journal of Botany* **101**, 1476–1485.

Silander OK, Weinreich DM, Wright KM, O'Keefe KJ, Rang CU, Turner PE, and Chao L (2005). Widespread genetic exchange among terrestrial bacteriophages. *Proceedings of the National Academy of Sciences USA* **102**, 19009–19014.

Sims GE and Kim S-H (2011). Whole-genome phylogeny of *Escherichia coli/Shigella* group by feature frequency profiles (FFPs). *Proceedings of the National Academy of Sciences USA* **108**, 8329–8334.

Sites JW, Jr and Marshall JC (2004). Operational criteria for delimiting species. *Annual Review of Ecology, Evolution and Systematics* **35**, 199–227.

Sithaldeen R, Bishop JM, and Ackermann RR (2009). Mitochondrial DNA analysis reveals Plio-Pleistocene diversification within the chacma baboon. *Molecular Phylogenetics and Evolution* **53**, 1042–1048.

Skoglund P and Jakobsson M (2011). Archaic human ancestry in East Asia. *Proceedings of the National Academy of Sciences USA* **108**, 18301–18306.

Slatkin M (1973). Gene flow and selection in a cline. *Genetics* **75**, 733–756.

Slatkin M (1985). Gene flow in natural populations. *Annual Review of Ecology and Systematics* **16**, 393–430.

Sloan DB, Nakabachi A, Richards S, Qu J, Murali SC, Gibbs RA, and Moran NA (2014). Parallel histories of horizontal gene transfer facilitated extreme reduction of endosymbiont genomes in sap-feeding insects. *Molecular Biology and Evolution* **31**, 857–871.

Sloop CM, Ayres DR, and Strong DR (2011). Spatial and temporal genetic structure in a hybrid cordgrass invasion. *Heredity* **106**, 547–556.

Slotman M, della Torre A, and Powell JR (2004). The genetics of inviability and male sterility in hybrids between *Anopheles gambiae* and *An. Arabiensis. Genetics* **167**, 275–287.

Slotman M, della Torre A, and Powell JR (2005). Female sterility in hybrids between *Anopheles gambiae* and *A. arabiensis*, and the causes of Haldane's rule. *Evolution* **59**, 1016–1026.

Smadja CM and Butlin RK (2011). A framework for comparing processes of speciation in the presence of gene flow. *Molecular Ecology* **20**, 5123–5140.

Small JK and Alexander EJ (1931). Botanical interpretation of the Iridaceous plants of the Gulf States. *New York Botanical Garden Contribution* **327**, 325–357.

Smith EB (1968). Pollen competition and relatedness in *Haplopappus* section *Isopappus. Botanical Gazette* **129**, 371–373.

Smith EB (1970). Pollen competition and relatedness in *Haplopappus* section *Isopappus* (Compositae). II. *American Journal of Botany* **57**, 874–880.

Smith TB, Kinnison MT, Strauss SY, Fuller TL, and Carroll SP (2014). Prescriptive evolution to conserve and manage biodiversity. *Annual Review of Ecology, Evolution, and Systematics* **45**, 1–22.

Smith GJD, Vijaykrishna D, Bahl J, Lycett SJ, Worobey M, Pybus OG, et al. (2009). Origins and evolutionary genomics of the 2009 swine-origin H1N1 influenza A epidemic. *Nature* **459**, 1122–1125.

Snow AA, Pilson D, Rieseberg LH, Paulsen MJ, Pleskac N, Reagon MR, et al. (2003). A Bt transgene reduces herbivory and enhances fecundity in wild sunflowers. *Ecological Applications* **13**, 279–286.

Sobel JM, Chen GF, Watt LR, and Schemske DW (2010). The biology of speciation. *Evolution* **64**, 295–315.

Solignac M and Monnerot M (1986). Race formation, speciation, and introgression within *Drosophila simulans, D. mauritiana*, and *D. sechellia* inferred from mitochondrial DNA analysis. *Evolution* **40**, 531–539.

Soltis DE, Albert VA, Leebens-Mack J, Bell CD, Paterson AH, Zheng C, et al. (2009). Polyploidy and angiosperm diversification. *American Journal of Botany* **96**, 336–348.

Soltis PS and Soltis DE (2009). The role of hybridization in plant speciation. *Annual Review of Plant Biology* **60**, 561–588.

Som C and Reyer H-U (2007). Hemiclonal reproduction slows down the speed of Muller's ratchet in the hybridogenetic frog *Rana esculenta. Journal of Evolutionary Biology* **20**, 650–660.

Song Y, Endepols S, Klemann N, Richter D, Matuschka F-R, Shih C-H, et al. (2011). Adaptive introgression of anticoagulant rodent poison resistance by hybridization between Old World mice. *Current Biology* **21**, 1296–1301.

Song B-H, Wang X-Q, Wang X-R, Ding K-Y, and Hong D-Y (2003). Cytoplasmic composition in *Pinus densata* and population establishment of the diploid hybrid pine. *Molecular Ecology* **12**, 2995–3001.

Song B-H, Wang X-Q, Wang X-R, Sun L-J, Hong D-Y, and Peng P-H (2002). Maternal lineages of *Pinus densata*, a diploid hybrid. *Molecular Ecology* **11**, 1057–1063.

Soria-Carrasco V, Gompert Z, Comeault AA, Farkas TE, Parchman TL, Johnston JS, et al. (2014). Stick insect genomes reveal natural selection's role in parallel speciation. *Science* **344**, 738–742.

Sotero-Caio C, Volleth M, Gollahon LS, Fu B, Cheng W, Ng BL, Yang F, and Baker RJ (2013). Chromosomal evolution among leaf-nosed nectarivorous bats—evidence from cross-species chromosome painting (Phyllostomidae, Chiroptera). *BMC Evolutionary Biology* **13**, 276.

Soto-Calderón ID, Clark NJ, Wildschutte JVH, DiMattio K, Jensen-Seaman MI, and Anthony NM (2014). Identification of species-specific nuclear insertions of mitochondrial DNA (numts) in gorillas and their potential as population genetic markers. *Molecular Phylogenetics and Evolution* **81**, 61–70.

Sousa V and Hey J (2013). Understanding the origin of species with genome-scale data: modelling gene flow. *Nature Reviews Genetics* **14**, 404–414.

Sousa-Santos C, Gante HF, Robalo J, Proença Cunha P, Martins A, Arruda M, *et al.* (2014). Evolutionary history and population genetics of a cyprinid fish (*Iberochondrostoma olisiponensis*) endangered by introgression from a more abundant relative. *Conservation Genetics* **15**, 665–677.

Spellman GM, Riddle B, and Klicka J (2007). Phylogeography of the mountain chickadee (*Poecile gambeli*): diversification, introgression, and expansion in response to Quaternary climate change. *Molecular Ecology* **16**, 1055–1068.

Spring J (1997). Vertebrate evolution by interspecific hybridisation—are we polyploid? *FEBS Letters* **400**, 2–8.

Stackebrandt E, Frederiksen W, Garrity GM, Grimont PAD, Kämpfer P, Maiden MCJ, *et al.* (2002). Report of the ad hoc committee for the re-evaluation of the species definition in bacteriology. *International Journal of Systematic and Evolutionary Microbiology* **52**, 1043–1047.

Stairs CW, Eme L, Brown MW, Mutsaers C, Susko E, Dellaire G, *et al.* (2014). A SUF Fe-S cluster biogenesis system in the mitochondrion-related organelles of the anaerobic protist *Pygsuia*. *Current Biology* **24**, 1176–1186.

Stathos A and Fishman L (2014). Chromosomal rearrangements directly cause underdominant F_1 pollen sterility in *Mimulus lewisii-Mimulus cardinalis* hybrids. *Evolution* **68**, 3109–3119.

Staubach F, Lorenc A, Messer PW, Tang K, Petrov DA, and Tautz D (2012). Genome patterns of selection and introgression of haplotypes in natural populations of the house mouse (*Mus musculus*). *PLoS Genetics* **8**, e1002891.

Steane DA, Nicolle D, Sansaloni CP, Petroli CD, Carling J, Kilian A, *et al.* (2011). Population genetic analysis and phylogeny reconstruction in *Eucalyptus* (Myrtaceae) using high-throughput, genome-wide genotyping. *Molecular Phylogenetics and Evolution* **59**, 206–224.

Stebbins GL Jr (1947). Types of polyploids: Their classification and significance *Advances in Genetics* **1**, 403–429.

Stebbins GL Jr (1950). *Variation and Evolution in Plants.* Columbia University Press, New York.

Stebbins GL Jr (1956). Cytogenetics and evolution of the grass family. *American Journal of Botany* **43**, 890–905.

Stebbins GL Jr (1959). The role of hybridization in evolution. *Proceedings of the American Philosophical Society* **103**, 231–251.

Stebbins GL Jr (1999). A brief summary of my ideas on evolution. *American Journal of Botany* **86**, 1207–1208.

Stebbins GL Jr and Daly K (1961). Changes in the variation pattern of a hybrid population of *Helianthus* over an eight-year period. *Evolution* **15**, 60–71.

Stebbins GL Jr, Matzke EB, and Epling C (1947). Hybridization in a population of *Quercus marilandica* and *Quercus ilicifolia*. *Evolution* **1**, 79–88.

Stegemann S, Keuthe M, Greiner S, and Bock R (2012). Horizontal transfer of chloroplast genomes between plant species. *Proceedings of the National Academy of Sciences USA* **109**, 2434–2438.

Stelkens RB, Brockhurst MA, Hurst GDD, Miller EL, and Grieg D (2014). The effect of hybrid transgression on environmental tolerance in experimental yeast crosses. *Journal of Evolutionary Biology* **27**, 2507–2519.

Stelkens RB and Seehausen O (2009). Phenotypic divergence but not genetic distance predicts assortative mating among species of a cichlid fish radiation. *Journal of Evolutionary Biology* **22**, 1679–1694.

Stelkens RB, Young KA, and Seehausen O (2009). The accumulation of reproductive incompatibilities in African cichlid fish. *Evolution* **64**, 617–633.

Stewart JE, Timmer LW, Lawrence CB, Pryor BM, and Peever TL (2014). Discord between morphological and phylogenetic species boundaries: Incomplete lineage sorting and recombination results in fuzzy species boundaries in an asexual fungal pathogen. *BMC Evolutionary Biology* **14**, 38.

Stöck M, Lampert KP, Möller D, Schlupp I, and Schartl M (2010). Monophyletic origin of multiple clonal lineages in an asexual fish (*Poecilia formosa*). *Molecular Ecology* **19**, 5204–5215.

Stöck M, Ustinova J, Lamatsch DK, Schartl M, Perrin N, and Moritz C (2009). A vertebrate reproductive system involving three ploidy levels: Hybrid origin of triploids in a contact zone of diploid and tetraploids palearctic green toads (*Bufo viridis* subgroup). *Evolution* **64**, 944–959.

Stölting KN, Nipper R, Lindtke D, Caseys C, Waeber S, Castiglione S, and Lexer C (2013). Genomic scan for single nucleotide polymorphisms reveals patterns of divergence and gene flow between ecologically divergent species. *Molecular Ecology* **22**, 842–855.

Stone AC, Griffiths RC, Zegura SL, and Hammer MF (2002). High levels of Y-chromosome nucleotide diversity in the genus *Pan*. *Proceedings of the National Academy of Sciences USA* **99**, 43–48.

Strasburg JL and Kearney M (2005). Phylogeography of sexual *Heteronotia binoei* (Gekkonidae) in the Australian arid zone: Climatic cycling and repetitive hybridization. *Molecular Ecology* **14**, 2755–2772.

Strasburg JL, Kearney M, Moritz C, and Templeton AR (2007). Combining phylogeography with distribution modeling: Multiple Pleistocene range expansions in a parthenogenetic gecko from the Australian arid zone. *PLoS ONE* **2**, e760.

Strasburg JL and Rieseberg LH (2008). Molecular demographic history of the annual sunflowers *Helianthus annuus* and *H. petiolaris*—Large effective population sizes and rates of long-term gene flow. *Evolution* **62**, 1936–1950.

Streisfeld MA, Young WN, and Sobel JM (2013). Divergent selection drives genetic differentiation in an R2R3-MYB transcription factor that contributes to incipient speciation in *Mimulus aurantiacus*. *PLoS Genetics* **9**, e1003385.

Stringer C (2014). Why we are not all multiregionalists now. *Trends in Ecology and Evolution* **29**, 248–251.

Stringer CB and Andrews P (1988). Genetic and fossil evidence for the origin of modern humans. *Science* **239**, 1263–1268.

Strong DR and Ayres DR (2013). Ecological and evolutionary misadventures of *Spartina*. *Annual Review of Ecology, Evolution and Systematics* **44**, 389–410.

Stump AD, Fitzpatrick MC, Lobo NF, Traore S, Sagnon NF, Costantini C, *et al.* (2005a). Centromere-proximal differentiation and speciation in *Anopheles gambiae*. *Proceedings of the National Academy of Sciences USA* **102**, 15930–15935.

Stump AD, Shoener JA, Costantini C, Sagnon NF, and Besansky NJ (2005b). Sex-linked differentiation between incipient species of *Anopheles gambiae*. *Genetics* **169**, 1509–1519.

Sturgill-Koszycki S and Swanson MS (2000). *Legionella pneumophila* replication vacuoles mature into acidic, endocytic organelles. *Journal of Experimental Medicine* **192**, 1261–1272.

Sturmbauer C, Salzburger W, Duftner N, Schelly R, and Koblmüller S (2010). Evolutionary history of the Lake Tanganyika cichlid tribe Lamprologini (Teleostei: Perciformes) derived from mitochondrial and nuclear DNA data. *Molecular Phylogenetics and Evolution* **57**, 266–284.

Suárez NM, Pestano J, and Brown RP (2014). Ecological divergence combined with ancient allopatry in lizard populations from a small volcanic island. *Molecular Ecology* **23**, 4799–4812.

Sullivan J, Demboski JR, Bell KC, Hird S, Sarver B, Reid N, and Good JM (2014). Divergence with gene flow within the recent chipmunk radiation (*Tamias*). *Heredity* **113**, 185–194.

Sullivan MB, Waterbury JB, and Chisholm SW (2003). Cyanophages infecting the oceanic cyanobacterium *Prochlorococcus*. *Nature* **424**, 1047–1051.

Sumner FB (1929a). The analysis of a concrete case of intergradation between two subspecies. *Proceedings of the National Academy of Sciences USA* **15**, 110–120.

Sumner FB (1929b). The analysis of a concrete case of intergradation between two subspecies. II. Additional data and interpretations. *Proceedings of the National Academy of Sciences USA* **15**, 481–493.

Sun Y, Abbott RJ, Li L, Li L, Zou J, and Liu J (2014). Evolutionary history of Purple cone spruce (*Picea purpurea*) in the Qinghai-Tibet Plateau: Homoploid hybrid origin and Pleistocene expansion. *Molecular Ecology* **23**, 343–359.

Sundqvist A-K, Björnerfeldt S, Leonard JA, Hailer F, Hedhammar Ä, Ellegren H, and Vilà C (2006). Unequal contribution of sexes in the origin of dog breeds. *Genetics* **172**, 1121–1128.

Sweigart AL (2010). The genetics of postmating, prezygotic reproductive isolation between *Drosophila virilis* and *D. americana*. *Genetics* **184**, 401–410.

Sweigart AL and Flagel LE (2015). Evidence of natural selection acting on a polymorphic hybrid incompatibility locus in *Mimulus*. *Genetics* **199**, 543–554.

Sweigart AL, Martin NH, and Willis JH (2008). Patterns of nucleotide variation and reproductive isolation between a *Mimulus* allotetraploid and its progenitor species. *Molecular Ecology* **17**, 2089–2100.

Sweigart AL and Willis JH (2003). Patterns of nucleotide diversity in two species of *Mimulus* are affected by mating system and asymmetric introgression. *Evolution* **57**, 2490–2506.

Swenson NG and Howard DJ (2005). Clustering of contact zones, hybrid zones, and phylogeographic breaks in North America. *American Naturalist* **166**, 581–591.

Swisher CC III, Rink WJ, Antón SC, Schwarcz HP, Curtis GH, Suprijo A, and Widiasmoro (1996). Latest *Homo erectus* of Java: Potential contemporaneity with *Homo sapiens* in Southeast Asia. *Science* **274**, 1870–1874.

Symonds VV, Soltis PS, and Soltis DE (2010). Dynamics of polyploid formation in *Tragopogon* (Asteraceae): Recurrent formation, gene flow, and population structure. *Evolution* **64**, 1984–2003.

Syvanen M (2012). Evolutionary implications of horizontal gene transfer. *Annual Review of Genetics* **46**, 341–358.

Szymura JM and Barton NH (1986). Genetic analysis of a hybrid zone between the fire-bellied toads, *Bombina bombina* and *B. variegata*, near Cracow in southern Poland. *Evolution* **40**, 1141–1159.

Tagliaro CH, Schneider MPC, Schneider H, Sampaio IC, and Stanhope MJ (1997). Marmoset phylogenetics, conservation perspectives, and evolution of the mtDNA control region. *Molecular Biology and Evolution* **14**, 674–684.

Takahashi K, Terai Y, Nishida M, and Okada N (2001). Phylogenetic relationships and ancient incomplete lineage sorting among cichlid fishes in Lake Tanganyika as revealed by analysis of the insertion of retroposons. *Molecular Biology and Evolution* **18**, 2057–2066.

Takahata N, Lee S-H, and Satta Y (2001). Testing multiregionality of modern human origins. *Molecular Biology and Evolution* **18**, 172–183.

Tan CC (1946). Genetics of sexual isolation between *Drosophila pseudoobscura* and *Drosophila persimilis*. *Genetics* **31**, 558–573.

Tang S, Okashah RA, Knapp SJ, Arnold ML, and Martin NH (2010). Transmission ratio distortion results in asymmetric introgression in Louisiana Iris. *BMC Plant Biology* **10**, 48.

Tao Y and Hartl DL (2003). Genetic dissection of hybrid incompatibilities between *Drosophila simulans* and *D. mauritiana*. III. Heterogeneous accumulation of hybrid incompatibilities, degree of dominance, and implications for Haldane's Rule. *Evolution* **57**, 2580–2598.

Tattersall I and Schwartz JH (1999). Hominids and hybrids: The place of Neanderthals in human evolution. *Proceedings of the National Academy of Sciences USA* **96**, 7117–7119.

Taubenberger JK and Morens DM (2006). 1918 influenza: The mother of all pandemics. *Emerging Infectious Diseases* **12**, 15–22.

Taylor SJ, Arnold ML, and Martin NH (2009). The genetic architecture of reproductive isolation in Louisiana Irises: Hybrid fitness in nature. *Evolution* **63**, 2581–2594.

Taylor SJ, AuBuchon KJ, and Martin NH (2012). Identification of floral visitors of *Iris nelsonii*. *Southeastern Naturalist* **11**, 141–144.

Taylor SA, Curry RL, White TA, Ferretti V, and Lovette I (2014a). Spatiotemporally consistent genomic signatures of reproductive isolation in a moving hybrid zone. *Evolution* **68**, 3066–3081.

Taylor DJ, Finston TL, and Hebert PDN (1998). Biogeography of a widespread freshwater crustacean: Pseudocongruence and cryptic endemism in the North American *Daphnia laevis* complex. *Evolution* **52**, 1648–1670.

Taylor SJ, Rojas LD, Ho SW, and Martin NH (2013). Genomic collinearity and the genetic architecture of floral differences between the homoploid hybrid species *Iris nelsonii* and one of its progenitors, *Iris hexagona*. *Heredity* **110**, 63–70.

Taylor SA, White TA, Hochachka WM, Ferretti V, Curry RL, and Lovette I (2014b). Climate-mediated movement of an avian hybrid zone. *Current Biology* **24**, 671–676.

Taylor SJ, Willard RW, Shaw JP, Dobson MC, and Martin NH (2011). Differential response of the homoploid hybrid species *Iris nelsonii* (Iridaceae) and its progenitors to abiotic habitat conditions. *American Journal of Botany* **98**, 1309–1316.

Tazzyman SJ and Bonhoeffer S (2014). Plasmids and evolutionary rescue by drug resistance. *Evolution* **68**, 2066–2078.

Templeton AR (1981). Mechanisms of speciation—a population genetic approach. *Annual Review of Ecology and Systematics* **12**, 23–48.

Templeton AR (1989). The meaning of species and speciation: a genetic perspective. In D Otte and JA Endler, eds., *Speciation and its Consequences*, pp. 3–27. Sinauer Associates, Inc., Sunderland, Massachusetts.

Templeton AR (2001). Using phylogeographic analyses of gene trees to test species status and processes. *Molecular Ecology* **10**, 779–791.

Templeton AR (2002). Out of Africa again and again. *Nature* **416**, 45–51.

Templeton AR (2004). Using haplotype trees for phylogeographic and species inferences in fish populations. *Environmental Biology of Fishes* **69**, 7–20.

Templeton AR (2005). Haplotype trees and modern human origins. *Yearbook of Physical Anthropology* **48**, 33–59.

Templeton AR (2007). Genetics and recent human evolution. *Evolution* **61**, 1507–1519.

Templeton AR (2009). Why does a method that fails continue to be used? The answer. *Evolution* **63**, 807–812.

Templeton AR (2010a). Coalescent-based, maximum likelihood inference in phylogeography. *Molecular Ecology* **19**, 431–435.

Templeton AR (2010b). Coherent and incoherent inference in phylogeography and human evolution. *Proceedings of the National Academy of Sciences USA* **107**, 6376–6381.

Thalmann O, Fischer A, Lankester F, Pääbo S, and Vigilant L (2007). The complex evolutionary history of gorillas: Insights from genomic data. *Molecular Biology and Evolution* **24**, 146–158.

Thalmann O, Shapiro B, Cui P, Schuenemann VJ, Sawyer SK, Greenfield DL, *et al.* (2013). Complete mitochondrial genomes of ancient canids suggest a European origin of domestic dogs. *Science* **342**, 871–874.

Theis A, Ronco F, Indermaur A, Salzburger W, and Egger B (2014). Adaptive divergence between lake and stream populations of an East African cichlid fish. *Molecular Ecology* **23**, 5304–5322.

Thielsch A, Brede N, Petrusek A, De Meester L, and Schwenk K (2009). Contribution of cyclic parthenogenesis and colonization history to population structure in *Daphnia*. *Molecular Ecology* **18**, 1616–1628.

Thiry E, Meurens F, Muylkens B, McVoy M, Gogev S, Thiry J, *et al.* (2005). Recombination in alphaherpesviruses. *Reviews in Medical Virology* **15**, 89–103.

This P, Lacombe T, and Thomas MR (2006). Historical origins and genetic diversity of wine grapes. *Trends in Genetics* **22**, 511–519.

Thompson JD, Gaudeul M, and Debussche M (2010). Conservation value of sites of hybridization in peripheral populations of rare plant species. *Conservation Biology* **24**, 236–245.

Thompson SL, Lamothe M, Meirmans PG, Périnet P, and Isabel N (2010). Repeated unidirectional introgression towards *Populus balsamifera* in contact zones of exotic and native poplars. *Molecular Ecology* **19**, 132–145.

Thulin C-G, Jaarola M, and Tegelström H (1997). The occurrence of mountain hare mitochondrial DNA in wild brown hares. *Molecular Ecology* **6**, 463–467.

Tillett BJ, Meekan MG, Broderick D, Field IC, Cliff G, and Ovenden JR (2012). Pleistocene isolation, secondary introgression and restricted contemporary gene flow in the pig-eye shark, *Carcharhinus amboinensis* across northern Australia. *Conservation Genetics* **13**, 99–115.

Ting N, Tosi AJ, Li Y, Zhang Y-P, and Disotell TR (2008). Phylogenetic incongruence between nuclear and mitochondrial markers in the Asian colobines and the evolution of the langurs and leaf monkeys. *Molecular Phylogenetics and Evolution* **46**, 466–474.

Toews DPL and Brelsford A (2012). The biogeography of mitochondrial and nuclear discordance in animals. *Molecular Ecology* **21**, 3907–3930.

Touchon M, Hoede C, Tenaillon O, Barbe V, Baeriswyl S, Bidet P, *et al.* (2009). Organised genome dynamics in the *Escherichia coli* species results in highly diverse adaptive paths. *PLoS Genetics* **5**, e1000344.

Travis SE, Marburger JE, Windels S, and Kubátová B (2010). Hybridization dynamics of invasive cattail (Typhaceae) stands in the Western Great Lakes Region of North America: A molecular analysis. *Journal of Ecology* **98**, 7–16.

Trelease W (1917). Naming American hybrid oaks. *Proceedings of the American Philosophical Society* **56**, 44–52.

Trier CN, Hermansen JS, Sætre G-P, and Bailey RI (2014). Evidence for mito-nuclear and sex-linked reproductive barriers between the hybrid Italian sparrow and its parent species. *PLoS Genetics* **10**, e1004075.

Trigo TC, Freitas TRO, Kunzler G, Cardoso L, Silva JCR, Johnson WE, *et al.* (2008). Inter-species hybridization among Neotropical cats of the genus *Leopardus*, and evidence for an introgressive hybrid zone between *L. geoffroyi* and *L. tigrinus* in southern Brazil. *Molecular Ecology* **17**, 4317–4333.

Trigo TC, Schneider A, de Oliveira TG, Lehugeur LM, Silveira L, Freitas TRO, and Eizirik E (2013). Molecular data reveal complex hybridization and a cryptic species of Neotropical wild cat. *Current Biology* **23**, 2528–2533.

Trigo TC, Tirelli FP, de Freitas TRO, and Eizirik E (2014) Comparative assessment of genetic and morphological variation at an extensive hybrid zone between two wild cats in southern Brazil. *PLoS ONE* **9**, e108469.

Trinkaus E (2007). European early modern humans and the fate of the Neandertals. *Proceedings of the National Academy of Sciences USA* **104**, 7367–7372.

Tripet F, Dolo G, and Lanzaro GC (2005). Multilevel analyses of genetic differentiation in *Anopheles gambiae* s.s. reveal patterns of gene flow important for malaria-fighting mosquito projects. *Genetics* **169**, 313–324.

Triplett JK, Clark LG, Fisher AE, and Wen J (2014). Independent allopolyploidization events preceded speciation in the temperate and tropical woody bamboos. *New Phytologist* **204**, 66–73.

Trucco F, Tatum T, Rayburn AL, and Tranel PJ (2009). Out of the swamp: unidirectional hybridization with weedy species may explain the prevalence of *Amaranthus tuberculatus* as a weed. *New Phytologist* **184**, 819–827.

Tucker AE, Ackerman MS, Eads BD, Xu S, and Lynch M (2013). Population-genomic insights into the evolutionary origin and fate of obligately asexual *Daphnia pulex*. *Proceedings of the National Academy of Sciences USA* **110**, 15740–15745.

Tung J, Charpentier MJE, Garfield DA, Altmann J, and Alberts SC (2008). Genetic evidence reveals temporal change in hybridization patterns in a wild baboon population. *Molecular Ecology* **17**, 1998–2011.

Turelli M, Lipkowitz JR, and Brandvain Y (2014). On the Coyne and Orr-ign of species: Effects of intrinsic postzygotic isolation, ecological differentiation, X chromosome size, and sympatry on *Drosophila* speciation. *Evolution* **68**, 1176–1187.

Turgeon J, Tayeh A, Facon B, Lombaert E, de Clercq P, Berkvens N, *et al.* (2011). Experimental evidence for the phenotypic impact of admixture between wild and biocontrol Asian ladybird (*Harmonia axyridis*) involved in the European invasion. *Journal of Evolutionary Biology* **24**, 1044–1052.

Turner BJ, Brett B-LH, and Miller RR (1980). Interspecific hybridization and the evolutionary origin of a gynogenetic fish, *Poecilia formosa*. *Evolution* **34**, 917–922.

Turner TL, Hahn MW, and Nuzhdin SV (2005). Genomic islands of speciation in *Anopheles gambiae*. *PLoS Biology* **3**, e285.

Turner LM and Harr B (2014). Genome-wide mapping in a house mouse hybrid zone reveals hybrid sterility loci and Dobzhansky-Muller interactions. *eLife* **3**, e02504.

Turner LM, Schwahn DJ, and Harr B (2012). Reduced male fertility is common but highly variable in form and severity in a natural house mouse hybrid zone. *Evolution* **66**, 443–458.

Turner GF, Seehausen O, Knight ME, Allender CJ, and Robinson RL (2001). How many species of cichlid fishes are there in African lakes? *Molecular Ecology* **10**, 793–806.

Turner LM, White MA, Tautz D, and Payseur BA (2014). Genomic networks of hybrid sterility. *PLoS Genetics* **10**, e1004162.

Unger GM, Vendramin GG, and Robledo-Arnuncio JJ (2014). Estimating exotic gene flow into native pine stands: Zygotic vs. gametic components. *Molecular Ecology* **23**, 5435–5447.

Ungerer MC, Baird SJE, Pan J, and Rieseberg LH (1998). Rapid hybrid speciation in wild sunflowers. *Proceedings of the National Academy of Sciences USA* **95**, 11757–11762.

United States Department of Agriculture (2012). Animal production and marketing issues—Glossary. <http://www.ers.usda.gov/topics/animal-products/animal-production-marketing-issues/glossary.aspx>

United States Department of Agriculture (2015). Livestock and meat domestic data. <http://ers.usda.gov/data-products/livestock-meat-domestic-data.aspx#26126>

Vallejo-Marin M and Lye GC (2013). Hybridisation and genetic diversity in introduced *Mimulus* (Phrymaceae). *Heredity* **110**, 111–122.

Van de Peer Y, Maere S, and Meyer A (2009). The evolutionary significance of ancient genome duplications. *Nature Reviews Genetics* **10**, 725–732.

Van der Niet T and Linder HP (2008). Dealing with incongruence in the quest for the species tree: A case study from the orchid genus *Satyrium*. *Molecular Phylogenetics and Evolution* **47**, 154–174.

van Dijk PJ (2003). Ecological and evolutionary opportunities of apomixis: Insights from *Taraxacum* and *Chondrilla*. *Philosophical Transactions of the Royal Society of London B* **358**, 1113–1121.

van Heerwaarden J, Doebley J, Briggs WH, Glaubitz JC, Goodman MM, Sanchez Gonzalez JS, and Ross-Ibarra J (2011). Genetic signals of origin, spread, and introgression in a large sample of maize landraces. *Proceedings of the National Academy of Sciences USA* **108**, 1088–1092.

Van Heuverswyn F and Peeters M (2007). The origins of HIV and implications for the global epidemic. *Current Infectious Disease Reports* **9**, 338–346.

Vanin S, Bhutani S, Montelli S, Menegazzi P, Green EW, Pegoraro M, *et al.* (2012). Unexpected features of *Drosophila* circadian behavioural rhythms under natural conditions. *Nature* **484**, 371–375.

van Rijssel JC, Hoogwater ES, Kishe-Machumu MA, van Reenen E, Spits KV, van der Stelt RC, *et al.* (2015). Fast adaptive responses in the oral jaw of Lake Victoria cichlids. *Evolution* **69**, 179–189.

Van Zandt PA and Mopper S (2002). Delayed and carryover effects of salinity on flowering in *Iris hexagona* (Iridaceae). *American Journal of Botany* **89**, 1847–1851.

Van Zandt PA and Mopper S (2004). The effects of maternal salinity and seed environment on germination and growth in *Iris hexagona*. *Evolutionary Ecology Research* **6**, 813–832.

Van Zandt PA, Tobler MA, Mouton E, Hasenstein KH, and Mopper S (2003). Positive and negative consequences of salinity stress for the growth and reproduction of the clonal plant, *Iris hexagona*. *Journal of Ecology* **91**, 837–846.

Veeramah KR and Hammer MF (2014). The impact of whole-genome sequencing on the reconstruction of human population history. *Nature Reviews Genetics* **15**, 149–162.

Velasquez-Manoff M (2014). Should you fear the pizzly bear? *New York Times Sunday Magazine*. August 14, 2014.

Verardi A, Lucchini V, and Randi E (2006). Detecting introgressive hybridization between free-ranging domestic dogs and wild wolves (*Canis lupus*) by admixture linkage disequilibrium analysis. *Molecular Ecology* **15**, 2845–2855.

Vergilino R, Markova S, Ventura M, Manca M, and Dufresne F (2011). Reticulate evolution of the *Daphnia pulex* complex as revealed by nuclear markers. *Molecular Ecology* **20**, 1191–1207.

Verhoeven KJF, Macel M, Wolfe LM, and Biere A (2011). Population admixture, biological invasions and the balance between local adaptation and inbreeding depression. *Proceedings of the Royal Society of London B* **278**, 2–8.

Verkaar ELC, Nijman IJ, Beeke M, Hanekamp E, and Lenstra JA (2004). Maternal and paternal lineages in crossbreeding bovine species. Has wisent a hybrid origin? *Molecular Biology and Evolution* **21**, 1165–1170.

Vernot B and Akey JM (2014). Resurrecting surviving Neandertal lineages from modern human genomes. *Science* **343**, 1017–1021.

Via S (2012). Divergence hitchhiking and the spread of genomic isolation during ecological speciation-with-gene-flow. *Philosophical Transactions of the Royal Society of London B: Biological Sciences* **367**, 451–460.

Vickery RK Jr (1959). Barriers to gene exchange within *Mimulus guttatus* (Scrophulariaceae). *Evolution* **13**, 300–310.

Vickery RK Jr (1964). Barriers to gene exchange between members of the *Mimulus guttatus* complex (Scrophulariaceae). *Evolution* **18**, 52–69.

Vigouroux Y, Glaubitz JC, Matsuoka Y, Goodman MM, Sánchez J, and Doebley J (2008). Populations structure and genetic diversity of New World maize races assessed by DNA microsatellites. *American Journal of Botany* **95**, 1240–1253.

Vijaykrishna D, Poon LLM, Zhu HC, Ma SK, Li OTW, Cheung CL, *et al.* (2010). Reassortment of pandemic H1N1/2009 influenza A virus in swine. *Science* **328**, 1529.

Vilaça ST, Vargas SM, Lara-Ruiz P, Molfetti E, Reis EC, Lôbo-Hajdu G, *et al.* (2012). Nuclear markers reveal a complex introgression pattern among marine turtle species on the Brazilian coast. *Molecular Ecology* **21**, 4300–4312.

Vilà C, Leonard JA, Götherström A, Marklund S, Sandberg K, Lidén K, *et al.* (2001). Widespread origins of domestic horse lineages. *Science* **291**, 474–477.

Vilà C, Walker C, Sundqvist A-K, Flagstad Ø, Andersone Z, Casulli A, *et al.* (2003). Combined use of maternal, paternal and bi-parental genetic markers for the identification of wolf-dog hybrids. *Heredity* **90**, 17–24.

Viosca P Jr (1935). The irises of southeastern Louisiana—A taxonomic and ecological interpretation. *Bulletin of the American Iris Society* **57**, 3–56.

Vogel JP, Andrews HL, Wong SK, and Isberg RR (1998). Conjugative transfer by the virulence system of *Legionella pneumophila*. *Science* **279**, 873–876.

vonHoldt BM, Pollinger JP, Lohmueller KE, Han E, Parker HG, Quignon P, *et al.* (2010). Genome-wide SNP and haplotype analyses reveal a rich history underlying dog domestication. *Nature* **464**, 898–902.

Vonlanthen P, Bittner D, Hudson AG, Young KA, Muller R, Lundsgaard-Hansen B, *et al.* (2012). Eutrophication causes speciation reversal in whitefish adaptive radiations. *Nature* **482**, 357–362.

von Seidlein L, Kim DR, Ali M, Lee H, Wang XY, Thiem VD, *et al.* (2006). A multicentre study of *Shigella* diarrhoea in six Asian countries: Disease burden, clinical manifestations, and microbiology. *PLoS Medicine* **3**, e353.

Vorburger C and Reyer H-U (2003). A genetic mechanism of species replacement in European waterfrogs? *Conservation Genetics* **4**, 141–155.

Wagner WH, Jr (1970). Biosystematics and evolutionary noise. *Taxon* **19**, 146–151.

Wagner CE, Harmon LJ, and Seehausen O (2012). Ecological opportunity and sexual selection together predict adaptive radiation. *Nature* **487**, 366–370.

Wagner CE, Keller I, Wittwer S, Selz OM, Mwaiko S, Greuter L, *et al.* (2013). Genome-wide RAD sequence data provide unprecedented resolution of species boundaries and relationships in the Lake Victoria cichlid adaptive radiation. *Molecular Ecology* **22**, 787–798.

Wagner PL and Waldor MK (2002). Bacteriophage control of bacterial virulence. *Infection and Immunity* **70**, 3985–3993.

Wakeley J (2008). Complex speciation of humans and chimpanzees. *Nature* **452**, E3–E4.

Wakeley J and Hey J (1997). Estimating ancestral population parameters. *Genetics* **145**, 847–855.

Waldor MK and Mekalanos JJ (1996). Lysogenic conversion by a filamentous phage encoding cholera toxin. *Science* **272**, 1910–1914.

Walker MJ, Gibert J, López MV, Lombardi AV, Pérez-Pérez A, Zapata J, *et al.* (2008). Late Neandertals in southeastern Iberia: Sima de las Palomas del Cabezo Gordo, Murcia, Spain. *Proceedings of the National Academy of Sciences USA* **105**, 20631–20636.

Walker EP, Warnick F, Hamlet SE, *et al.* (1975). *Mammals of the World, Third Edition*. Johns Hopkins University Press, Baltimore.

Wall JD, Lohmueller KE, and Plagnol V (2009). Detecting ancient admixture and estimating demographic parameters in multiple human populations. *Molecular Biology and Evolution* **26**, 1823–1827.

Wall JD, Yang MA, Jay F, Kim SK, Durand EY, Stevison LS, *et al.* (2013). Higher levels of Neanderthal ancestry in East Asians than in Europeans. *Genetics* **194**, 199–209.

Wallander E (2008). Systematics of *Fraxinus* (Oleaceae) and evolution of dioecy. *Plant Systematics and Evolution* **273**, 25–49.

Walton C, Handley JM, Collins FH, Baimai V, Harbach RE, Deesin V, and Butlin RK (2001). Genetic population structure and introgression in *Anopheles dirus* mosquitoes in south-east Asia. *Molecular Ecology* **10**, 569–580.

Walton C, Handley JM, Tun-Lin W, Collins FH, Harbach RE, Baimai V, and Butlin RK (2000). Population structure and population history of *Anopheles dirus* mosquitoes in south-east Asia. *Molecular Biology and Evolution* **17**, 962–974.

Wang H, Feng Z, Shu Y, Yu H, Zhou L, Zu R, *et al.* (2008). Probable limited person-to-person transmission of highly pathogenic avian influenza A (H5N1) virus in China. *Lancet* **371**, 1427–1434.

Wang R-L and Hey J (1996). The speciation history of *Drosophila pseudoobscura* and close relatives: Inferences from DNA sequence variation at the Period locus. *Genetics* **144**, 1113–1126.

Wang B, Mao J-F, Gao J, Zhao W, and Wang X-R (2011). Colonization of the Tibetan Plateau by the homoploid hybrid pine *Pinus densata*. *Molecular Ecology* **20**, 3796–3811.

Wang Y-Q and Su B (2004). Molecular evolution of *microcephalin*, a gene determining human brain size. *Human Molecular Genetics* **13**, 1131–1137.

Wang X-R and Szmidt AE (1994). Hybridization and chloroplast DNA variation in a *Pinus* species complex from Asia. *Evolution* **48**, 1020–1031.

Wang X-R, Szmidt AE, and Savolainen O (2001). Genetic composition and diploid hybrid speciation of a high mountain pine, Pinus densata, native to the Tibetan plateau. *Genetics* **159**, 337–346.

Wang R-L, Wakeley J, and Hey J (1997). Gene flow and natural selection in the origin of *Drosophila pseudoobscura* and close relatives. *Genetics* **147**, 1091–1106.

Wang B and Wang X-R (2014). Mitochondrial DNA capture and divergence in Pinus provide new insights into the evolution of the genus. *Molecular Phylogenetics and Evolution* **80**, 20–30.

Ward TJ, Bielawski JP, Davis SK, Templeton JW, and Derr JN (1999). Identification of domestic cattle hybrids in wild cattle and bison species: A general approach using mtDNA markers and the parametric bootstrap. *Animal Conservation* **2**, 51–57.

Wargo AR and Kurath G (2011). In vivo fitness associated with high virulence in a vertebrate virus is a complex trait regulated by host entry, replication, and shedding. *Journal of Virology* **85**, 3959–3967.

Warmuth V, Eriksson A, Bower MA, Barker G, Barrett E, Hanks BK, *et al.* (2012). Reconstructing the origin and spread of horse domestication in the Eurasian steppe. *Proceedings of the National Academy of Sciences USA* **109**, 8202–8206.

Waser NM, Vickery RK Jr, and Price MV (1982). Patterns of seed dispersal and population differentiation in *Mimulus guttatus*. *Evolution* **36**, 753–761.

Watanabe T, Kiso M, Fukuyama S, Nakajima N, Imai M, Yamada S, *et al.* (2013). Characterization of H7N9 influenza A viruses isolated from humans. *Nature* **501**, 551–555.

Waterbury JB and Valois FW (1993). Resistance to co-occurring phages enables marine *Synechococcus* communities to coexist with cyanophages abundant in seawater. *Applied Environmental Microbiology* **59**, 3393–3399.

Wayne RK and Jenks SM (1991). Mitochondrial DNA analysis implying extensive hybridization of the endangered red wolf *Canis rufus*. *Nature* **351**, 565–568.

Wayne RK and Ostrander EA (2007). Lessons learned from the dog genome. *Trends in Genetics* **23**, 557–567.

Webb GC, White MJD, Contreras N, and Cheney J (1978). Cytogenetics of the parthenogenetic grasshopper *Warramaba* (formerly *Moraba*) *virgo* and its bisexual relatives. IV. Chromosome banding studies. *Chromosoma* **67**, 309–339.

Weetman D, Wilding CS, Steen K, Pinto J, and Donnelly MJ (2012). Gene flow-dependent genomic divergence between *Anopheles gambiae* M and S forms. *Molecular Biology and Evolution* **29**, 279–291.

Wegmann D and Excoffier L (2010). Bayesian inference of the demographic history of chimpanzees. *Molecular Biology and Evolution* **27**, 1425–1435.

Wei F, Coe E, Nelson W, Bharti AK, Engler F, Butler E, *et al.* (2007). Physical and genetic structure of the maize genome reflects its complex evolutionary history. *PLoS Genetics* **3**, e123.

Wei W, Davis RE, Jomantiene R, and Zhao Y (2008). Ancient, recurrent phage attacks and recombination shaped dynamic sequence-variable mosaics at the root of phytoplasma genome evolution. *Proceedings of the National Academy of Sciences USA* **105**, 11827–11832.

Weider LJ, Hobæk A, Hebert PDN, and Crease TJ (1999). Holarctic phylogeography of an asexual species complex—II. Allozymic variation and clonal structure in Arctic *Daphnia*. *Molecular Ecology* **8**, 1–13.

Weiss RA (2006). The discovery of endogenous retroviruses. *Retrovirology* **3**, 67.

Weller SG, Sakai AK, Culley TM, Duong L, and Danielson RE (2014). Segregation of male-sterility alleles across a species boundary. *Journal of Evolutionary Biology* **27**, 429–436.

Wendel JF (1989). New World tetraploid cottons contain Old World cytoplasm. *Proceedings of the National Academy of Sciences USA* **86**, 4132–4136.

Wendel JF (1995). Cotton—*Gossypium* (Malvaceae). In J Smartt and NW Simmonds, eds., *Evolution of Crop Plants*, Second Edition, pp. 358–366. Longman Scientific and Technical, Essex, England.

Wendel JF, Schnabel A, and Seelanan T (1995). An unusual ribosomal DNA sequence from *Gossypium gossypioides* reveals ancient, cryptic, intergenomic introgression. *Molecular Phylogenetics and Evolution* **4**, 298–313.

Wesselingh RA and Arnold ML (2000). Nectar production in Louisiana Iris hybrids. *International Journal of Plant Sciences* **161**, 245–251.

White GB (1971). Chromosomal evidence for natural interspecific hybridization by mosquitoes of the *Anopheles gambiae* complex. *Nature* **231**, 184–185.

White MJD (1978). *Modes of Speciation*. WH Freeman and Company, San Francisco.

White MJD (1980). Meiotic mechanisms in a parthenogenetic grasshopper species and its hybrids with related bisexual species. *Genetica* **52/53**, 379–383.

White S (2011). From globalized pig breeds to capitalist pigs: A study in animal cultures and evolutionary history. *Environmental History* **16**, 94–120.

White MJD, Cheney J, and Key KHL (1963). A parthenogenetic species of grasshopper with complex structural heterozygosity (Orthoptera: Acridoidea). *Australian Journal of Zoology* **11**, 1–19.

White MJD and Contreras N (1982). Cytogenetics of the parthenogenetic grasshopper *Warramaba virgo* and its bisexual relatives. VIII. Karyotypes and C-banding patterns in the clones of *W. virgo*. *Cytogenetics and Cell Genetics* **34**, 168–177.

White MJD, Contreras N, Cheney J, and Webb GC (1977). Cytogenetics of the parthenogenetic grasshopper *Warramaba* (formerly *Moraba*) *virgo* and its bisexual relatives. II. Hybridization studies. *Chromosoma* **61**, 127–148.

White MJD, Dennis ES, Honeycutt RL, Contreras N, and Peacock WJ (1982). Cytogenetics of the parthenogenetic grasshopper *Warramaba virgo* and its bisexual relatives. IX. The ribosomal RNA cistrons. *Chromosoma* **85**, 181–199.

White MJD, Webb GC, and Contreras N (1980). Cytogenetics of the parthenogenetic grasshopper *Warramaba* (formerly *Moraba*) *virgo* and its bisexual relatives. VI. DNA replication patterns of the chromosomes. *Chromosoma* **81**, 213–248.

Whiteley AR, Fitzpatrick SW, Funk WC, and Tallmon DA (2015). Genetic rescue to the rescue. *Trends in Ecology and Evolution* **30**, 42–49.

Whitney KD, Randell RA, and Rieseberg LH (2006). Adaptive introgression of herbivore resistance traits in the weedy sunflower *Helianthus annuus*. *American Naturalist* **167**, 794–807.

Whitney KD, Randell RA, and Rieseberg LH (2010). Adaptive introgression of abiotic tolerance traits in the sunflower *Helianthus annuus*. *New Phytologist* **187**, 230–239.

Whittall JB, Carlson ML, Beardsley PM, Meinke RJ, and Liston A (2006). The *Mimulus moschatus* alliance (Phrymaceae): Molecular and morphological phylogenetics and their conservation implications. *Systematic Botany* **31**, 380–397.

Wildman DE, Bergman TJ, al-Aghbari A, Sterner KN, Newman TK, Phillips-Conroy JE, *et al.* (2004). Mitochondrial evidence for the origin of hamadryas baboons. *Molecular Phylogenetics and Evolution* **32**, 287–296.

Willett CS (2008). Significant variation for fitness impacts of ETS loci in hybrids between populations of *Tigriopus californicus*. *Journal of Heredity* **99**, 56–65.

Willett CS and Berkowitz JN (2007). Viability effects and not meiotic drive cause dramatic departures from Mendelian inheritance for malic enzyme in hybrids of *Tigriopus californicus* populations. *Journal of Evolutionary Biology* **20**, 1196–1205.

Williams JH Jr and Arnold ML (2001). Sources of genetic structure in the woody perennial Betula occidentalis. *International Journal of Plant Sciences* **162**, 1097–1109.

Williams JH Jr, Friedman WE, and Arnold ML (1999). Developmental selection within the angiosperm style: using gamete DNA to visualize interspecific pollen competition. *Proceedings of the National Academy of Sciences USA* **96**, 9201–9206.

Williams EG, Kaul V, Rouse JL, and Palser BF (1986). Overgrowth of pollen tubes in embryo sacs of *Rhododendron* following interspecific pollinations. *Australian Journal of Botany* **34**, 413–423.

Williams EG and Rouse JL (1988). Disparate style lengths contribute to isolation of species in *Rhododendron*. *Australian Journal of Botany* **36**, 183–191.

Willis SC, Farias IP, and Ortí G (2014). Testing mitochondrial capture and deep coalescence in Amazonian cichlid fishes (Cichlidae: *Cichla*). *Evolution* **68**, 256–268.

Willis BL, van Oppen MJH, Miller DJ, Vollmer SV, and Ayre DJ (2006). The role of hybridization in the evolution of reef corals. *Annual Review of Ecology, Evolution and Systematics* **37**, 489–517.

Wilson WA, Harrington SE, Woodman WL, Lee M, Sorrells ME, and McCouch SR (1999). Inferences on the genome structure of progenitor maize through comparative analysis of rice, maize and the domesticated Panicoids. *Genetics* **153**, 453–473.

Winker K, McCracken KG, Gibson DD, and Peters JL (2013). Heteropatric speciation in a duck, *Anas crecca*. *Molecular Ecology* **22**, 5922–5935.

Winkler KA, Pamminger-Lahnsteiner B, Wanzenböck J, and Weiss S (2011). Hybridization and restricted gene flow between native and introduced stocks of Alpine whitefish (*Coregonus* sp.) across multiple environments. *Molecular Ecology* **20**, 456–472.

Winter CB and Porter AH (2010). AFLP linkage map of hybridizing swallowtail butterflies, *Papilio glaucus* and *Papilio canadensis*. *Journal of Heredity* **101**, 83–90.

Wisecaver JH and Hackett JD (2014). The impact of automated filtering of BLAST-determined homologs in the phylogenetic detection of horizontal gene transfer from a transcriptome assembly. *Molecular Phylogenetics and Evolution* **71**, 184–192.

Wisecaver JH, Slot JC, and Rokas A (2014). The evolution of fungal metabolic pathways. *PLoS Genetics* **10**, e1004816.

Wisniewski-Dyé F, Borziak K, Khalsa-Moyers G, Alexandre G, Sukharnikov LO, Wuichet K, *et al.* (2011). *Azospirillum* genomes reveal transition of bacteria from aquatic to terrestrial environments. *PLoS Genetics* **7**, e1002430.

Witt KE, Judd K, Kitchen A, Grier C, Kohler TA, Ortman SG, *et al.* (2015). DNA analysis of ancient dogs of the Americas: Identifying possible founding haplotypes and reconstructing population histories. *Journal of Human Evolution* **79**, 105–118.

Witter MS and Carr GD (1988). Adaptive radiation and genetic differentiation in the Hawaiian silversword alliance (Compositae: Madiinae). *Evolution* **42**, 1278–1287.

Woese CR (1987). Bacterial evolution. *Microbiological Reviews* **51**, 221–271.

Wolf JBW, Bayer T, Haubold B, Schilhabel M, Rosenstiel P, and Tautz D (2010). Nucleotide divergence vs. gene expression differentiation: comparative transcriptome sequencing in natural isolates from the carrion crow and its hybrid zone with the hooded crow. *Molecular Ecology* **19 (Supplement 1)**, 162–175.

Wolf DE, Takebayashi N, and Rieseberg LH (2001). Predicting the risk of extinction through hybridization. *Conservation Biology* **15**, 1039–1053.

Wolfe LM, Blair AC, and Penna BM (2007). Does intraspecific hybridization contribute to the evolution of invasiveness?: An experimental test. *Biological Invasions* **9**, 515–521.

Wolinska J, Bittner K, Ebert D, and Spaak P (2006). The coexistence of hybrid and parental *Daphnia*: The role of parasites. *Proceedings of the Royal Society of London B* **273**, 1977–1983.

Wolpoff MH (2009). How Neandertals inform human variation. *American Journal of Physical Anthropology* **139**, 91–102.

Won Y-J and Hey J (2005). Divergence population genetics of chimpanzees. *Molecular Biology and Evolution* **22**, 297–307.

Won Y-J, Sivasundar A, Wang Y, and Hey J (2005). On the origin of Lake Malawi cichlid species: A population genetic analysis of divergence. *Proceedings of the National Academy of Sciences USA* **102**, 6581–6586.

Wong GK-S, Liu B, Wang J, Zhang Y, Yang X, Zhang Z, et al. (2004). A genetic variation map for chicken with 2.8 million single-nucleotide polymorphisms. *Nature* **432**, 717–722.

Woodruff DS and Gould SJ (1987). Fifty years of interspecific hybridization: Genetics and morphometrics of a controlled experiment on the land snail *Cerion* in the Florida Keys. *Evolution* **41**, 1022–1045.

Woolhouse MEJ, Haydon DT, and Antia R (2005). Emerging pathogens: The epidemiology and evolution of species jumps. *Trends in Ecology and Evolution* **20**, 238–244.

Worden AZ, Lee J-H, Mock T, Rouzé P, Simmons MP, Aerts AL, et al. (2009). Green evolution and dynamic adaptations revealed by genomes of the marine Picoeukaryotes *Micromonas*. *Science* **324**, 268–272.

World Health Organization (2003). Influenza—report by the Secretariat. 4 pp.

World Health Organization (2014). HIV/AIDS Fact sheet N°360. <http://www.who.int/mediacentre/factsheets/fs360/en/>

Worobey M, Han G-Z, and Rambault A (2014). Genesis and pathogenesis of the 1918 pandemic H1N1 influenza A virus. *Proceedings of the National Academy of Sciences USA* **111**, 8107–8112.

Wright, S. 1931. Evolution in mendelian populations. *Genetics* **16**, 97–159.

Wright S (1951). The genetical structure of populations. *Annals of Eugenics* **15**, 323–354.

Wright D, Rubin C-J, Martinez Barrio A, Schütz K, Kerje S, Brändström H, et al. (2010). The genetic architecture of domestication in the chicken: Effects of pleiotropy and linkage. *Molecular Ecology* **19**, 5140–5156.

Wu C-I (2001). The genic view of the process of speciation. *Journal of Evolutionary Biology* **14**, 851–865.

Wu CA and Campbell DR (2005). Cytoplasmic and nuclear markers reveal contrasting patterns of spatial genetic structure in a natural *Ipomopsis* hybrid zone. *Molecular Ecology* **14**, 781–791.

Wu CA and Campbell DR (2006). Environmental stressors differentially affect leaf ecophysiological responses in two *Ipomopsis* species and their hybrids. *Oecologia* **148**, 202–212.

Wu CA and Campbell DR (2007). Leaf physiology reflects environmental differences and cytoplasmic background in *Ipomopsis* (Polemoniaceae) hybrids. *American Journal of Botany* **94**, 1804–1812.

Wu Y, Xia L, Zhang Q, Yang Q, and Meng X (2011). Bidirectional introgressive hybridization between *Lepus capensis* and *Lepus yarkandensis*. *Molecular Phylogenetics and Evolution* **59**, 545–555.

Xi Z, Wang Y, Bradley RK, Sugumaran M, Marx CJ, Rest JS, and Davis CC (2013). Massive mitochondrial gene transfer in a parasitic flowering plant clade. *PLoS Genetics* **9**, e1003265.

Xiang H, Gao J, Yu B, Zhou H, Cai D, Zhang Y, et al. (2014). Early Holocene chicken domestication in northern China. *Proceedings of the National Academy of Sciences USA* **111**, 17564–17569.

Xie J-B, Du Z, Bai L, Tian C, Zhang Y, Xie J-Y, et al. (2014). Comparative genomic analysis of N2-fixing and non-N2-fixing *Paenibacillus* spp.: Organization, evolution and expression of the nitrogen fixation genes. *PLoS Genetics* **10**, e1004231.

Xing F, Mao J-F, Meng J, Dai J, Zhao W, Liu H, et al. (2014). Needle morphological evidence of the homoploid hybrid origin of *Pinus densata* based on analysis of artificial hybrids and the putative parents, *Pinus tabuliformis* and *Pinus yunnanensis*. *Ecology and Evolution* **4**, 1890–1902.

Xu C, Bai Y, Lin X, Zhao N, Hu L, Gong Z, Wendel JF, and Liu B (2014). Genome-wide disruption of gene expression in allopolyploids but not hybrids of rice subspecies. *Molecular Biology and Evolution* **31**, 1066–1076.

Xu YC, Fang SG, and Li ZK (2007). Sustainability of the South China tiger: Implications of inbreeding depression and introgression. *Conservation Genetics* **8**, 1199–1207.

Xu S, Innes DJ, Lynch M, and Cristescu ME (2013). The role of hybridization in the origin and spread of asexuality in *Daphnia*. *Molecular Ecology* **22**, 4549–4561.

Xu B, Zhi N, Hu G, Wan Z, Zheng X, Liu X, et al. (2013). Hybrid DNA virus in Chinese patients with seronegative hepatitis discovered by deep sequencing. *Proceedings of the National Academy of Sciences USA* **110**, 10264–10269.

Yakimowski SB and Rieseberg LH (2014). The role of homoploid hybridization in evolution: A century of studies synthesizing genetics and ecology. *American Journal of Botany* **101**, 1247–1258.

Yamamichi M, Gojobori J, and Innan H (2012). An autosomal analysis gives no genetic evidence for complex speciation of humans and chimpanzees. *Molecular Biology and Evolution* **29**, 145–156.

Yang MA, Harris K, and Slatkin M (2014). The projection of a test genome onto a reference population and applications to humans and archaic hominins. *Genetics* **198**, 1655–1670.

Yang MA, Malaspinas A-S, Durand EY, and Slatkin M (2012). Ancient structure in Africa unlikely to explain Neanderthal and non-African genetic similarity. *Molecular Biology and Evolution* **29**, 2987–2995.

Yatabe Y, Kane NC, Scotti-Saintagne C, and Rieseberg LH (2007). Rampant gene exchange across a strong reproductive barrier between the annual sunflowers, *Helianthus annuus* and *H. petiolaris*. *Genetics* **175**, 1883–1893.

Yau MM and Taylor EB (2013). Environmental and anthropogenic correlates of hybridization between westslope cutthroat trout (*Oncorhynchus clarkii lewisi*) and introduced rainbow trout (*O. mykiss*). *Conservation Genetics* **14**, 885–900.

Yawson AE, Weetman D, Wilson MD, and Donnelly MJ (2007). Ecological zones rather than molecular forms predict genetic differentiation in the malaria vector *Anopheles gambiae* s.s. in Ghana. *Genetics* **175**, 751–761.

Yi X, Liang Y, Huerta-Sánchez E, Jin X, Cuo ZXP, Pool JE, *et al.* (2010). Sequencing of 50 human exomes reveals adaptation to high altitude. *Science* **329**, 75–78.

Yoo M-J, Liu X, Pires JC, Soltis PS, and Soltis DE (2014). Nonadditive gene expression in polyploids. *Annual Review of Genetics* **48**, 485–517.

Yoshida K, Makino T, Yamaguchi K, Shigenobu S, Hasebe M, Kawata M, *et al.* (2014). Sex chromosome turnover contributes to genomic divergence between incipient stickleback species. *PLoS Genetics* **10**, e1004223.

Yoshida S, Maruyama S, Nozaki H, and Shirasu K (2010). Horizontal gene transfer by the parasitic plant *Striga hermonthica*. *Science* **328**, 1128.

Yotova V, Lefebvre J-F, Moreau C, Gbeha E, Hovhannesyan K, Bourgeois S, *et al.* (2011). An X-linked haplotype of Neandertal origin is present among all non-African populations. *Molecular Biology and Evolution* **28**, 1957–1962.

Yu Y, Barnett RM, and Nakhleh L (2013). Parsimonious inference of hybridization in the presence of incomplete lineage sorting. *Systematic Biology* **62**, 738–751.

Yu Y, Degnan JH, and Nakhleh L (2012). The probability of a gene tree topology within a phylogenetic network with applications to hybridization detection. *PLoS Genetics* **8**, e1002660.

Yu Y, Dong J, Liu KJ, and Nakhleh L (2014). Maximum likelihood inference of reticulate evolutionary histories. *Proceedings of the National Academy of Sciences USA* **111**, 16448–16453.

Yu N, Jensen-Seaman MI, Chemnick L, Kidd JR, Deinard AS, Ryder O, *et al.* (2003). Low nucleotide diversity in chimpanzees and bonobos. *Genetics* **164**, 1511–1518.

Yu VL, Plouffe JF, Pastoris MC, Stout JE, Schousboe M, Widmer A, *et al.* (2002). Distribution of *Legionella* species and serogroups isolated by culture in patients with sporadic community-acquired legionellosis: An international collaborative survey. *Journal of Infectious Diseases* **186**, 127–128.

Yu Y, Than C, Degnan JH, and Nakhleh L (2011). Coalescent histories on phylogenetic networks and detection of hybridization despite incomplete lineage sorting. *Systematic Biology* **60**, 138–149.

Yuan Y-W, Sagawa JM, Young RC, Christensen BJ, and Bradshaw HD Jr (2013). Genetic dissection of a major anthocyanin QTL contributing to pollinator-mediated reproductive isolation between sister species of *Mimulus*. *Genetics* **194**, 255–263.

Zeng Y-F, Liao W-J, Petit RJ, and Zhang D-Y (2010). Exploring species limits in two closely related Chinese oaks. *PLoS ONE* **5**, e15529.

Zeng Y-F, Liao W-J, Petit RJ, and Zhang D-Y (2011). Geographic variation in the structure of oak hybrid zones provides insights into the dynamics of speciation. *Molecular Ecology* **20**, 4995–5011.

Zeng L-W and Singh RS (1993). The genetic basis of Haldane's Rule and the nature of asymmetric hybrid male sterility among *Drosophila simulans*, *Drosophila mauritiana* and *Drosophila sechellia*. *Genetics* **134**, 251–260.

Zeyland J, Wolko L, Lipiński D, Woźniak A, Nowak A, Szalata M, *et al.* (2012). Tracking of wisent–bison–yak mitochondrial evolution. *Journal of Applied Genetics* **53**, 317–322.

Zhang W, Kunte K, and Kronforst MR (2013). Genome-wide characterization of adaptation and speciation in tiger swallowtail butterflies using de novo transcriptome assemblies. *Genome Biology and Evolution* **5**, 1233–1245.

Zhang J, Wang X, and Podlaha O (2004). Testing the chromosomal speciation hypothesis for humans and chimpanzees. *Genome Research* **14**, 845–851.

Zhang J-J, Ye Q-G, Yao X-H, and Huang H-W (2010). Spontaneous interspecific hybridization and patterns of pollen dispersal in ex situ populations of a tree species (*Sinojackia xylocarpa*) that is extinct in the wild. *Conservation Biology* **24**, 246–255.

Zhao W, Meng J, Wang B, Zhang L, Xu Y, Zeng Q-Y, *et al.* (2014). Weak crossability barrier but strong juvenile selection supports ecological speciation of the hybrid pine *Pinus densata* on the Tibetan plateau. *Evolution* **68**, 3120–3133.

Zhaxybayeva O and Gogarten JP (2004). Cladogenesis, coalescence and the evolution of the three domains of life. *Trends in Genetics* **20**, 182–187.

Zhaxybayeva O, Lapierre P, and Gogarten JP (2004). Genome mosaicism and organismal lineages. *Trends in Genetics* **20**, 254–260.

Zhou R, Moshgabadi N, and Adams KL (2011). Extensive changes to alternative splicing patterns following allopolyploidy in natural and resynthesized polyploids. *Proceedings of the National Academy of Sciences USA* **108**, 16122–16127.

Zhou X, Wang B, Pan Q, Zhang J, Kumar S, Sun X, *et al.* (2014). Whole-genome sequencing of the snub-nosed monkey provides insights into folivory and evolutionary history. *Nature Genetics* **46**, 1303–1310.

Zhou W-W, Wen Y, Fu J, Xu Y-B, Jin J-Q, Ding L, *et al.* (2012). Speciation in the *Rana chensinensis* species complex and its relationship to the uplift of the Qinghai–Tibetan plateau. *Molecular Ecology* **21**, 960–973.

Zilhão J, d'Errico F, Bordes J-G, Lenoble A, Texier J-P, and Rigaud J-P (2006). Analysis of Aurignacian interstratification at the Châtelperronian-type site and implications for the behavioral modernity of Neandertals. *Proceedings of the National Academy of Sciences USA* **103**, 12643–12648.

Zimmer JT (1941). Studies of Peruvian birds. Number XXXVII. The genera *Sublegatus, Phaeomyias, Camptostoma, Xanthomyias, Phyllomyias,* and *Tyranniscus. American Museum Novitates* **1109**, 1–25.

Zimmer EA, Martin SL, Beverley SM, Kan YW, and Wilson AC (1980). Rapid duplication and loss of genes coding for the α chains of hemoglobin. *Proceedings of the National Academy of Sciences USA* **77**, 2158–2162.

Zinner D, Arnold ML, and Roos C (2009a). Is the new primate genus *Rungwecebus* a baboon?. *PLoS ONE* **4**, e4859.

Zinner D, Arnold ML, and Roos C (2011). The strange blood: Natural hybridization in primates. *Evolutionary Anthropology* **20**, 96–103.

Zinner D, Groeneveld LF, Keller C, and Roos C (2009b). Mitochondrial phylogeography of baboons (*Papio* spp.)—Indication for introgressive hybridization? *BMC Evolutionary Biology* **9**, 83.

Zinner D and Roos C (2014). So what is a species anyway? A primatological perspective. *Evolutionary Anthropology* **23**, 21–23.

Zinner D, Wertheimer J, Liedigk R, Groeneveld LF, and Roos C (2013). Baboon phylogeny as inferred from complete mitochondrial genomes. *American Journal of Physical Anthropology* **150**, 133–140.

Zou J-B, Peng X-L, Li L, Liu J-Q, Miehe G, and Opgenoorth L (2012). Molecular phylogeography and evolutionary history of *Picea likiangensis* in the Qinghai-Tibetan Plateau inferred from mitochondrial and chloroplast DNA sequence variation. *Journal of Systematics and Evolution* **50**, 341–350.

Zozomová-Lihová J, Mandáková T, Kovaříková A, Mühlhausen A, Mummenhoff K, Lysak MA, and Kovařík A (2014). When fathers are instant losers: Homogenization of rDNA loci in recently formed *Cardamine* x *schulzii* trigenomic allopolyploid. *New Phytologist* **203**, 1096–1108.

Index

Notes: Page numbers in *italics* refer to Figures and Tables. Page numbers in **bold** refer to Glossary entries.
To save space in the index, the following abbreviations have been used:

HGT - horizontal gene transfer
ILS - incomplete lineage sorting
PSC - phylogenetic species concept

A

ABBA/BABA 37
  Zimmerus ILS 38
Achtman, M 73–4
Ackermann, R 42
adaptive homoploid hybrids 18–19
adaptive introgression
  Heliconius adaptive radiations 122
  human evolution 161–3
adaptive radiation 104–22
  cichlids 120–1, *121*
  definition **177**
  Hawaiian silversword
    assemblage 127–9
  Heliconius 121–2, *122*
  Lord Howe Island flora 129
  plants 123–9
    see also individual species
adaptive trait transfer 11–16
  canine coat colour gene 12–13
  definition **177**
  Geospiza (Darwin's finches) 15
  Iris 13–15, *14*
  Mus (mouse) 55
  Senecio 15–16
Adh 84
Allendorf, FW 132
Allonemobius, contemporary hybrid
  zones 53–4
allopatric divergence 2
allopatric speciation 4
  definition **177**
allopolyploid speciation 2, 26, *26*,
  114–19
  asexual species 116–19
  definition **177**

Hawaiian silversword
  assemblage 128–9
  plants 126–7
  sexual species 115–16
  see also individual species
allopolyploidy 26
Alloteropsis 22
allozyme analysis 52–3
Alouatta (howler monkeys)
  evolution 146–7, *147*
Alpine lake whitefish (*Corgeonus*),
  ILS 38, *39*
Alternaria, PSC 29–30
Alves, PC 105–7
Ambystoma (salamander), hybrid
  zone studies 9
Anas (duck), divergence-with-
  introgression 108–9, *109*, *110*
animals, hybrid fitness 80–8
Anopheles (malaria mosquito)
  hybrid invariability 75–6
  introgressive hybridization 75–6,
    76
  mosaic genomes 76
antibacterial resistance, *Shigella* 95–96
apples (*Malus*) evolution *164*, 164–5
Arabidopsis thaliana, allopolyploid
  hybrid speciation 126
Arroyo-García R 171
Artibeus, homoploid hybrid
  speciation 109, *110*
asexual species
  allopolyploid speciation 116–19
    see also individual species
  homoploid hybrid speciation
    109–14

asymmetric mtDNA
  introgression 112, *112*
Australian gekkonid lizard
  (*Heteronotia*), asexual
  allopolyploid speciation
  116–18, *117*
Australian grasshopper (*Warramaba*),
  asexual homoploid hybrid
  speciation 109–12
Australopithecus, introgression
  evidence 159
autopolyploidy definition **177**
autosomal loci
  Oryctolagus (rabbit) introgressive
    hybridization 107–8
  Pan (chimpanzee) evolution 155–6
  Ursus introgressive
    hybridization *133*, 133–4

B

baboon *(Papio)* evolution 149–51,
  150
Bacillus subtilis, HGT 94
backcrossing, *Canis* (dog)
  evolution 169–70
bacteria
  animal HGT 21
  intragenomic divergence 56
  reproductive isolation 72–4
    see also individual species
bacteriophages 100–2
  HGT 101
  photosynthesis gene
    sequences 101–2
  recombination 101
Baltrus, DA 92